Recent Advances in
NANOTECHNOLOGY

Recent Advances in
NANOTECHNOLOGY

Edited By
Changhong Ke, PhD
Professor of Mechanical Engineering,
Binghamton University, State University of New York, U.S.A.

Apple Academic Press

TORONTO   NEW YORK

CRC Press
Taylor & Francis Group
6000 Broken Sound Parkway NW, Suite 300
Boca Raton, FL 33487-2742

Apple Academic Press, Inc
3333 Mistwell Crescent
Oakville, ON L6L 0A2
Canada

© 2012 by Apple Academic Press, Inc.
Exclusive worldwide distribution by CRC Press an imprint of Taylor & Francis Group, an Informa business

First issued in paperback 2021

No claim to original U.S. Government works

Version Date: 20120530

ISBN 13: 978-1-77463-191-1 (pbk)
ISBN 13: 978-1-926692-73-9 (hbk)

Visit the Taylor & Francis Web site at
http://www.taylorandfrancis.com

and the CRC Press Web site at
http://www.crcpress.com

For information about Apple Academic Press product
http://www.appleacademicpress.com

Preface

Nanotechnology deals with the smallest building blocks of matter on the atomic and molecular scale and the development of devices and materials using these particles that are between one and 100 nanometers. Particles of this size are governed by the laws of quantum physics, not the Newtonian rules that govern the visible world around us. Studies in nanotechnology, therefore, must involve studies across disciplines, including quantum physics, engineering, biology and chemistry, among others.

Nanotechnology has grown exponentially over the last few decades with more and more national and international funding going toward research. Nobel laureate and physics professor Richard Feynman is credited with sparking the nanotechnology revolution during a talk at Caltech in the 1950s that was titled, "There's Plenty of Room at the Bottom", encouraging researchers to look at the smaller end of the spectrum. Feynman and others offered awards for research in the field. In 2000, then-President Bill Clinton announced the National Nanotechnology Initiative that furthered the expansion of the field.

Nanotechnology has been used to improve upon current technologies as well as to create groundbreaking advances. As the research reported in the following pages will show, nanotechnology is being used to improve the delivery of anti-cancer drugs, for early detection and identification of diseased tissue, to increase the efficiency of photovoltaic cells, and to improve the effectiveness of biosensors, among other uses. Researchers have also worked to improve the processes by which nanodevices are made by developing refined methods of creating nanocrystalline metals. As advances in nanotechnology continue, studies are being conducted to test the safety of these materials among environmental systems, including soil, water and organisms.

In the article "Nanotechnology: Looking as We Leap," Ernie Hood surveys the field providing a view from the top, looking at general directions in the field and laying out suggestions for future research. Hood defines the world of nano thus: "Nanoparticles behaves like neither solids, liquids, nor gases and exist in the topsy-turvy world of quantum physics, which governs those denizens small enough to have escaped the laws of Newtonian physics. This allows them to perform their almost magical feats of conductivity, reactivity, and optical sensitivity, among others." He then describes four phases that he expects nanotechnology to move through as the science develops. The first phase would involve simple particles designed to perform single tasks. After some accomplishment in that arena, active nanostructures would be developed for commercial purposes with nanoscale building blocks available for more complex projects, then nanosystems comprised of thousands of components would be the next step. Finally, Hood foresees nanotechnology developing "molecular" nanodevices, which are systems-within-systems that will perform highly complicated procedures. At the time that Hood was writing, research in nanotechnology included projects that would improve water filtration systems and clean energy production, clean up toxic wastes, reduce pollution and waste in manufacturing processes, and even a more effective sunscreen. While current research moves forward in these areas and dozens of others, other projects are focused on the safety of nanotechnology itself. Early studies have shown that nanoparticles can be absorbed into the body causing toxic reactions, although more research needs to focus on whether nanoparticles can be metabolized and naturally processed out of the body. Further research is also looking at what happens to nanoparticles in soil and water systems to further predict what might happen to people or other organisms that are exposed. The excitement over the possible advances that nanotech promises across a wide variety of fields should be tampered with an air of caution, Hood says, however, that this shouldn't keep researchers from moving forward.

Several research teams reported on the use of nanotechnology in cancer research. In "Cancer Nanotechnology", the US Dept of Health and Human Resources recognizes that nanotechnology provides a unique opportunity to change the way we diagnose, image, and treat cancer. In keeping with their goal to eliminate death and suffering from cancer by 2015, the department is promoting a Cancer Nanotechnology Plan, which will help to coordinate interdisciplinary research into the diagnosis, treatment, and prevention of cancer, and the National Nanotechnology Characterization Laboratory, which will provide resources and support for researchers. The department has identified six key challenge areas of research: prevention and control, early detection and proteomics, imaging diagnostics, multifunctional therapeutics, quality of life enhancement in cancer care, and interdisciplinary training. Nanoscale devices are already proving to be successful in delivering therapies directly to the cells where they are needed. As nanotechnology improves and research across fields refines the processes by which nanoparticles can be used, we will find more and more efficient ways of using this technology for preventing and fighting cancer on the cellular level.

Yiqan Du and partners report on research that is an example of such an experiment, using nanoparticles for focused treatment of cancer. Thermotherapy and arsenic are used against a variety of cancerous tumors; however, these treatments have limitations including toxicity of arsenic for healthy tissues. In an effort to reduce the limitations that these treatments imposed, researchers Du, Zhang, Liu and Lai experimented with a complex of nanoparticles with Fe_3O_4 encapsulated within As_2O_3 in order to increase the tissue-specific delivery. The team found that the spherical As_2O_3/Fe_3O_4 nanoparticles could become part of a process to eliminate tumors in a minimally invasive way, or in preoperative treatment because of their ability to reduce tumor volume and intraoperative bleeding.

The studies by Raje Chouhan and A. K. Bajpai and by Yan Xiao et al. exhibited further research of the use of nanoparticles in targeting cancerous cells for treatment while leaving healthy cells untouched. Chouhan and Bajpai found that nanocarriers can be designed to improve the targeted delivery of anticancer therapy to malignant cells, avoiding harmful contact with healthy cells. They developed poly-2-hydroxyethyl methacrylate (PHEMA) particles that could deliver the anticancer drug doxorubicin to targeted cells and release it in a controlled manner. Xiao et al. experimented with single-walled carbon nanotubes (SWNT), an inorganic nanomaterial that affords researchers advantages over organic materials, being particularly useful in the development of biosensors and drug delivery transport. Their research showed that SWNTs have properties that make them useful for the detection and destruction of cancer cells. Another key direction of research incorporated the use of nanoparticles into biosensors. Biosensors are used to detect specific biological components within samples, such as DNA, RNA, enzymes or toxins. Biosensors rely on molecular affinity between particles, such as enzyme-substrate or antibody-antigen complementary pairs. In order to increase the efficiency of these devices, researchers have engineered nanoparticles that allow biosensors to detect smaller molecules. Wildgruber et al. developed biosensors that identify the presence of monocytes in human blood as a potential signal of certain diseases, including coronary artery disease, sepsis, and viral infection. The team custom engineered magnetic nanoparticles to find monocytes in *ex vivo* studies. Zhang et al. surveyed the use of nanomaterials in biosensors, including gold nanoparticles, carbon nanotubes, magnetic nanoparticles, and quantum dots, in order to determine if nanomaterial-based biosensors could provide improved results in disease diagnosis and food safety testing. Research is showing that the addition of nanomaterials to biosensors can improve the sensitivity and specificity of their detection abilities, increasing the sensitivity to small molecules in samples such as DNA, RNA, proteins, glucose, and pesticides. The eventual direction of these devices will be toward single molecule detection, although the technology is not quite there yet.

In "Nanotechnology: A Tool for Improved Performance on Electrochemical Screen-Printed (Bio)Sensors," Elena Jubete and her partners compared ways that nanotechnology has improved the screen-printing processes that are used for the fabrication of disposable electrodes for biosensor applications. Screen-printing is accomplished by squeezing a paste through a screen onto a substrate. The process becomes more complicated based on the materials that are used in the paste and substrate. To make a biosensor, the biological material, such as an enzyme or antibody, is mixed in with the paste or added as a layer on top, but these components are delicate and can be unstable. Introducing nanoscale materials such as carbon nanotubes into the manufacture of screen-printed biosensors added more stability to the biological materials, increased the reaction rate, and allowed for more detection capabilities.

Other research topics focused on refining the methods that produce nanoparticles that are used for microdevices to improve efficiency and operation of current technologies, such as magnetic resonance imaging and photovoltaic cells. J. Michl discusses the design process for creating the first crystalline material for singlet fission that can be used in photovoltaic cells. Michl set out to find a way to create sensitizer dyes that would increase the efficiency of photovoltaic cells and found that it is not possible at the moment, because his design model found a contradictory requirement for singlet fission to occur. Michl identified the two key hurdles as, "One of these is thermodynamic and deals with the arrangement of the electronic levels of the chromophore and the other is kinetic and deals with the strength of coupling between the two or more chromophores that need to be present in a singlet fission sensitizer". Further directions for future research are suggested at the end of his study.

In another study, the team of Huang et al. used superparamagnetic iron oxide nanoparticles (SPIOs) as a contrast agent in magnetic resonance imaging to test their ability to load more efficiently and with less cytotoxicity than the formulas that were previously available. The researchers formulated cationic lipid nanoparticles containing SPIOs that could be used for in vivo imaging. The lipid-coated SPIOs showed efficient cellular uptake with high loading efficiency, low toxicity, and long-term imaging capability.

The team of David Gill et al. set out with the goal of creating bulk nanocrystalline metal in order to produce materials that exhibit high levels of hardness and strength and more homogenous properties for small nanoscale devices. They experimented with a variety of fabrication methods including large strain extrusion machining, modulation assisted machining, cold spray processing, and shock consolidation. The cold spray process and shock compaction proved to be the most effective means of creating bulk nanocrystalline metal that would exhibit the strength, hardness, and resiliency needed. Next steps would include testing these materials in practical applications in order to determine their efficiency.

In "The Origin of Nanoscopic Grooving on Vesicle Walls in Submarine Basaltic Glass: Implications for Nanotechnology," the researchers French and Muehlenbachs look at the internal features of submarine pillow basalts in order to determine new possibilities for replicating such structures in the fabrication of nanodevices. The researchers study examples of nanoscopic grooving in glass in order to develop new methods of nanolithography that can generate complex patterns. New directions in nanolithography and the creation of nanodevices may result from their findings.

As the field continues to expand and grow, as Ernie Hood cautioned, it is important for it to grow smartly, with an eye toward detrimental effects from research or use of nanotechnology materials in the environment. There are a variety of organizations that have formed with the specific goal of providing oversight to the field, including the Woodrow Wilson International Center for Scholar's Project on Emerging Nanotechnologies and the Center for Responsible Nanotechnologies. Opportunities for collaborative research abound, with the National Nanotechnology Initiative taking a lead in bringing stakeholders together. Advances in nanotechnology will certainly be a part of a bright future, helping to improve the quality of life around the world.

— Changhong Ke, PhD

List of Contributors

C. Thomas Avedisian
Sibley School of Mechanical and Aerospace Engineering, Cornell University, Ithaca, NY, USA.

A. K. Bajpai
Bose Memorial Research Laboratory, Department of Chemistry, Government Autonomous Science College, Jabalpur (MP)-482001, India.

Richard E. Cavicchi
Chemical Science and Technology Laboratory, National Institute of Standards and Technology (NIST), Gaithersburg, MD, USA.

Fu-Hsiung Chang
Institute of Biochemistry and Molecular Biology, College of Medicine, National Taiwan University, Taipei, Taiwan.

Karen Chang
Department of Life Science, College of Science and Engineering, Fu-Jen Catholic University, Taipei, Taiwan.

Po-Yuan Chang
Department of Internal Medicine, National Taiwan University Hospital, Taipei, Taiwan.

Zee-Fen Chang
Institute of Biochemistry and Molecular Biology, College of Medicine, National Taiwan University, Taipei, Taiwan.

Chao-Yu Chen
Institute of Biochemistry and Molecular Biology, College of Medicine, National Taiwan University, Taipei, Taiwan.

Jyh-Horng Chen
Department of Electrical Engineering, National Taiwan University, Taipei, Taiwan.

Raje Chouhan
Bose Memorial Research Laboratory, Department of Chemistry, Government Autonomous Science College, Jabalpur (MP)-482001, India.

Aleksey Chudnovskiy
Center for Systems Biology, Massachusetts General Hospital and Harvard Medical School, Boston, MA, USA.

Kevin Croce
Cardiovascular Division, Department of Medicine, Brigham and Women's Hospital, Boston, MA, USA.
Center for Excellence in Vascular Biology, Brigham and Women's Hospital, Boston, MA, USA.

Daxiang Cui
Shanghai Jiao Tong University, Shanghai, 200240, P.R. China.

Y. Du
Department of Pathology and Pathophysiology, School of Basic Medical Sciences, Southeast University, Nanjing, P.R. China.

Martin Etzrodt
Center for Systems Biology, Massachusetts General Hospital and Harvard Medical School, Boston, MA, USA.

D. Anthony Fredenburg
Georgia Institute of Technology, Atlanta, GA, USA.

Jason E. French
Department of Earth and Atmospheric Sciences, University of Alberta, 1-26 Earth Science Building, Edmonton, AB, T6G 2E3, Canada.

Xiugong Gao
Research and Development, Translabion, Clarksburg, MD, USA.

David D. Gill
Ktech Corporation, Albuquerque, NM, USA.

Hans Grande
New Materials Department, Sensors and Photonics Unit, Centre for Electrochemical Technologies (CIDETEC), Paseo Miramón 196, 20009 Donostia-San Sebastián, Spain.

Qin Guo
Shanghai Jiao Tong University, Shanghai, 200240, P.R. China.

Aaron C. Hall
Ktech Corporation, Albuquerque, NM, USA.

Huixin He
Department of Chemistry, Rutgers University, Newark, NJ, USA.

R. David Holbrook
Chemical Science and Technology Laboratory, National Institute of Standards and Technology (NIST), Gaithersburg, MD, USA.

Huey-Chung Huang
Institute of Biochemistry and Molecular Biology, College of Medicine, National Taiwan University, Taipei, Taiwan.

Elena Jubete
New Materials Department, Sensors and Photonics Unit, Centre for Electrochemical Technologies (CIDETEC), Paseo Miramón 196, 20009 Donostia-San Sebastián, Spain.

R. Lai
Department of Pathology and Pathophysiology, School of Basic Medical Sciences, Southeast University, Nanjing, P.R. China.

Hakho Lee
Center for Systems Biology, Massachusetts General Hospital and Harvard Medical School, Boston, MA, USA.

Peter Libby
Cardiovascular Division, Department of Medicine, Brigham and Women's Hospital, Boston, MA, USA.
Center for Excellence in Vascular Biology, Brigham and Women's Hospital, Boston, MA, USA.

Chung-Wu Lin
Department of Pathology, College of Medicine, National Taiwan University, Taipei, Taiwan.

H. Liu
Department of Pathology and Pathophysiology, School of Basic Medical Sciences, Southeast University, Nanjing, P.R. China.

Oscar A. Loaiza
New Materials Department, Sensors and Photonics Unit, Centre for Electrochemical Technologies (CIDETEC), Paseo Miramón 196, 20009 Donostia-San Sebastián, Spain.

J. Michl
University of Colorado Boulder, Colorado, USA.

Somenath Mitra
Department of Chemistry and Environmental Science, New Jersey Institute of Technology, Newark, NJ, USA.

Chung-Yuan Mou
Department of Chemistry, College of Science, National Taiwan University, Taipei, Taiwan.

Karlis Muehlenbachs
Department of Earth and Atmospheric Sciences, University of Alberta, 1-26 Earth Science Building, Edmonton, AB, T6G 2E3, Canada.

Matthias Nahrendorf
Center for Systems Biology, Massachusetts General Hospital and Harvard Medical School, Boston, MA, USA.

Estibalitz Ochoteco
New Materials Department, Sensors and Photonics Unit, Centre for Electrochemical Technologies (CIDETEC), Paseo Miramón 196, 20009 Donostia-San Sebastián, Spain.

Mikael J. Pittet
Center for Systems Biology, Massachusetts General Hospital and Harvard Medical School, Boston, MA, USA.

Jose A. Pomposo
New Materials Department, Sensors and Photonics Unit, Centre for Electrochemical Technologies (CIDETEC), Paseo Miramón 196, 20009 Donostia-San Sebastián, Spain.

Javier Rodríguez
New Materials Department, Sensors and Photonics Unit, Centre for Electrochemical Technologies (CIDETEC), Paseo Miramón 196, 20009 Donostia-San Sebastián, Spain.

Timothy J. Roemer
Ktech Corporation, Albuquerque, NM, USA.

Christopher J. Saldana
Purdue University, West Lafayette, IN, USA.

Ronak Savla
Department of Chemistry, Rutgers University, Newark, NJ, USA.

Sudhir Srivastava
Division of Cancer Prevention, National Cancer Institute (NCI), Bethesda, MD, USA.

Filip K. Swirski
Center for Systems Biology, Massachusetts General Hospital and Harvard Medical School, Boston, MA, USA.

Oleh Taratula
Department of Chemistry, Rutgers University, Newark, NJ, USA.

Stephen Treado
Building Environment Division, National Institute of Standards and Technology (NIST), Gaithersburg, MD, USA.

Aaron Urbas
Chemical Science and Technology Laboratory, National Institute of Standards and Technology (NIST), Gaithersburg, MD, USA.

Paul D. Wagner
Division of Cancer Prevention, National Cancer Institute (NCI), Bethesda, MD, USA.

Ralph Weissleder
Center for Systems Biology, Massachusetts General Hospital and Harvard Medical School, Boston, MA, USA.
Department of Systems Biology, Harvard Medical School, Boston, MA, USA.

Moritz Wildgruber
Center for Systems Biology, Massachusetts General Hospital and Harvard Medical School, Boston, MA, USA.
Department of Radiology, Klinikum Rechts der Isar, Technische Universität München, Munich, Germany.

Yan Xiao
Chemical Science and Technology Laboratory, National Institute of Standards and Technology (NIST), Gaithersburg, MD, USA.

Tae-Jong Yoon
Center for Systems Biology, Massachusetts General Hospital and Harvard Medical School, Boston, MA, USA.

D. Zhang
Department of Pathology and Pathophysiology, School of Basic Medical Sciences, Southeast University, Nanjing, P.R. China.

Xueqing Zhang
Shanghai Jiao Tong University, Shanghai, 200240, P.R. China.

List of Abbreviations

AA2P	Ascorbic acid 2-phosphate
AChE	Acetylcholinesterase
ADH	Alcohol dehydrogenase
AFM	Atomic force microscopy
ALP	Alkaline phosphatase
AOX	Alcohol oxidase
APPL	Acute promyelocytic leukemia
APT	Aptamer
ARE	Antioxidant response element
AuNPs	Gold nanoparticles
BF	Bright field
BPO	Benzoyl peroxide
CAD	Coronary artery disease
CD44	Cluster of differentiation 44
ChE	Cholinesterase
CHO	Choline oxidase
ChOx	Cholesterol oxidase
CLIO	Cross-linked iron oxide
CNTs	Carbon nanotubes
CT	Computed tomography
DF	Dark field
DLS	Dynamic light scattering
DMR	Diagnostic magnetic resonance
DOTAP	1,2-Dioleoyl-3-(trimethylammonium) propane
DSC	Differential scanning calorimetry
DSDP	Deep Sea Drilling Project
EBSD	Electron backscattered diffraction
ECAP	Equal channel angular pressing
ECM	Electro-chemical machining
EDC	1-Ethyl-3-(3-dimethylaminopropyl) carbodiimide
EDM	Electro-discharge machining
EELS	Electron energy loss
EGDMA	Ethyleneglycol dimethacrylate
EOS	Equation of state
EPA	Environmental Protection Agency

EthD-1	Ethidium homodimer-1
FABS	Force amplified biological sensor
FCS	Fetal calf serum
FIA	Flow injection analysis
FIB	Focused ion beam
FRET	Fluorescence resonance energy transfer
FSC	Fractional solid content
FSC/SSC	Forward scatter/side scatter
FTIR	Fourier transform infrared
GMOs	Genetically modified organisms
GNRs	Gold nanorods
GOX	Glucose oxidase
HAMA	Human anti-mouse IgG antibodies
Hb	Hemoglobin
HCG	Human chorionic gonadotrophin
HCMV	Human cytomegalovirus
HEMA	2-Hydroxyethyl methacrylate
HPLC	High-pressure liquid chromatography
HQ	Hydroquinone
HRP	Horseradish peroxidase
HS-CdS	Hollow nanospheres CdS
HUVEC	Human umbilical vein endothelial cell
IM	Mass inhibition
ISC	Intersystem crossing
ITO	Indium tin oxide
IV	Volume inhibition
LBL	Layer-by-layer
LDH	Lactate dehydrogenase
LDV	Laser doppler velocimetry
LEQDs	Lipid-enclosed quantum dots
LN_2	Liquid nitrogen
LSEM	Large strain extrusion machining
L-SPIOs	Lipid-coated Superparamagnetic iron oxide nanoparticles
LTG	Lamotrigine
MAM	Modulation assisted machining
MB	Methylene blue
MCMS	Magnetic chitosan microsphere
MCSFR	Macrophage colony stimulating factor
MDNS	Magnetic dextran microsphere

MDR	Multidrug resistant
MF	Magnetic fluid
MFH	Magnetic fluid hyperthermia
MFI	Mean fluorescent intensity
MMP-9	Matrix metalloproteinase-9
MMPs	Magnetic microparticles
MNP	Magnetic nanoparticles
MORB	Mid-ocean ridge basaltic
MOS	Mitochondrial oxidative stress
MOSFETs	Metal-oxide-semiconductor field-effect transistors
MRI	Magnetic resonance imaging
MRS	Magnetic relaxation switches
MT	Methimazole
MWCNTs	Multiwalled carbon nanotubes
nc	Nanocrystalline
NCI	National Cancer Institute
NHS	N-Hydroxysuccinimide
NIEHS	National Institute of Environmental Health Sciences
NIOSH	National Institute for Occupational Safety and Health
NIR	Near-infrared
NIRF	Near-infrared fluorescent
NNI	National nanotechnology initiative
NSF	National Science Foundation
NTP	National Toxicology Program
OA	Over-ageing
OPH	Organophosporus hydrolase
OPs	Organophosphate pesticides
PA	Peak-ageing
PB	Prusian blue
PBS	Phosphate-buffered saline
PE-co-GMA	Poly (ethylene-co-glycidyl methacrylate)
PEG	Polyethylene glycol
PEG-DSPE	Polyethylene-glycol-2000-1,2-distearyl-3-sn-phosphatidyl-ethanolamine
PIV[9]	Particle image velocimetry
POD	Peroxidase
PSA	Prostate specific antigen
PSD	Power spectral density
P-V	Pressure-volume

PVA	Polyvinyl alcohol
QD	Quantum dots
RBM	Radial breathing mode
RC	Resistor-capacitor
Rhodamine-DOPE	Rhodamine-dioleoyl-phosphatidylethanolamine
RT	Room temperature
RTM	Relative tumor mass
RTV	Relative tumor volume
SAD	Selected-area diffraction
SEI	Secondary electron imaging
SEM	Scanning electron microscopy
SF	Singlet fission
SOI	Silicon-on-insulator
SORS	Spatially offset Raman spectroscopy
SPD	Severe plastic deformation
SPE	Screen printed electrode
SPIOs	Superparamagnetic iron oxide nanoparticles
SQUID	Superconducting quantum interference device
ST	Solution treated
STAR	Science to achieve results program
SWNTs	Single-walled carbon nanotubes
TEM	Transmission electron microscopy
THG	Third harmonic generation
TRUS	Transrectal ultrasound
UFG	Ultra-fine grained
VEGF-C	Vascular endothelial growth factor-C

Contents

Chapter 1

Nanotechnology: Introduction

Ernie Hood

INTRODUCTION

Since 1989, when IBM researchers whimsically demonstrated a scientific breakthrough by constructing a 35-atom depiction of the company's logo, the ability to manipulate individual atoms has spawned a tidal wave of research and development at the nanoscale (from the Greek word for "dwarf"). Nanomaterials are defined as having at least one dimension of 100 nanometers or less—about the size of your average virus. Nanotechnology—the creation, manipulation, and application of materials at the nanoscale—involves the ability to engineer, control, and exploit the unique chemical, physical, and electrical properties that emerge from the infinitesimally tiny man-made particles.

Nanoparticles behave like neither solids, liquids, nor gases, and exist in the topsy-turvy world of quantum physics, which governs those denizens small enough to have escaped the laws of Newtonian physics. This allows them to perform their almost magical feats of conductivity, reactivity, and optical sensitivity, among others.

"That's why nanomaterials are useful and interesting and so hot right now," says Kristen Kulinowski, executive director for education and policy at the Rice University Center for Biological and Environmental Nanotechnology (CBEN). "Being in this quantum regime enables new properties to emerge that are not possible or not exhibited by those same chemicals when they're much smaller or much larger. These include different colors, electronic properties, magnetic properties, mechanical properties—depending on the particle, any or all of these can be altered at the nanoscale. That's the power of nanotech."

Many observers not normally given to hyperbole are calling nanotechnology "the next Industrial Revolution." The National Nanotechnology Initiative (NNI), the interagency consortium overseeing the federal government's widespread and well-funded nanotechnology activities, has predicted the field will be worth $1 trillion to the US economy alone by 2015—or sooner. Clearly, nanotechnology is poised to become a major factor in the world's economy and part of our everyday lives in the near future. The science of the very small is going to be very big, very soon.

THE SPRINGBOARD

The first swells presaging the approaching nanotechnology tidal wave have already reached the shore. Engineered nanoparticles are already being produced, sold, and used commercially in products such as sporting goods, tires, and stain-resistant clothing. Engineered nanomaterials designed to provide nontoxic, noncorrosive, and nonflammable neutralization of chemical spills or chemical warfare agents are currently on the market. Even sunscreens have gone nano—some now contain nanoscale titanium dioxide or zinc oxide particles that, unlike their larger, opaque white incarnations, are transparent, while still blocking ultraviolet rays effectively. Fullerenes, which are used in commercial products from semiconductors to coatings for bowling balls, are being produced by the ton at a Mitsubishi plant in Japan.

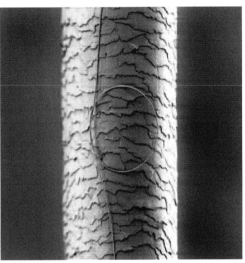

Figure 1. For comparison's sake. A micrograph shows a nanowire curled into a loop in front of a human hair. Nanowires can be as slender as 50 nanometers, about 1000th the width of a hair.

Within a few years, experts say, these initial market forays will seem as quaint as eight-track tapes. According to Mihail Roco, senior advisor on nanotechnology to the National Science Foundation (NSF) and coordinator of the NNI, nanotechnology will have four generations, or phases of development. We were already in the first, consisting of "passive" nanostructures—simple particles designed to perform one task. Roco predicts the second phase will start in 2005, with the appearance of commercial prototypes of "active" nanostructures such as special actuators, drug delivery devices, and new types of transistors and sensors.

As evidence of progress toward this second phase, a team of Northwestern University chemists led by Chad Mirkin recently announced that they have discovered ways to precisely construct nanoscale building blocks that assemble into flat or curved structures. The ability to create unusual nanostructures such as bundles, sheets, and tubes holds promise for new and powerful drug delivery systems, electronic circuits, catalysts, and light-harvesting materials.

By 2010, Roco says, the third generation will arrive, featuring nanosystems with thousands of interacting components. And a few years after that, the first "molecular" nanodevices will appear, devices that will be composed of systems within systems operating much like a cell.

As manufacturing methods are perfected and scaled up, nanotechnology is expected to soon pervade, and often revolutionize, virtually every sector of industrial activity, from electronics to warfare, from medicine to agriculture, from the energy we use to drive our cars and light our homes to the water we drink and the food we eat. Nanotechnology is today's version of the space race, and countries around the globe are enthusiastically pouring billions of dollars into support of research, development, and commercialization.

Figure 2. Small learning curve. Self-assembly of gold polymer nanorods results in a curved structure. The ability to control the size and curvature of nanostructures could aid in applications in drug delivery and electronics.

In terms of the environment and human health, nanotechnology presents the same conundrum as past major technological advances: there may be enormous benefits in terms of benign applications, but there are inherent risks as well. What will happen when nanomaterials and nanoparticles get into our soil, water, and air, as they most assuredly will, whether deliberately or accidentally? What will happen when they inevitably get into our bodies, whether through environmental exposures or targeted applications? The answers to those vital questions remain largely unanswered, although some early findings are less than reassuring, as evidenced by a recent study implicating fullerenes in oxidative stress in the brains of large-mouth bass ("Fullerenes and Fish Brains: Nanomaterials Cause Oxidative Stress," EHP 112:A568 (2004)).

Questions of another sort also need to be answered. Is anyone looking at these health and safety issues? And can enough solid, reliable risk assessment knowledge be gained in time to ensure that the public will—or even that it should—be comfortable with the proliferation of the technology? Will the paradigm shift smoothly into the nano world, or will issues of safety and trust surround nanotechnology with controversy that may hinder its potential, as has happened in the past with such achievements as genetically modified organisms (GMOs)?

Kulinowski expresses the nearly universal sentiments of the field's advocates: "We think nanotechnology has enormous potential to benefit society in a whole variety of sectors and applications, from the next cancer treatment, to environmental applications, to energy—you name it. So we don't want to see that potential limited or eliminated by real or perceived risk factors associated with engineered nanomaterials." To ensure that nanotechnology flourishes responsibly and with strong public support, Kulinowski says, advocates believe it's very important to gather risk data so that questions can be answered and problems addressed early on in the trajectory of the technology development.

Sean Murdock, executive director of the NanoBusiness Alliance, a nanotechnology trade association, thinks it is possible to avoid past mistakes of rolling out a new technology too far ahead of health and safety information. "The risks are there, they're real, but they're manageable," he says. "And on balance, with the right processes in place, we're going to be able to deal with all of those risks, we're going to mitigate those risks, and we're going to realize the upside of the potential."

NANOMEDICINE: A TINY DOSE FOR HEALTH

One of the most promising applications of nanotechnology, known as nanomedicine, involves the development of nanoscale tools and machines designed to monitor health, deliver drugs, cure diseases, and repair damaged tissues, all within the molecular factories of living cells and organelles. The NIH Roadmap for Medical Research—the agency's master plan to accelerate the pace of discovery and speed the application of new knowledge to biomedical prevention strategies, diagnostics, and treatments—contains a significant nanomedicine initiative that will begin with the establishment of 3–4 Nanomedicine Development Centers. These multidisciplinary facilities will serve as the intellectual and technological centerpiece of the endeavor. Funding for the centers of $6 million/year will begin in September 2005.

Today, the initiative's long-term goals sound like scenarios straight out of Isaac Asimov's Fantastic Voyage: nanobots that can search out and destroy cancer cells before they can form tumors … nanomachines that can remove and replace broken parts of cells … molecule-sized implanted pumps that can deliver precisely targeted doses of drugs when and where they're needed … even "smart" nanosensors that can detect pathology or perturbation in any or every cell in the body, and instantly communicate that information to doctors. Science fiction may soon become science fact—these and many other nanomedicine innovations are currently in development, and the NIH predicts that its nanomedicine initiative will start yielding medical benefits in as soon as 10 years. Roco also foresees that fully half of all drug discovery and delivery technology will be based on nanotechnology by 2015.

Experts predict that nanosensors will also provide significantly improved tools to determine both internal and external exposures in real time, assess risk, link exposure to disease etiology, characterize gene-environment interactions, and ultimately improve public health. The National Institute of Environmental Health Sciences

(NIEHS), through its extramural grants and Superfund Basic Research Program, is funding the research and development behind many of these expected innovations.

For example, with a small business innovation research grant, the institute is supporting Platypus Technologies of Madison, Wisconsin, in its work on smart nanosensors designed to act as personal dosimeters for real-time and cumulative exposure to toxic compounds. Combining scaled-down photo optics and nanomaterials to form a uniquely sensitive platform for exposure detection, the initial prototype device is intended to detect even very low exposures to organophosphate pesticides. The sensor, expected to be available commercially within 2 years, is small, lightweight, passive, inexpensive, and easily operated—one immediate application will be monitoring the chemical environments of children.

Platypus CEO Barbara Israel elaborates: "Our product is 'tunable' for different anticipated concentration ranges and monitoring time periods. Therefore, it can be applied to monitor workers for occupational exposure to toxic compounds during manufacturing, as well as to the monitoring of field exposure of agricultural workers." The company is also developing sensors that will immediately respond to the ambient presence of very low concentrations of other toxic agents, and expects that units will be networked by the thousands in security systems at facilities such as airports and train stations, as well as having industrial applications.

"This technology is going to revolutionize how we do business"—the business of environmental health science, that is—according to William Suk, director of the Center for Risk and Integrated Sciences within the NIEHS Division of Extramural Research and Training. Suk oversees many of the institute's extramural grants involving nanotechnology. "One of the real potentials of this technology is to truly be able to understand gene-environment interactions, to be able to take the "omics" revolution and scale it down in such a way that you have a comprehensive global approach to understanding how things fit together," he says. "We're really looking at the use of these technologies in systems biology, to understand how systems communicate—how cells communicate amongst themselves, and within themselves, and with other cell systems within our body. It's all connected."

A wide variety of extraordinarily sophisticated nanobiosensors fitting Suk's vision are well along in development among institute grantees. For example, neurotoxicologist Martin Philbert at the University of Michigan is perfecting a sensor that measures and identifies chemical perturbations within the mitochondria of neurons, and may eventually allow intervention or prevention of such cellular disturbances. Roger Tsien, a professor of pharmacology, biochemistry, and chemistry at the University of California, San Diego, is developing toxicity sensors that can indicate exposures and the perturbations they cause at the genomic level in real time. Kenneth Turtletaub, a scientist at Lawrence Livermore National Laboratory, uses an accelerator mass spectrometer to look at nanostructures for biomarkers of exposure to carcinogenic chemicals, characterizing perturbations at the atomic level. According to Suk, these and other nanodevices will be making major contributions to the field of environmental health within the next 5 years. When nanotechnology achieves its full impact, he says, toxicogenomics will evolve beyond its infancy and begin to fulfill its promise of significant improvements in public health.

SMALL IMPROVEMENTS IN A BIG WORLD

Although research and development of environmental applications is still a relatively narrow area of nanotechnology work, it is growing rapidly, and nanomaterials promise just as dazzling an array of benefits here as they do in other fields. Nanotechnology will be applied to both ends of the environmental spectrum, to clean up existing pollution and to decrease or prevent its generation. It is also expected to contribute to significant leaps forward in the near future in environmental monitoring and environmental health science.

The Science to Achieve Results (STAR) program of the US Environmental Protection Agency (EPA), administered by the agency's National Center for Environmental Research, was an early investor in and promoter of environmental applications of nanotechnology. Beginning in 2001, the agency devoted a small discretionary portion of its grant-making budget to nanotechnology. "We decided to do applications with respect to the environment first," says Barbara Karn, who oversees the nanotechnology aspect of the program. "We wanted to make a case for the new technology being useful for the legacy issues of EPA."

Contaminated soil and groundwater are among the most prominent of those legacy issues, and there has been considerable progress in nanotechnology-based remediation methods. Environmental engineer Wei-xian Zhang of Lehigh University, a STAR grantee who also receives funding from the NSF, has been working since 1996 to develop a remediation method using nanoscale metallic particles, particularly iron nanoparticles, which he has found to be powerful reductants. "If any contaminant can be degraded or transformed by reduction," he says, "you can use the iron nanoparticles." He has been field-testing the method since 2000, both in pilot studies and at several industrial sites contaminated with such toxicants as polychlorinated biphenyls, DDT, and dioxin, and the results have been encouraging.

Zhang's nanoremediation offers several potential advantages over existing methods. The implementation is very simple-the nanoparticles are suspended in a slurry and basically pumped directly into the heart of a contaminated site. By comparison, current methods often involve digging up the soil and treating it. "You can inject (the nanoparticles) in some difficult situations, for example, under a runway, under a building, or other sites where typical engineering methods may not be feasible," says Zhang.

Nanomaterials have a large proportion of surface atoms, and the surface of any material is where reactions happen. Because of nanoparticles' huge surface area and thus very high surface activity, workers can potentially use much less material. The amount of surface area also allows a fast reaction with less time for intermediates to form—a boon in biodegradation, where the intermediate products are sometimes more toxic than the parent compound. Finally, Zhang's method is also much faster. "Because of the higher activity, it takes much less time to achieve remediation goals than conventional technology, which, using biological processes, can take years," he says. With the iron nanoparticles, in most cases the team saw contaminants neutralized into benign compounds in a few days.

Zhang is currently focusing on scaling up production of the iron nanoparticles to make them more cost-competitive, and plans to establish a business based upon his techniques. His is just one of dozens of nanoremediation methods being developed, but is probably the closest to large-scale deployment-he expects that within a year or two, there will be tens to hundreds of projects using the metallic nanoparticle technology. And this type of "passive" application is only the beginning.

"In the future," says Zhang, "we'll have more sophisticated devices that can function not only as a treatment device, but also as a sensor with detection functions and communication capability that you can put into the ground and get feedback on different environmental parameters." That type of device will give remediators the ability to determine when a treatment has been adequately completed, currently a problematic determination to make. Similar nanosensors that will allow real-time in situ detection and analysis of pollutants are being developed for environmental monitoring purposes.

The environmental benefits portended by nanotechnology go farther still. Improvements in membrane technology afforded by nanomaterials, for example, will allow greatly enhanced water filtration, desalination, and treatment of wastewater through finer and "smarter" selective filtration. The technology that is expected to be proliferated is also anticipated to be very simple and very inexpensive. These developments are expected to eventually go a long way toward ameliorating the shortages of clean, plentiful, low-cost drinking water that plague many areas of the world.

Murdock says nanotechnology is also likely to help prevent a great deal of pollution in the future by affording the opportunity to "reinvent the energy infrastructure that powers the economy, which ultimately has been driving many of the issues that environmentalists … have been worrying about over the past few decades." Nanoscale materials and devices could result in game-changing breakthroughs in energy production through advances in hydrogen and solar energy, and could even beget vast improvements in the efficiency and cleanliness of carbon-based energy. There is serious talk, for example, that nanotechnology could make it possible to sustainably expand the use of coal in energy production, using a nanocatalyst that turns coal directly into cleaner-burning diesel fuel and gasoline.

On the other hand, nano-based lighting is already a reality-traffic lights across the country now use tiny light-emitting diode displays that remain in service longer and use less energy than bulbs. The NNI has projected that widespread proliferation of the technology for home and office lighting could cut US energy consumption by as much as 10%, dropping carbon emissions by up to 200 million tons annually.

With their extremely high reactivity, nanomaterials may also enable "green" chemistry and "exact" manufacturing, in which chemicals and other products are manufactured from the bottom up, atom by atom. This development would allow the creation of less-toxic products while reducing or eliminating both hazardous waste and the need for large quantities of toxic raw materials-so-called source reduction. The green chemistry concept applies to the production of nanoparticles themselves. University of Oregon chemist James Hutchinson recently patented a more benign (and faster and cheaper) method for producing gold nanoparticles, which are particularly important in the semiconductor industry.

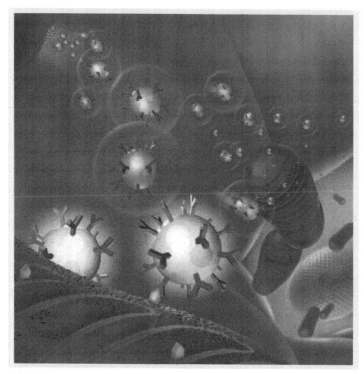

Figure 3. Probing insights. Nanoprobes studded with molecules that bind ions such as zinc, calcium, and potassium are injected into cells to reveal the patterns of ion exchange that make cells function. Computer models are used to interpret the fluorescent signatures probes emit when they capture a target ion.

Karn is excited by this and similar developments: "We really have such an opportunity here with this new technology, to make it without waste, to make the particles in an environmentally friendly way, so that we don't have to worry about the emissions [and] we don't have to worry about the cleanup afterwards."

A YELLOW LIGHT

The same properties that confer such incredible utility to engineered nanoparticles are those that raise concerns about the nature of their interactions with biological systems: their size, their shapes, their high reactivity, how they are coated, and other unique characteristics could prove to be harmful in some physiologic circumstances. Several recent studies have appeared in the literature showing that some nanomaterials are not inherently benign. Some can travel readily through the body, deposit in organ systems, and penetrate individual cells, and could trigger inflammatory responses similar to those seen with ambient nanoparticles-better known in environmental science as ultrafine particles-which are known to often be far more toxic than their larger counterparts. The primary difference between ambient and engineered nanoparticles is that the former have widely varying shapes, sizes, and compositions, whereas the latter are single, uniform compounds.

University of Rochester environmental toxicologist Gunter Oberdšrster has shown in rodent studies, published in June 2004 in Inhalation Toxicology, that inhaled nanoparticles accumulate in the nasal passages, lungs, and brains of rats. And in the January 2004 issue of Toxicological Sciences, National Aeronautics and Space Administration scientist Chiu-Wing Lam recently reported that a suspension of carbon nanotubes (one of the most widely used and researched engineered nanoparticles) placed directly into mouse lungs caused granulomas, unusual lesions that can interfere with oxygen absorption. David Warheit, a DuPont researcher, conducted a similar experiment in rats, reported in the same issue of Toxicological Sciences, and discovered immune cells gathering around clumps of nanotubes in the animals' lungs. At the highest dose, 15% of the rats essentially suffocated due to the clumping of the nanotubes having blocked bronchial passages. Although Lam's and Warheit's studies did not reflect potential real-world exposures, their results were nonetheless troubling, showing at least that nanotubes are biologically active and possibly toxic.

A study published in the July 2004 issue of EHP documenting oxidative stress (a sign of inflammation) in the brains of largemouth bass exposed to aqueous fullerenes has received perhaps the most attention and raised the most warnings of any nanomaterial health implication experiment to date. Eva Oberdšrster (Gunter's daughter), an environmental toxicologist at Southern Methodist University, describes herself as "shocked" at the amount of mainstream national press coverage the study has received. She is quick to stress that although some reports have described "brain damage" or even "severe brain damage" in the fish, she has actually characterized her findings as "significant damage in the brain, which is very different from brain damage." After 48 hr exposure to fairly high doses of fullerenes, the fish probably had the same effect as a very bad headache, she says, but they did survive the exposure. As to the inflammation, Oberdšrster says it could have been an appropriate response to a foreign stressor or a symptom of real physiologic damage. She plans to study this issue further in gene microarray experiments designed to more thoroughly characterize the inflammatory response involved, and to see whether the fish might actually metabolize and excrete the particles.

Oberdšrster has described the findings as "a yellow light, not a red one," and explains further that there are some indications from the inhalation and fish studies that there is a potential for nanoparticles to react with tissues and create inflammation. "So the next step then is to look at it in a broader spectrum before we bring all these products out into the market, to make sure that they are safe so that consumers are protected," she says.

Kulinowski feels that the early studies raise more questions than answers, and she cautions against overinterpreting individual studies. She is optimistic, however, that with technological advances, the potential negative impacts of engineered nanoparticles can be minimized or eliminated altogether. "The good news I see is that with the control we have over engineered nanoparticles, we may be able to engineer them to confer the benefits, but not the risks, not the hazards." Again, it's all about the surface, she says: "If we can control surface properties of nanoparticles, we may be able to tune out the toxicity.... It's like sliding a dimmer switch on a lamp-you can just tune it right down to pretty much beyond our capacity to measure it."

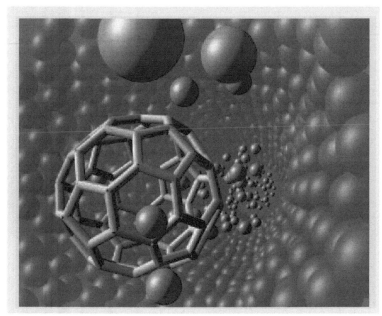

Figure 4. Shapes and sizes. A visualization of a nanohydraulic piston consists of common nanotechnology components including a carbon nanotube, helium atoms, and a "buckyball" molecule.

BIG ISSUES FOR NANOTECHNOLOGY

Nanotechnology accomplishments could soon affect every person on the planet. But the opinion is virtually unanimous, among advocates and skeptics alike, that the full realization of nanotechnology's potential benefits is threatened by ongoing concerns about the potentially negative effects nanomaterials could have on human health and the environment. Applications are being vigorously pursued. The question is, will knowledge of the implications keep pace?

In August 2002, the Action Group on Erosion, Technology, and Concentration (ETC Group), a Canadian environmental activist group that played a key role in the battle against acceptance of GMOs in the 1990s, called for a worldwide moratorium on research and commercialization of engineered nanomaterials until there are protocols in place to protect workers, including lab workers. They cited a dearth of research data about the potential negative implications of nanomaterials, and the lack of specific regulatory oversight or established best practices in the handling of nanoparticles in either the laboratory or the manufacturing setting.

Perhaps coincidentally, in the 2 years since the ETC Group's moratorium call there has been a palpable upsurge in research and bureaucratic activity regarding those missing pieces of the puzzle. All of the stakeholders, casting a weather eye at the GMO experience, apparently agree that the bountiful benefits of nanotechnology cannot be harvested without full and transparent characterization of the risks they could pose to human health and the environment.

"We're much better off over the long haul if we make sure that we address concerns and issues proactively," says Murdock. "That doesn't mean we should be hypersensitive and shy away from exploring new areas, because fundamentally we only make progress through exploration. But it needs to be balanced and tempered with a continuous examination of the implications."

Roco, who has been instrumental in the NNI's ongoing attention to both safety implications and the potential societal impacts of nanotechnology worldwide, agrees that the time for responsible risk assessment is now: "This is no longer something you do after the fact, after you do the other research, but has to be done from the beginning, to be an integral part of the research. You have to look at the whole cycle of activity, not only at the first phase when you create something."

The CBEN has been investigating the environmental fate of nanoparticles since its inception in 2001 as one of six Nanoscale Science and Engineering Centers established by the NSF, and Kulinowski has noted the recent explosion in interest and funding in nanotech environmental health and safety research. "We have seen tremendous movement on this issue over the last year and a half, from the point where we almost felt like we were calling out into the darkness, to where people are now moving forward independently," she says. "Most encouraging has been the federal government's response. We've also seen a tremendous response from industry … that gives us hope that as we move toward commercialization of nanotechnology products, these questions will be addressed early on in the development, before or when products come to market."

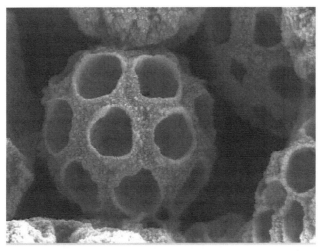

Figure 5. Tiny beach umbrellas. A titanium dioxide microsphere (approximately 1–50 microns in diameter) with closed-packed spherical inclusions functions in sunscreens as a small "photonic crystallite," scattering light very effectively.

Critics and even some participants maintain that funding of implications research is still inadequate in proportion to the nearly $1 billion the government is currently investing in nanotechnology development. But efforts to better understand the implications of nanotechnology are clearly gaining momentum. Important new research initiatives are getting under way, and coordination and collaboration are increasing among the federal regulatory agencies and research organizations involved.

The NNI is the central locus, with a number of agencies participating. Representatives from several of those agencies, including the NIEHS, the EPA, and the National Institute for Occupational Safety and Health (NIOSH), have formed a working group on the environmental and health impacts of nanotechnology. The group meets monthly to share knowledge, coordinate activities, identify research gaps and goals, and address urgent issues such as regulation and nomenclature.

Two further major research initiatives designed to establish fundamental knowledge about the toxicologic properties of engineered nanomaterials are in their initial stages. Both will contribute significantly to the knowledge base and allow more rational risk assessment in the future.

The first of these major initiatives arose from the CBEN's 2003 nomination of nanomaterials for study by the National Toxicology Program (NTP). The NTP, which is headquartered at the NIEHS, has embarked upon a research program involving safety studies of representative manufactured nanomaterials. "The aim of our program," says Nigel Walker, lead scientist of the investigation, "is actually to help guide the nanomaterial industry in identifying the key parameters that lead to biocompatibility of nanomaterials, versus toxicity of nanomaterials, so that we can avoid having problems such as the genetically modified food situation, where the industry and the technology got ahead of the biocompatibility issues."

The NTP program will focus initially on studies of single-walled carbon nanotubes, titanium dioxide, quantum dots (fluorescent semiconductor nanocrystals used in imaging equipment), and fullerenes. Because the most likely route of exposure to those nanomaterials as they are used today is through the skin, several studies will concentrate on dermal toxicity. Other exposure routes will be examined as well, however, all looking at general, acute, subchronic, and chronic levels of exposure.

One of the broad goals of the NTP initiative is to create models of nanomaterial chemical, physical, and pharmacokinetic properties that can be used to help evaluate new engineered nanostructures as they come along. According to John Bucher, deputy director of the NIEHS Environmental Toxicology Program, the purpose of this initiative is not to prevent or understand the toxicity of every material that can be manufactured under the "nano" rubric. Instead, he says, "What we're trying to do is understand some of the fundamental properties of nanomaterials-how they move, what kind of toxicities they have, what kinds of organ systems are generally targeted, what the effects of surface coatings are... . We're not trying to make the world safe from nanotechnology, nor do we believe that the world is necessarily at great risk from nanomaterials at this moment, or potentially even in the future. But the total absence of any information makes this an area that we just have to pursue."

The NTP, in association with the University of Florida, is also planning a workshop for November 2004 designed to bring together scientists from the toxicology community, environmental engineers, and representatives of the pharmaceutical and chemical industries. The workshop will focus on questions about how best to assess exposure to nanomaterials and evaluate their toxicity and safety.

Walker thinks these efforts are timed perfectly. "If we'd tried to do this two or three years ago, we may actually have been targeting things that weren't important," he says. "You don't want to be too early on the curve, but then you don't want to be too late. This is about the right time ... and we are being very open about how things are moving along, because the NTP is completely open, and all the data is ultimately the public's."

The second major initiative-research on occupational health risks associated with manufacturing and using nanomaterials-is being spearheaded by NIOSH. The institute recently organized a Nanotechnology Research Center to coordinate, track, and measure outcomes, and disseminate the output of nanotechnology-related activities throughout the institute.

The NIOSH has also undertaken a 5-year multidisciplinary initiative known as the NIOSH Nanotechnology and Health and Safety Research Program. As with the NTP's efforts, the idea is to characterize risks early in the industry's development, and the workplace is the most likely location of exposures at present. "There is some concern that these materials are of unknown effect, and there is interest in getting generalized industrial hygiene, generalized control measures, and best work practices involved early on," says researcher Vincent Castranova, who is principal investigator for the program's coordinating project as well as a separate study exploring particle surface area as a dose metric. "Normally, interest in these elements has come after proof of disease outcome. This is one case where the concerns are sufficient to cause the industry and the governmental agencies to try to get good work practices and prevention measures up front, before we know full health outcomes."

Another NIOSH scientist, Andrew Maynard, is investigating methods of characterizing and monitoring airborne nanoparticles. "Part of my project," says Maynard, "is developing and using the characterization techniques, so that we can understand very precisely the chemical and physical nature of the particles, and also the concentration of the particles being used in these experiments. Also, we will look at how we can effectively monitor exposures in the workplace, so that we can have simple, robust, inexpensive techniques that people can use in the workplace."

Although dermal exposure to nanomaterials is occurring as a result of their use in sunscreens and some cosmetics, inhalation is suspected to be the most likely route of exposure in the workplace, so other projects in the program will focus on pulmonary toxicity questions, particularly with respect to carbon nanotubes. Those questions will be tricky, again due to the unique attributes of nanomaterials, they are technically ultrafine particles, but can they be judged the same way?

"This is one of the big areas of debate at the moment," says Maynard. "To what extent do you treat engineered nanoparticles as just another ultrafine particle? It's fair to say that most of our concerns over nanomaterials are being driven by our experiences with ultrafine particles, which are substantially more inflammatory and toxic than fine particles."

"Another issue that is unresolved is that these nanoparticles tend to aggregate, and the aggregates often tend not to be under a hundred nanometers in diameter," adds Castranova. "So do they then behave as fine particles rather than ultrafine? That depends on whether they disaggregate either in handling or once they're in the lung, which is unknown. Their ability to enter the lungs, to cross the air-blood barrier, or to cause inflammation would be affected by (disaggregating)." The NIOSH is helping to organize the First International Symposium on Occupational Health Implications of Nanomaterials, which will convene in the UK in October to discuss these issues.

THE RIPPLE EFFECT

The NTP and NIOSH initiatives are the major new programs in the works, but a great deal of activity is continuing or beginning in other circles as well. The NNI is expanding its support of implications research, and recently held a landmark international meeting that brought together the leaders of nanotechnology programs in 25 countries and the European Union. The International Dialog on Responsible Research and Development of Nanotechnology took place June 17–18, 2004 in Arlington, Virginia, and was designed to help develop a global vision of how the technology can be fostered with the appropriate attention to and respect for concerns about the societal issues and environmental, health, and safety implications.

Roco, who called the meeting "a historic event," proposed the establishment of an ongoing international organization dedicated to responsible nanotechnology development. Participants agreed to form a "preparatory group" charged with exploring possible actions, mechanisms, timing, institutional frameworks, and principles involved in constructing a permanent institution designed to ensure international dialogue, cooperation, and coordination in nanotechnology research and development.

The ETC Group executive director Pat Mooney reacted favorably to the gathering as well: "That's the first time we've had an international meeting like this, and I think it's a very encouraging sign."

Encouraging signs of commitment and progress were also evident at two other landmark events held earlier this year. In March, the NIEHS held a workshop called Technologies for Improved Risk Stratification and Disease Prevention that brought together a panel of experts to formulate specific recommendations on how the institute should incorporate nanotechnology into its research agenda in the coming years. Participants embraced the idea that the NIEHS should lead the way in developing a single small-scale platform to detect individual chemical exposures, eliminate toxicants from the system, and intervene to reverse any harmful effects that might have been initiated by the exposure. Then, in May, a one-day public discussion was held by the Institute of Medicine's Roundtable on Environmental Health Sciences, Research, and Medicine, in which experts and members of the public explored the issues raised by nanotechnology from a public health perspective. The discussion illuminated potential public health benefits while acknowledging recent toxicological concerns. Events such as these serve to inform the scientific community and the public alike, encouraging the responsible development of the technology.

Recognizing the enormous opportunities at hand, the chemical industry is also placing a high priority on nanotechnology implications research. A consortium called the Chemical Industry Vision 2020 Technology Partnership, in cooperation with the NNI and the US Department of Energy Office of Energy Efficiency and Renewable Energy, released a comprehensive white paper in 2003 titled "Chemical Industry R&D Roadmap for Nanomaterials by Design: From Fundamentals to Function." This document calls for an unprecedented level of cooperation and collaboration among US chemical companies to foster the long-term success of the nanochemical industry, and stresses that environment, safety, and health knowledge will be an essential component. "The anticipated growth in nanoparticle utilization warrants parallel efforts in hazard identification, exposure evaluation, and risk assessment," the chapter states. "Chemical companies are prepared to serve a major role in this process as leaders in characterizing materials, identifying their potential risks, and providing guidelines for their safe and effective utilization."

The EPA's STAR program is planning to award new grants soon in nanotechnology implications research, and the CBEN is continuing its work on what it calls "the wet-dry interface"-the interactions between engineered nanomaterials and systems that are active in aqueous or water-based environments, including ecosystems and living beings. "We have several research projects we would characterize as implications research," says Kulinowski, "looking at what happens when nanomaterials get into the soil or into a water supply." By understanding how nanoparticles (which are typically not soluble in water, hence the "dry" side) interact with aqueous environments (the "wet" side), the researchers hope to create technologies that will improve human health and the environment, such as biocompatible nanoparticles or nanostructured catalysts that will break down organic pollutants. The wet-dry interface also plays a major role in determining the environmental fate and transport of nanomaterials.

Regulatory agencies such as the EPA, the FDA, and the Occupational Safety and Health Administration are all participating in the NNI, following the progress of the research carefully, and building their own knowledge bases with an eye toward the eventual development and implementation of nano-specific regulatory frameworks within their purviews. At present the consensus seems to be that existing regulations are sufficiently robust to appropriately address concerns related to nanomaterials, but as risks and hazards are characterized in more detail, that stance could change.

Even the ETC Group, although it has not rescinded its call for a moratorium, seems encouraged by recent progress. "We do feel like we have had a reasonable response-as reasonable as to be expected-from the governments," says Mooney, "and that there is work under way to try to correct the problems that we have identified." Mooney says that as individual nations put nano-specific laboratory protocols into place, his group will no longer call for a moratorium in those countries.

It appears that all of this research activity is reaching critical mass at just the right time. The nanotechnology bullet train has left the station with the power to take us to some magical places we've barely even dreamed of. Although public distrust of the technology could potentially derail the train, many passengers are hoping that increased understanding of both its potential benefits and dangers will keep it on track and allow the journey toward discovery to continue.

KEYWORDS

- **Biological and environmental nanotechnology**
- **Environmental protection agency**
- **Genetically modified organisms**
- **National toxicology program**

Chapter 2

Cancer Nanotechnology

US Department of Health and Human Services

NANOTECHNOLOGY AND CANCER

To help meet the goal of eliminating death and suffering from cancer by 2015, the National Cancer Institute (NCI) is engaged in efforts to harness the power of nanotechnology to radically change the way we diagnose, image, and treat cancer. Already, NCI programs have supported research on novel nanodevices capable of one or more clinically important functions, including detecting cancer at its earliest stages, pinpointing its location within the body, delivering anticancer drugs specifically to malignant cells, and determining if these drugs are killing malignant cells. As these nanodevices are evaluated in clinical trials, researchers envision that nanotechnology will serve as multifunctional tools that will not only be used with any number of diagnostic and therapeutic agents, but will change the very foundations of cancer diagnosis, treatment, and prevention.

The advent of nanotechnology in cancer research could not have come at a more opportune time. The vast knowledge of cancer genomics and proteomics emerging as a result of the Human Genome Project is providing critically important details of how cancer develops, which, in turn, creates new opportunities to attack the molecular underpinnings of cancer. However, scientists lack the technological innovations to turn promising molecular discoveries into benefits for cancer patients. It is here that nanotechnology can play a pivotal role, providing the technological power and tools that will enable those developing new diagnostics, therapeutics, and preventives to keep pace with today's explosion in knowledge.

To harness the potential of nanotechnology in cancer, NCI is seeking broad scientific input to provide direction to research and engineering applications. In doing so, NCI will develop a Cancer Nanotechnology Plan. Drafted with input from experts in both cancer research and nanotechnology, the Plan will guide NCI in supporting the interdisciplinary efforts needed to turn the promise of nanotechnology and the postgenomics revolution in knowledge into dramatic gains in our ability to diagnose, treat, and prevent cancer. Though this quest is near its beginning, the following pages highlight some of the significant advances that have already occurred from bridging the interface between modern molecular biology and nanotechnology.

DEVELOPING A CANCER NANOTECHNOLOGY PLAN

The NCI's Cancer Nanotechnology Plan will provide critical support for the field though extramural projects, intramural programs, and a new Nanotechnology Standardization Laboratory. This latter facility will develop important standards for nanotechnological constructs and devices that will enable researchers to develop cross-functional platforms that will serve multiple purposes. The laboratory will be a centralized characterization laboratory capable of generating technical data that will assist researchers in choosing which of the many promising nanoscale devices they might want to use for a particular clinical or research application. In addition, this new laboratory will facilitate the development of data to support regulatory sciences for the translation of nanotechnology into clinical applications.

The six major challenge areas of emphasis include:

Prevention and Control of Cancer
- Developing nanoscale devices that can deliver cancer prevention agents.
- Designing multicomponent anticancer vaccines using nanoscale delivery vehicles.

Early Detection and Proteomics
- Creating implantable, biofouling-indifferent molecular sensors that can detect cancer-associated biomarkers that can be collected for *ex vivo* analysis or analyzed *in situ*, with the results being transmitted via wireless technology to the physician.
- Developing "smart" collection platforms for simultaneous mass spectroscopic analysis of multiple cancer-associated markers.

Imaging Diagnostics
- Designing "smart" injectable, targeted contrast agents that improve the resolution of cancer to the single cell level.
- Engineering nanoscale devices capable of addressing the biological and evolutionary diversity of the multiple cancer cells that make up a tumor within an individual.

Multifunctional Therapeutics
- Developing nanoscale devices that integrate diagnostic and therapeutic functions.
- Creating "smart" therapeutic devices that can control the spatial and temporal release of therapeutic agents while monitoring the effectiveness of these agents.

Quality of Life Enhancement in Cancer Care
- Designing nanoscale devices that can optimally deliver medications for treating conditions that may arise over time with chronic anticancer therapy, including pain, nausea, loss of appetite, depression, and difficulty breathing.

Interdisciplinary Training
- Coordinating efforts to provide cross-training in molecular and systems biology to nanotechnology engineers and in nanotechnology to cancer researchers.
- Creating new interdisciplinary coursework/degree programs to train a new generation of researchers skilled in both cancer biology and nanotechnology.

WHAT IS NANOTECHNOLOGY?

Nanotechnology refers to the interactions of cellular and molecular components and engineered materials—typically clusters of atoms, molecules, and molecular fragments—at the most elemental level of biology. Such nanoscale objects—typically, though not exclusively, with dimensions smaller than 100 nanometers—can be useful by themselves or as part of larger devices containing multiple nanoscale objects. At the nanoscale, the physical, chemical, and biological properties of materials differ fundamentally and often unexpectedly from those of the corresponding bulk material because the quantum mechanical properties of atomic interactions are influenced by material variations on the nanometer scale. In fact, by creating nanometer-scale structures, it is possible to control fundamental characteristics of a material, including its melting point, magnetic properties, and even color, without changing the material's chemical composition.

Nanoscale devices and nanoscale components of larger devices are of the same size as biological entities. They are smaller than human cells (10,000–20,000 nanometers in diameter) and organelles and similar in size to large biological macromolecules such as enzymes and receptors—hemoglobin, for example, is approximately 5 nm in diameter, while the lipid bilayer surrounding cells is on the order of 6 nm thick. Nanoscale devices smaller than 50 nanometers can easily enter most cells, while those smaller than 20 nanometers can transit out of blood vessels. As a result, nanoscale devices can readily interact with biomolecules on both the cell surface and within the cell, often in ways that do not alter the behavior and biochemical properties of those molecules. From a scientific viewpoint, the actual construction and characterization of nanoscale devices may contribute to understanding carcinogenesis.

Noninvasive access to the interior of a living cell affords the opportunity for unprecedented gains on both clinical and basic research frontiers. The ability to simultaneously interact with multiple critical proteins and nucleic acids at the molecular scale should provide better understanding of the complex regulatory and signaling networks that govern the behavior of cells in their normal state and as they undergo malignant transformation. Nanotechnology provides a platform for integrating efforts in proteomics with other scientific investigations into the molecular nature of cancer by giving researchers the opportunity to simultaneously measure gene and protein expression, recognize specific protein structures and structural domains, and follow protein transport among different cellular compartments. Similarly, nanoscale devices are already proving that they can deliver therapeutic agents that can act where they are likely to be most effective, that is, within the cell or even within specific organelles. Yet despite their small size, nanoscale devices can also hold tens of thousands of small molecules, such as a contrast agent or a multicomponent diagnostic system capable of assaying a cell's metabolic state, creating the opportunity for unmatched sensitivity in detecting cancer in its earliest stages. For example, current approaches may link a monoclonal antibody to a single molecule of magnetic resonance imaging (MRI) contrast agent, requiring that many hundreds or thousands of this construct reach and bind to a targeted cancer cell in order to create a strong enough signal to be detected via MRI. Now imagine the same cancer-homing monoclonal antibody attached to a nanoparticle that contains tens of thousands of the same contrast agent—if even one such construct reaches and binds to a cancer cell, it would be detectable.

NANOTECHNOLOGY AND DIAGNOSTICS

Today, cancer-related nanotechnology research is proceeding on two main fronts: laboratory-based diagnostics and *in vivo* diagnostics and therapeutics. Nanoscale devices designed for laboratory use rely on many of the methods developed to construct computer chips. For example, 1–2 nanometer-wide wires built on a micron-scale silicon grid can be coated with monoclonal antibodies directed against various tumor markers. With minimal sample preparation, substrate binding to even a small number of antibodies produces a measurable change in the device's conductivity, leading to a 100-fold increase in sensitivity over current diagnostic techniques.

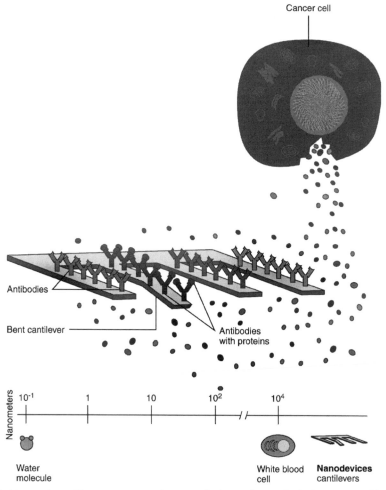

Figure 1. Nanoscale cantilevers, constructed as part of a larger diagnostic device, can provide rapid and sensitive detection of cancer-related molecules.

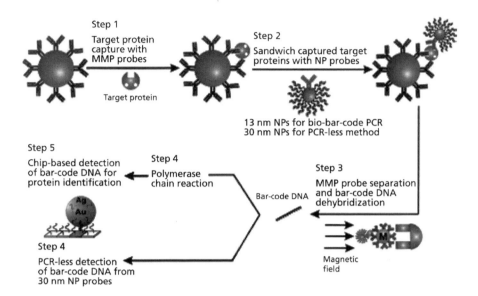

Figure 2. The DNA-coated gold nanoparticles (NPs) form the basis of a system that also uses larger magnetic microparticles (MMPs) to detect attomolar (10–18) concentrations of serum proteins. In this case, a monoclonal antibody to prostate specific antigen (PSA) is attached to the MMP, creating a reagent to capture free PSA. A second antibody to PSA, attached to the NPs, is then added, creating a "sandwich" of the captured protein and two particles that is easily separated using a magnetic field.

Nanoscale cantilevers, microscopic, flexible beams resembling a row of diving boards, are built using semiconductor lithographic techniques. These can be coated with molecules capable of binding specific substrates—DNA complementary to a specific gene sequence, for example. Such micron-sized devices, comprising many nanometer-sized cantilevers, can detect single molecules of DNA or protein.

Researchers have also been developing a wide variety of nanoscale particles to serve as diagnostic platform devices. For example, DNA-labeled magnetic nanobeads have the potential to serve as a versatile foundation for detecting virtually any protein or nucleic acid with far more sensitivity than is possible with conventional methods now in use. If this proves to be a general property of such systems, nanoparticle-based diagnostics could provide the means of turning even the rarest biomarkers into useful diagnostic or prognostic indicators.

Quantum dots, nanoscale crystals of a semiconductor material such as cadmium selenide, are another promising nanoscale tool for laboratory diagnostics. A product of the quest to develop new methods for harvesting solar energy, these coated nanoscale semiconductor crystals act as molecular light sources whose color depends solely on particle size. When linked to an antibody or other molecule capable of binding to a substance of interest, quantum dots act like a beacon that lights up when binding occurs.

Figure 3. The color of a quantum dot depends on its size. These quantum dots emit across the entire visible spectrum even though all are whose color depends solely irradiated with white light.

Because of the multitude of colors with which they can emit light, quantum dots can be combined to create assays capable of detecting multiple substances simultaneously. In one demonstration, researchers were able to simultaneously measure levels of the breast cancer marker Her-2, actin, microfibril proteins, and nuclear antigens.

NANOTECHNOLOGY AND CANCER THERAPY

Nanoscale devices have the potential to radically change cancer therapy for the better and to dramatically increase the number of highly effective therapeutic agents. Nanoscale constructs, for example, should serve as customizable, targeted drug delivery vehicles capable of ferrying large doses of chemotherapeutic agents or therapeutic genes into malignant cells while sparing healthy cells, which would greatly reduce or eliminate the often unpalatable side effects that accompany many current cancer therapies. Already, research has shown that nanoscale delivery devices, such as dendrimers (spherical, branched polymers), silica-coated micelles, ceramic nanoparticles, and cross-linked liposomes, can be targeted to cancer cells. This is done by attaching monoclonal antibodies or cell-surface receptor ligands that bind specifically to molecules found on the surfaces of cancer cells, such as the high-affinity folate receptor and luteinizing hormone releasing hormone (LH-RH), or molecules unique to endothelial cells that become co-opted by malignant cells, such as the integrin avb3. Once they reach their target, the nanoparticles are rapidly taken into cells. As efforts in proteomics and genomics uncover other molecules unique to cancer cells, targeted nanoparticles could become the method of choice for delivering anticancer drugs directly to tumor cells and their supporting endothelial cells. Eventually, it should be possible to mix and match anticancer drugs with any one of a number of nanotechnology-based delivery vehicles and targeting agents, giving researchers the opportunity to fine-tune therapeutic properties without needing to discover new bioactive molecules.

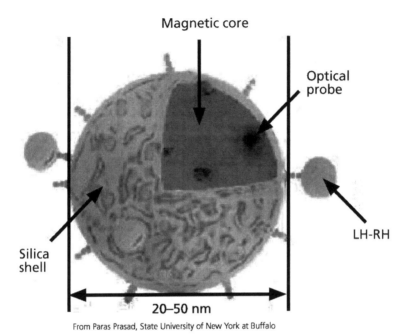

From Paras Prasad, State University of New York at Buffalo

Figure 4. Multifunctional nanoparticles can be targeted to cancer cells using receptor ligands.

On an equally unconventional front, efforts are focused on constructing robust "smart" nanostructures that will eventually be capable of detecting malignant cells *in vivo*, pinpointing their location in the body, killing the cells, and reporting back that their payload has done its job. The operative principles driving these current efforts are modularity and multifunctionality, that is, creating functional building blocks that can be snapped together and modified to meet the particular demands of a given clinical situation. A good example from the biological world is a virus capsule, made from a limited set of proteins, each with a specific chemical functionality, that comes together to create a multifunctional nanodelivery vehicle for genetic material.

In fact, at least one research group is using the empty RNA virus capsules from cowpea mosaic virus and flockhouse virus as potential nanodevices. The premise is that 60 copies of coat protein that assemble into a functional virus capsule offer a wide range of chemical functionality that could be put to use to attach homing molecules—such as monoclonal antibodies or cancer cell-specific receptor antagonists, and reporter molecules—such as MRI contrast agents, to the capsule surface, and to load therapeutic agents inside the capsule.

While such work with naturally existing nanostructures is promising, chemists and engineers have already made substantial progress turning synthetic materials into multifunctional nanodevices. Dendrimers, 1- to 10-nanometer spherical polymers of uniform molecular weight made from branched monomers, are proving particularly adept at providing multifunctional modularity. In one elegant demonstration, investigators attached folate—which targets the high-affinity folate receptor found on some malignant cells, the indicator fluorescein, and either of the anticancer drugs methotrexate or paclitaxel to a single dendrimer. Both *in vitro* and *in vivo* experiments showed that this nanodevice delivered its therapeutic payload specifically to folate receptor-positive cells while simultaneously labeling these cells for fluorescent detection. Subsequent work, in which a fluorescent indicator of cell death was linked to the dendrimer, provided evidence that the therapeutic compound was not only delivered to its target cell but also produced the desired effect. Already, some dendrimer-based constructs are making their way toward clinical trials for treating a variety of cancers.

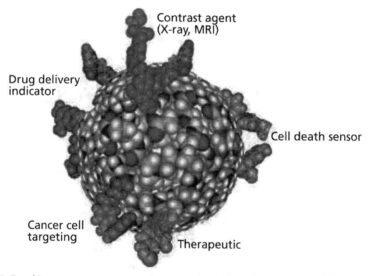

Figure 5. Dendrimers can serve as versatile nanoscale platforms for creating multifunctional devices capable of detecting cancer and delivery drugs.

Such multifunctional nanodevices, sometimes referred to as nanoclinics, may also enable new types of therapeutic approaches or broader application of existing approaches to killing malignant cells. For example, silica-coated lipid micelles containing LH-RH as a targeting agent have been used to deliver iron oxide particles to LH-RH receptor-positive cancer cells.

Once these so-called nanoclinics have been taken up by the target cell, they can not only be imaged using MRI, but can also be turned into molecular-scale thermal scalpels: applying a rapidly oscillating magnetic field causes the entrapped Fe_2O_3 molecules to become hot enough to kill the cell. The critical factor operating here is that nanoparticles can entrap 10,000 or more Fe_2O_3 molecules, providing both enhanced sensitivity for detection and enough thermal mass to destroy the cell.

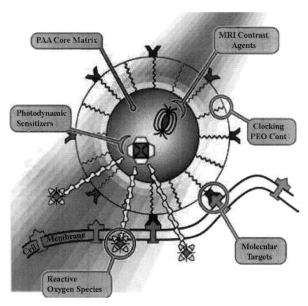

Figure 6. "Smart" dynamic nanoplatforms have the potential to change the way cancer is diagnosed, treated, and prevented. The outside of such "nanoclinics" could be decorated with a tumor-homing monoclonal antibody and coated with polyethylene glycol (PEG) to shield the device from immune system detection. The polymer matrix of such particles could be loaded with contrast agents, which would provide enhanced sensitivity for pinpointing tumor location within the body, and various types of therapeutic agents, such as reactive oxygen-generating photodynamic sensitizers that would be activated once the particle detected a malignant cell.

Photosensitizers used in photodynamic therapy, in which light is used to generate reactive oxygen locally within tumors, have also been entrapped in targeted nanodevices. The next step in this work is to also entrap a light-generating system, such as the luciferin-luciferase pair, in such a way as to trigger light production only after the nanoparticles have been taken up by a targeted cell. If successful, such an approach would greatly extend the usefulness of photodynamic therapy to include treatment of tumors deep within the body.

Such multifunctional nanodevices hold out the possibility of radically changing the practice of oncology, perhaps providing the means to survey the body for the first signs of cancer and deliver effective therapeutics during the earliest stages of the disease. Certainly, researchers envision a day when a smart nanodevice will be able to fingerprint a particular cancer and dispense the correct drug at the proper time in a malignant cell's life cycle, making individualized medicine a reality at the cellular level.

With the focus on modularity and multifunctionality, one goal is to create and characterize platform technologies that can be mixed and matched with new targeting agents that will come from large-scale proteomics programs already in action and therapeutics both old and new. Accomplishing this goal, however, will require that engineers and biologists work hand in hand to combine the best of both of their worlds in the fight against cancer.

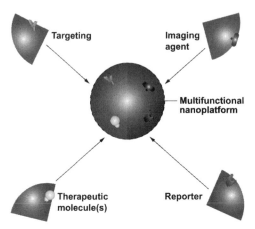

Figure 7. An important aspect of biomedical nanotechnology research is that most systems are being designed as general platforms that can be used to create a diverse set of multifunctional diagnostic and therapeutic devices.

A POWERFUL RESEARCH ENABLER

Nanotechnology is providing a critical bridge between the physical sciences and engineering, on the one hand, and modern molecular biology on the other. Materials scientists, for example, are learning the principles of the nanoscale world by studying the behavior of biomolecules and biomolecular assemblies. In return, engineers are creating a host of nanoscale tools that are required to develop the systems biology models of malignancy needed to better diagnose, treat, and ultimately prevent cancer. In particular, biomedical nanotechnology is benefiting from the combined efforts of scientists from a wide range of disciplines, in both the physical and biological sciences, who together are producing many different types and sizes of nanoscale devices, each with its own useful characteristics.

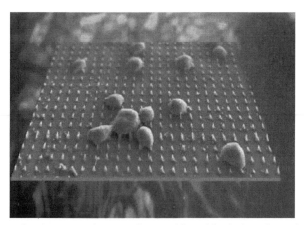

Figure 8. An array of carbon nanotubes provides an addressable platform for probing intact, living cells.

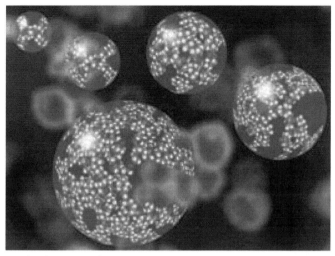

Figure 9. Nanotechnology research is generating a variety of constructs, giving cancer researchers great flexibility in their efforts to change the paradigm of cancer diagnosis, treatment, and prevention. Shown here are two such structures. On the left are highly stable nanoparticles containing a cross-linked hydrophilic shell and a hydrophobic interior. On the right are spherical dendrimers, which are of rigorously defined size based on the number of monomer layers used to create a particular dendrimer. Like most of the other nanoparticles being developed, these are easily manipulated, affording researchers the opportunity to add a variety of molecules to the surfaces and interiors of the nanoparticles.

NATIONAL NANOTECHNOLOGY CHARACTERIZATION LABORATORY FOR CANCER RESEARCH

As part of its cancer nanotechnology program, the NCI is establishing a national resource laboratory at its Frederick facility that will provide critical infrastructure support to this rapidly developing field. The National Nanotechnology Characterization Laboratory (NCL) will fill a major resource gap in biomedical nanotechnology by providing nanodevice assessment and standardization capabilities that many experts have identified as a critical requirement to rapidly integrating nanotechnology into the clinical realm.

In creating the NCL, NCI will provide a stable and standardized environment for intramural and extramural researchers to bring their nanodevices and nanomaterials for assessment and development. The laboratory will also serve as an incubator in which private and public sector research efforts could be coordinated, yielding new partnerships that would accelerate the translation of basic research into clinical advances.

The basic functions carried out by the NCL will include:

- The GMP synthesis of sizable quantities of a variety of nanoparticles and nanodevices;
- Characterization of nanoparticles and devices;
- Functionalization of nanoparticles;

- Development of tools and methods for characterizing both native and functionalized nanoparticles;
- Creation of reference standards and release specifications; and
- Facilitation of testing and analysis protocol development that will speed the regulatory review of novel diagnostics, therapeutics, and prevention strategies that use nanoscale devices.

In addition, the laboratory will have the capability of performing preclinical toxicology, pharmacology, and efficacy testing of nanodevices created both by NCI intramural and extramural efforts as well as by the private sector.

KEYWORDS

- **Cancer Nanotechnology Plan**
- **Luteinizing hormone releasing hormone**
- **Magnetic resonance imaging**
- **Nanodevices**
- **Nanoscale**

Chapter 3

Recent Advances Nanotechnology Applied to Biosensors

Xueqing Zhang, Qin Guo, and Daxiang Cui

INTRODUCTION

In recent years there has been great progress the application of nanomaterials in biosensors. The importance of these to the fundamental development of biosensors has been recognized. In particular, nanomaterials such as gold nanoparticles (GNPs), carbon nanotubes (CNTs), magnetic nanoparticles (MNPs), and quantum dots (QDs) have been being actively investigated for their applications in biosensors, which have become a new interdisciplinary frontier between biological detection and material science. Here we review some of the main advances in this field over the past few years, explore the application prospects, and discuss the issues, approaches, and challenges, with the aim of stimulating a broader interest in developing nanomaterial-based biosensors and improving their applications in disease diagnosis and food safety examination.

A biosensor is a device incorporating a biological sensing element either intimately connected to or integrated within a transducer. Specific molecular recognition is a fundamental prerequisite, based on affinity between complementary structures such as enzyme-substrate, antibody-antigen and receptor-hormone, and this property in biosensor is used for the production of concentration–proportional signals. Biosensor's selectivity and specificity highly depend on biological recognition systems connected to a suitable transducer [1-3].

In recent years, with the development of nanotechnology, a lot of novel nanomaterials are being fabricated, their novel properties are being gradually discovered, and the applications of nanomaterials in biosensors have also advanced greatly. For example, nanomaterials-based biosensors, which represent the integration of material science, molecular engineering, chemistry, and biotechnology, can markedly improve the sensitivity and specificity of biomolecule detection, hold the capability of detecting or manipulating atoms and molecules, and have great potential in applications such as biomolecular recognition, pathogenic diagnosis, and environment monitoring [4-6].

Here we review some of the main advances in this field over the past few years, explore the application prospects, and discuss the issues, approaches, and challenges, with the aim of stimulating a broader interest in developing nanomaterials-based biosensor technology.

THE USE OF NANOMATERIALS IN BIOSENSORS

To date, modern materials science has reached a high degree of sophistication. As a result of continuous progress in synthesizing and controlling materials on the submicron and nanometer scales, novel advanced functional materials with tailored properties can be created. When scaled down to a nanoscale, most materials exhibit novel properties that cannot be extrapolated from their bulk behavior. The interdisciplinary boundary between materials science and biology has become a fertile ground for new scientific and technological development. For the fabrication of an efficient biosensor, the selection of substrate for dispersing the sensing material decides the sensor performance. Various kinds of nanomaterials, such as GNPs [7], CNTs [8], MNPs [9], and QDs [10], are being gradually applied to biosensors because of their unique physical, chemical, mechanical, magnetic, and optical properties, and markedly enhance the sensitivity, and specificity of detection.

The Use of Gold Nanoparticles in Biosensors

The GNPs show a strong absorption band in the visible region due to the collective oscillations of metal conduction band electrons in strong resonance with visible frequencies of light, which is called surface plasmon resonance (SPR). There are several parameters that influence the SPR frequency. For example, the size and shape of nanoparicles, surface charges, dielectric constant of surrounding medium and so forth. By changing the shape of GNPs from spherical to rod, the new SPR spectrum will present two absorption bands: a weaker short-wavelength in the visible region due to the transverse electronic oscillation and a stronger long-wavelength band in NIR due to the longitudinal oscillation of electrons. The change of aspect ratio can greatly affect the absorption spectrum of gold nanorods (GNRs) [11]. In the same vein, increasing the aspect ratio can lead to longitudinal SPR absorption band redshifts. Different GNP structures shows different properties. In comparison with a gold nanoparticle-conjugating probe, the gold nanowire-functionalized probe could avoid the leakage of biomolecules from the composite film, and enhanced the stability of the sensor [12, 13]. This interesting phenomenon will be enormously beneficial in practical applications such as biosensors.

It is well known that well-dispersed solutions of GNPs display a red color, while aggregated GNPs appear a blue color. Based on this phenomenon, Jena et al. [14] established a GNPs-based biosensor to quantitatively detect the polyionic drugs such as protamine and heparin. As shown in Figure 1, the degree of aggregation and de-aggregation of GNPs is proportional to the concentration of added protamine and heparin.

Non-crosslinking GNP aggregation can also be applied for enzymatic activity sensing and potentially inhibitor screening [15]. Wei et al. [16] described a simple and sensitive aptamer-based colorimetric sensing of alpha-thrombin protein using unmodified 13 nm GNP probes, as shown in Figure 2. This method's advantage lies in that the general steps such as surface modification and separation can be avoided, which ensures the original conformation of the aptamer while interacting with its target, thereby leading to high binding affinity and sensitive detection.

The GNPs in biosensors can also provide a biocompatible microenvironment for biomolecules, greatly increasing the amount of immobilized biomolecules on the electrode surface, and thus improving the sensitivity of the biosensor [17, 18]. The glassy carbon electrode (GCE) was widely used in biosensor, and GNP modified GCEs showed much better electrochemical stability and sensitivity. The GNPs and methylene blue (MB) could be assembled via a layer-by-layer (LBL) technique into films on the GCE modified for detection of human chorionic gonadotrophin (HCG) [19]. Due to the high surface area of the nanoparticles for loading anti-HCG, this immunosensor can be used to detect the HCG concentration in human urine or blood samples.

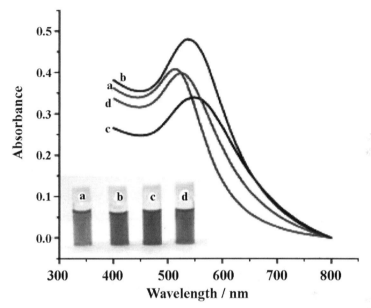

Figure 1. Absorption spectra illustrating the protamine-induced aggregation and heparin-driven de-aggregation of AuNPs. (a) The AuNPs alone; (b, c) after the addition of protamine: (b) 0.7 µg/ml and (c) 1.6 µg/ml; (d) after the addition of heparin (10.2 µg/ml). Inset shows the corresponding colorimetric response [14].

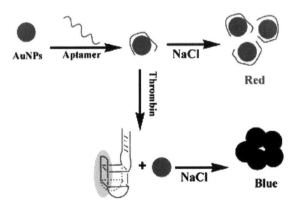

Figure 2. The AuNPs colorimetric strategy for thrombin detection [16].

For the detection of reduction of H_2O_2, GNP-modified electrodes also showed much wider pH adaptive range and larger response currents [20]. Due to the large specific surface area and good biocompatibility of GNPs, horseradish peroxidase (HRP) can be adsorbed onto a GNP layer for the detection of H_2O_2 without loss of biological activity [21]. Shi et al. [22] confirmed that this kind of HRP-GNP biosensor exhibited long-term stability and good reproducibility.

The GNPs/CNTs multilayers can also provide a suitable microenvironment to retain enzyme activity and amplify the electrochemical signal of the product of the enzymatic reaction [23]. For example, GNPs/CNTs nanohybrids were covered on the surface of a GCE, which formed an effective antibody immobilization matrix and gave the immobilized biomolecules high stability and bioactivity. The approach provided a linear response range between 0.125 and 80 ng/ml with a detection limit of 40 pg/ml. As shown in Figure 3, because of the advantages of GNPs and CNTs, the hybrid composite has more potential applications for electrochemical sensor, which could be easily extended to other protein detection schemes and DNA analysis [24]. For example, Wang et al. [25] described the fabrication of ZrO_2/Au nano-composite films through a combination of sol–gel procedure and electroless plating, the organophosphate pesticides (OPs) can be strongly adsorbed on the ZrO_2/Au film electrode surface, which provides an effective quantitative method for OPs analysis.

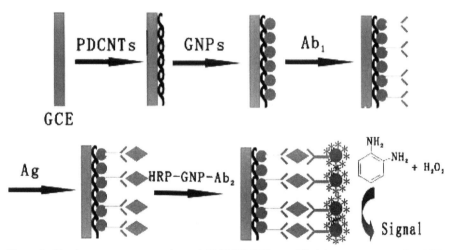

Figure 3. The immunoassay procedure of GNPs/PDCNTs modified immunosensor using HRP–GNPs–Ab2 conjugates as label [24].

The GNR modified electrode layer shows a better analytical response than GNPs [26]. The GNR-based immunosensors have advantages such as simplicity, being label free, low sample volume, reusability, and being more suitable for lab-on-chip devices over GNPs. The GNRs are sensitive to the dielectric constant of the surrounding medium due to SPR, therefore a slight change of the local refractive index around GNRs will result in an observable plasmon resonance frequency shift. Irudayaraj and

Yu fabricated different aspect ratios of GNRs with targeted antibodies to detect three targets (goat anti-human IgG1 Fab, rabbit antimouse IgG1 Fab, rabbit anti-sheep IgG (H+L)). Results showed that GNRs can be used for a multiplexing detection device of various targets. In another study, they examined the quantification of the plasmonic binding events and estimation of ligand binding kinetics tethered to GNRs via a mathematical method. The GNRs sensors were found to be highly specific and sensitive with a dynamic response in the range between 10^{-9} and 10^{-6} M. For higher-target affinity pair, one can expect to reach femtomolar levels limit of detection. This is promising for developing sensitive and precise sensors for biological molecule interactions. Chilkoti and his co-workers have miniaturized the biosensor to the dimensions of a single GNR [27]. Based on a proof-of-concept experiment with streptavidin and biotin, they tracked the wavelength shift using a dark-field microspectroscopy system. The GNRs binding 1 nm of streptavidin could bring about a 0.59 nm mean wavelength shift. Furthermore, they also indicated that the current optical setup could reliably measure wavelength shifts as small as 0.3 nm. Frasch and co-workers have set single molecules DNA detection in spin by linking F1-ATPase motors and GNRs [28]. The biosensor overcomes the defects inherent to polymerase chain reaction (PCR) or ligase chain reaction (LCR), is faster and reaches zeptomol concentrations, which is greatly superior to traditional fluorescence-based DNA detection systems which have only about a five picomolar detection limit.

The Use of CNTs in Biosensors

Since Iijima discovered CNTs in 1991, CNTs have attracted enormous interest due to their many novel properties such as unique mechanical, physical, chemical properties. The CNTs have great potential in applications such as nanoelectronics, biomedical engineering, and biosensing and bioanalysis [5, 29, 30]. For example, polymer-CNTs composites can achieve high electrical conductivity and good mechanical properties, which offer the exciting possibility of developing ultrasensitive, electrochemical biosensors. As shown in Figures 4 and 5, amperometric biosensors [31] was constructed by incorporation of single-walled carbon nanotubes (SWNT) modified with enzyme into redox polymer hydrogels. First, an enzyme was incubated in a SWNT solution, then cross-linked within a poly [(vinylpyridine)Os(bipyridyl)(2)Cl$^{2+/3+}$] polymer film, and finally formed into composite films. The redox polymer films incorporated with glucose oxidase modified SWNTs resulted in a 2- to 10-fold increase in the oxidation and reduction peak currents during cyclic voltammetry, while the glucose electrooxidation current was increased 3-fold to close to 1 mA/cm^2 for glucose sensors. Similar effects were also observed when SWNTs were modified with HRP prior to incorporation into redox hydrogels.

Conductive polymer-based nano-composite has been utilized as a MEMS sensing material via a one-step, selective on-chip deposition process at room temperature [32]. For example, the doped PPy-MWCNT is confirmed to be sensitive to glucose concentrations up to 20 m, which covers the physiologically important 0–20 mm range for diabetics, so they can be used for diagnosis of diabetes [33, 34]. So far, these electrochemical sensors such as enzyme-based biosensors, DNA sensors, and immunosensors have been developed based on polymer-CNT composites, and can be used to diagnose different kinds of diseases quickly [35, 36].

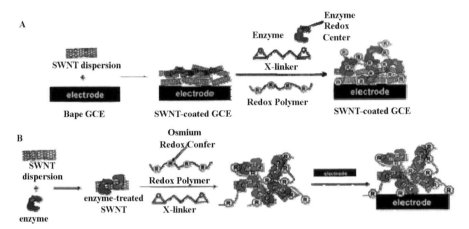

Figure 4. Schematic of the construction of type A and type B sensors. (A) Fabrication of type A sensors in which a film of SWNTs was first cast onto a bare glassy carbon electrode and allowed to dry, before an alquot of the redox hydrogel was cast on top of the SWNT-coated electrode. (B) Fabrication of type B sensors in which SWNTs were first incubated with an enzyme solution before they were incorporated into the redox hydrogel. An aliquot of the redox hydrogel solution containing the enzyme-modified SWNTs was then cast on top of a bare glassy carbon electrode [31].

The bionanocomposite layer of multiwalled carbon nanotubes (MWNTs) in chitosan (CHIT) can be used in the detection of DNA [34]. The biocomponent, represented by double-stranded herring sperm DNA, was immobilized on this composite using LBL coverage to form a robust film. The ssDNA probes could be immobilized on the surface of GCE modified with MWNTs/ZnO/CHIT composite film [37]. The sensor can effectively discriminate different DNA sequences related to PAT gene in the transgenic corn, with a detection limit of 2.8 mol/l of target molecules.

Carbon nanofibers are found to be an effective strategy for building a biosensor platform [38]. Bai et al. [39] found that the synergistic effects of MWNTs and ZnO improved the performance of the biosensors formed. They reported an amperometric biosensor for hydrogen peroxide, which was developed based on adsorption of HRP at the GCE modified with ZnO nanoflowers produced by electrodeposition onto MWNTs film. Zhang et al. described a controllable LBL self-assembly modification technique of GCE with MWNTs and introduce a controllable direct immobilization of acetylcholinesterase (AChE) on the modified electrode. By the activity decreasing of immobilized AChE caused by pesticides, the composition of pesticides can be determined [40-43].

Our group also just reported a highly selective, ultrasensitive, fluorescent detection method for DNA and antigen based on self-assembly of multi-walled CNT and CdSe QDs via oligonucleotide hybridization; its principle is shown in Figure 6 [44]. The MWNTs and QDs, their surfaces are functionalized with oligonucleotide (ASODN) or antibody (Ab), can be assembled into nanohybrid structures upon the addition of a target complementary oligonucleotide or antigen (Ag). As shown in Figure 6, nano-material building blocks that vary in chemical composition, size or shape are arranged in space on the basis of their interactions with complementary linking oligonucleotide for potential application in biosensors. We show how this oligonucleotide directed assembly strategy could be used to prepare binary (two-component) assembly materials comprising two different shaped oligonucleotide-functionalized nanomaterials. Importantly, the proof-of-concept demonstrations reported herein suggest that this strategy could be extended easily to a wide variety of multicomponent systems.

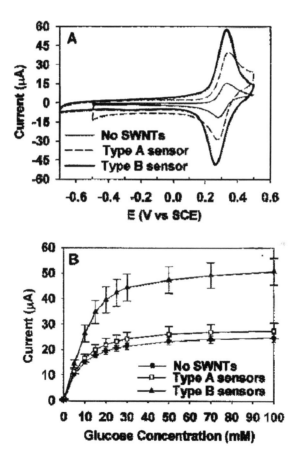

Figure 5. Electrochemical characterization of glucose oxidase sensors. (A) Cyclic voltammograms of a GCE modified with the redox hydrogel alone (-); a GCE modified first with a film of SWNT and then coated with the redox hydrogel (----) (type A sensor); (III) a GCE modified with a redox hydrogel containing GOX-treated SWNTs (-) (type B sensor). Scan rate 50 mV/s. (B) Glucose calibration curves for the three types of sensors described in (A). T = 25C, E = 0.5 V vs SCE. Values are mean ±SEM [31].

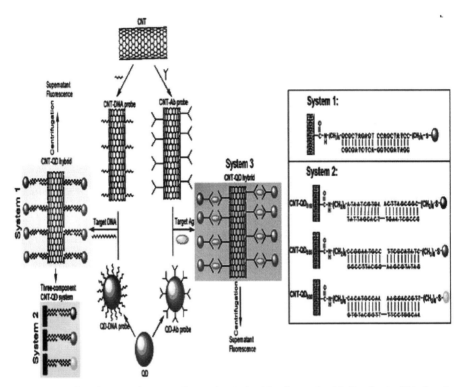

Figure 6. Surface functionalization of CNT (or QD) with oligonucleotide/Angibody (Ab), forming CNT-DNA (or -Ab) probe and QD-DNA (or -Ab) probe, and subsequent addition of target oligonucleotide (or Antigen) to form CNT-QD assembly. The unbound QD probe was obtained by simple centrifugation separation and the supernatant fluorescence intensity of QDs was monitored by spectrofluorometer. (System 1) Formation of CNT-QD hybrid in the presence of complementary DNA target; (System 2) Three-component CNT-QD system with the purpose to detect three different DNA target simultaneously; (System 3) CNT-QD protein detection system based on antigen-antibody immunoreaction [44].

The Use of Magnetic Nanoparticales in Biosensor

The MNPs, because of their special magnetic properties, have been widely explored in applications such as hyperthermia [45], magnetic resonance imaging (MRI) contrast agent [46], tissue repair [47], immunoassay [48], drug/gene delivery [49], cell separation [50], GMR-sensor [51], and so forth. Zhang et al. [52] prepared a new kind of magnetic dextran microsphere (MDMS) by suspension crosslinking using iron nanoparticles and dextran. The HRP was then immobilized on a MDMS-modified GCE. On the basis of the immobilized HRP-modified electrode with hydroquinone (HQ) as mediator, an amperometric H_2O_2 biosensor was fabricated. Lai et al. [53] prepared a magnetic chitosan microsphere (MCMS) using carbon-coated MNPs and chitosan. Hemoglobin (Hb) was successfully immobilized on the surface of MCMS modified GCE with the cross-linking of glutaraldehyde.

Janssen et al. [54] demonstrated that a rotating magnetic field can be used to apply a controlled torque on superparamagnetic beads which leads to a tunable bead rotation frequency in fluid and develop a quantitative model, based on results from a comprehensive set of experiments. This control of torque and rotation will enable novel functional assays in bead-based biosensors.

The amperometric biosensor was based on the reaction of alkaline phosphatase (ALP) with the substrate ascorbic acid 2-phosphate (AA2P), where the Fe_3O_4 nanoparticles have led to the enhancement of the biosensor response with an improved linear response range. This biosensor was applied to the determination of the herbicide 2, 4-dichlorophenoxyacetic acid (2, 4-D) [55].

In fact, a wide variety of methods have been developed for sensing and enumerating individual micron-scale magnetic particles [56]. Direct detection of magnetic particle labels includes Maxwell bridge, Frequency-dependent magnetometer, Superconducting quantum interference device (SQUID), and methods of magnetoresistance. Indirect detection includes Micro-cantilever-based force amplified biological sensor (FABS) and magnetic relaxation switches (MRS). Two examples follow.

Recently, a highly sensitive, giant magnetoresistance-spin valve (GMR-SV) biosensing device with high linearity and very low hysteresis was fabricated by photolithography [57]. The signal from even one drop of human blood and nanoparticles in distilled water was sufficient for their detection and analysis.

For the immunomagnetic detection and quantification of the pathogen *Escherichia coli* (*E. coli*) O157:H7, a giant magnetoresistive multilayer structure implemented as sensing film consists of 20 [Cu5.10 nm/Co2.47 nm] with a magnetoresistance of 3.20% at 235 Oe and a sensitivity up to 0.06 Ω/Oe between 150 and 230 Oe. Silicon nitride has been selected as optimum sensor surface coating. In order to guide the biological samples, a microfluidic network made of SU-8 photoresist and 3D stereolithographic techniques have been included [58, 59].

The Use of QDs in Biosensors

The QDs have been subject to intensive investigations because of their unique photoluminescent properties and potential applications [60-62]. So far, several methods have been developed to synthesize water-soluble QDs for use in biologically relevant studies. For example, QDs have been used successfully in cellular imaging [63], immunoassays [64], DNA hybridization [65], biosensor, and optical barcoding [66]. The QDs also have been used to study the interaction between protein molecules or detect the dynamic course of signal transduction in live cells by fluorescence resonance energy transfer (FRET) [67, 68]. These synthesized QDs have significant advantages over traditional fluorescent dyes, including better stability, stronger fluorescent intensity, and different colors, which are adjusted by controlling the size of the dots [64]. Therefore, QDs provide a new functional platform for bioanalytical sciences and biomedical engineering.

For example, CdTe QDs led to an increased effective surface area for immobilization of enzyme and their electrocatalytic activity promoted electron transfer reactions and catalyzed the electro-oxidation of thiocholine, thus amplifying the detection sensitivity [69]. As shown in Figure 7, Deng et al. [70] reported that green and orange CdTe QDs can be used as pH-sensitive fluorescent probes, which could monitor the proton

(H+) flux driven by ATP synthesis for dual simultaneous and independent detection of viruses on the basis of antibody—antigen reactions.

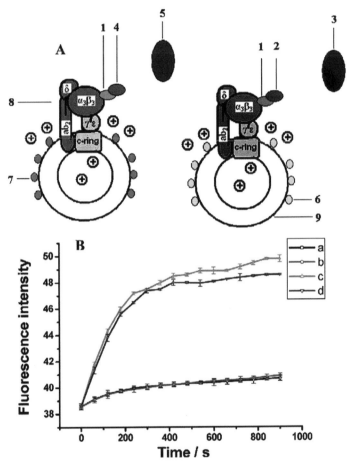

Figure 7. (a) Basic design of QD biosensors based on F0F1-ATPase: (1) antibody of β¬subunit; (2) the antibody of MHV68; (3) MHV68; (4) the antibody of H9 avian influenza virus; (5) H9 avian influenza virus; (6) CdTe QDs with emission wavelength at 585 nm; (7) CdTe QDs with emission wavelength at 535 nm; (8) F0F1-ATPase within chromatophores; and (9) chromatophores. (b) Changes of fluorescence intensity of QD biosensors with and without viruses. Curve a: The changes of fluorescence intensity of orange QD biosensors without MHV68 when the ADP is added to initialize reaction. Curve b: The changes of fluorescence intensity of green QD biosensors without H9 avian influenza virus when the ADP is added to initialize reaction. Curve c: The changes of fluorescence intensity of orange QD biosensors with capturing MHV68 when the ADP is added to initialize reaction. Curve d: The changes of fluorescence intensity of green QD biosensors with capturing H9 avian influenza virus when the ADP is added to initialize reaction [70].

The Use of Other Nanomaterials in Biosensors

Aside from GNPs, CNTs, MNP, and QDs, there are still many other nanomaterials such as metals, metal-oxides [71, 72], polymers and other compounds [73-75], which could be used in biosensors. For example, hollow nanospheres CdS (HS-CdS) [76] were first used to study the direct electrochemical behavior of Hb and the construction of nitrite biosensors. The HS-CdS nanostructure provides a microenvironment around the protein to retain the enzymatic bioactivity.

Metal nanoparticles [77], for example, nano-Cu, with great surface area and high surface energy, are used as electron-conductors and show good catalytic ability to the reduction of H_2O_2 [78]. Platinum nanoparticles have also been widely used in biosensors.

Nanoscale metal-oxides have also been widely used in immobilization of proteins and enzymes for bioanalytical applications. For example, metal-oxide-based semiconducting nanowires or nanotubes play an important role on electric, optical, electrochemical, and magnetic transducers [79]. Cheng et al. [80] reported a nano-TiO_2 based biosensor for the detection of lactate dehydrogenase (LDH). Waxberry-like nanoscale ZnO balls, as shown in Figure 8, can act as excellent materials for immobilization of enzymes and the rapid electron transfer agent for the fabrication of efficient biosensors due to the wide direct band gap [81, 82]. The porous structure can greatly enhances the active surface area available for protein binding, provide a protective microenvironment for the enzymes to retain their enzymatic stability and activity [83].

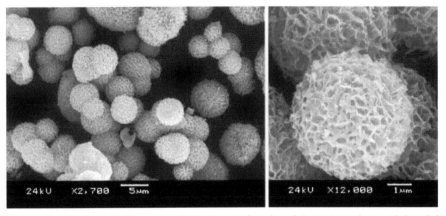

Figure 8. The SEM images of as-prepared porous nanosheet-based ZnO microsphere with low (left) and high magnification (right) [83].

Surface functionalized silicon nano-channels with the enzyme urease could detect and quantify urea concentration [84]. The differential conductance of silicon nano-channels can be tuned for optimum performance using the source drain bias voltage, and is sensitive to urea at low concentration. Zhang et al. [85] used silicon-on-insulator (SOI) substrate to fabricate the planar type patch clamp ion-channel biosensor, which is suitable for the high throughput screening. The channel current showing the desensitization unique to TRPV1 is measured successfully.

Poly (ethylene-co-glycidyl methacrylate) (PE-co-GMA) nanofibers with abundant active epoxy groups on their surfaces were fabricated through a novel manufacturing process [85-86]. The results demonstrated that the PE-co-GMA nanofibers prepared could be a promising candidate as solid support materials for potential biosensor applications.

POTENTIAL APPLICATION OF NANOMATERIALS-BASED BIOSENSORS

Although few sensors based on nanomaterials work at all in commercial applications, however, nanomaterial-based biosensors exhibit fascinating prospects. Compared with traditional biosensors, nanomaterial-based biosensors have marked advantages such as enhanced detection sensitivity and specificity, and possess great potential in applications such as the detection of DNA, RNA, proteins, glucose [87], pesticides [88], and other small molecules from clinical samples, food industrial samples, as well as environmental monitoring.

Nanomaterials-based Biosensors for the Detection of Glucose

The glucose biosensor has been widely used as a clinical indicator of diabetes. Nanoscale materials such as GNPs, CNTs, MNPs [89], Pt nanoparticles [90], QDs, and so forth play an important role in glucose sensor performance, fibrous morphology, and wrapping of PDDA over MWCNTs result in a high loading of GOx into the electrospun matrix [91]. Platinum nanoparticles could be electrodeposited on MWNTs matrix in a simple and robust way. The immobilization of glucose oxidase onto Pt/MWNTs electrode surfaces also could be carried out by chitosan-SiO_2 gel [92]. The resulting biosensors could be used to determine the glucose levels of serum samples with high sensivity.

Nanomaterials-based Biosensors for the Detection of DNA and Protein

The ssDNA–CNTs probes might be used as optical biosensors to detect specific kinds of DNA oligonucleotides [93]. The MWNTs/ZnO/CHIT composite film modified GCE can be used to immobilize ssDNA probes to effectively discriminate different DNA sequences [94, 95]. A biosensor for the detection of deep DNA damage is designed employing the bionanocomposite layer of MWNT in chitosan deposited on a SPCE [96]. The biocomponent represented by double-stranded herring sperm DNA was immobilized on this composite using LBL coverage to form a robust film. The GNPs can also be used to recognize DNA sequences by the interactions of DNA and chemical materials [97]. And for single-stranded DNA, GNPs functionalized with alkanethiol-capped LNA/DNA chimeras in a tail-to-tail hybridization mode could perform excellent [98], and these probes show remarkable discrimination between a complementary target and one containing a single-base mismatch. Nano-SiO_2/p-aminothiophenol (PATP) film was fabricated by self-assembly and electrodeposition methods and was successfully applied to the detection of the PAT gene sequences by a label-free EIS method [99]. Maki et al. [100] reported the first nanowire field effect transistor based biosensor which achieves simple and ultra-sensitive electronic DNA methylation detection and avoids complicated bisulfite treatment and PCR amplification. Similarly, using protein–ligand (antigen) interaction properties, protein-nanoparticles based biosensors can realize the ultra-sensitive detection of special protein molecules.

Nanomaterials-based Biosensors for the Detection of Other Molecules

Liposome-based biosensors have successfully monitored the organophosphorus pesticides such as dichlorvos and paraoxon at very low levels [101]. The nano-sized liposomes provide a suitable environment for the effective stabilization of AChE and they can be utilized as fluorescent biosensors. Porins embedded into the lipid membrane allow for the free substrate and pesticide transport into the liposomes. Pesticide concentrations down to 10^{-10} M can be monitored.

By flow injection analysis (FIA), Zhang et al. [102] developed a method for the detection of *E. coli* using bismuth nanofilm modified GCE. Seo et al. [103] constructed a biochip sensor system, consisting of two Ti contact pads and a 150 nm wide Ti nanowell device on $LiNbO_3$ substrate. When the bacteria were resistant to the phages (uninfected bacteria), small voltage fluctuations were observed in the nanowell displaying a power spectral density (PSD) of 1/f shape. Medley et al. [104] developed a colorimetric assay for the direct detection of diseased cells. This assay uses aptamer-conjugated GNPs to combine the selectivity and affinity of aptamers and the spectroscopic advantages of GNPs. Samples with diseased cells present exhibited a distinct color change while non-target samples did not change the color.

Mitochondrial oxidative stress (MOS) has been hypothesized as one of the earliest insults in diabetes. Some data support the hypothesis that the induction of MOS is more sensitive to hyperglycemia than the induction of the antioxidant response element (ARE). An ARE-GFP vector constructed with nanoparticles was successfully delivered to the eyes by using sub-retinal injection [105]. These data support the use of nanoparticle-delivered biosensors for monitoring the oxidative status of tissues *in vivo*.

Li et al. [106] reported an electrochemical aptamer biosensor for the detection of adenosine based on impedance spectroscopy measurement, which gives not only a label-free but also a reusable platform to make the detection of small molecules simple and convenient. For this method did not rely on the molecule size or the conformational change of the aptamer, it may possess the potential of wider application for different target molecules.

CHALLENGES AND PROSPECTS

In recent years, applications of nanomaterials in biosensors provides novel opportunities for developing a new generation of biosensor technologies. Nanomaterials can improve mechanical, electrochemical, optical, and magnetic properties of biosensors, nanomaterial-based biosensors are developing towards single molecule biosensors and high throughput biosensor arrays [107]. However, like any emerging field, they face many challenges. Biological molecules possess special structures and functions, and determining how to fully use the structure and function of nanomaterials and biomolecules to fabricate single molecule multifunctional nanocomposites, nanofilms, and nanoelectrodes, is still a great challenge. The mechanism of interaction between biomolecules and nanomaterials is also not clarified very well yet. How to use these laws and principles of an optimized biosystem for fabricating novel multifunctional or homogenous nanofilms or modifying electrodes is also a great challenge. The processing, characterization, interface problems, availability of high quality nanomaterials, tailoring of nanomateriala, and the mechanisms governing the behavior of these nanoscale composites on the surface of electrodes are also great challenges for the presently existing techniques. For example, how to align nanomaterials such as CNTs in a polymer matrix along identical direction is a great challenge. How to enhance the signal to noise ratio, how to enhance transduction and amplification of the signals, are also great challenges. Future work should concentrate on further clarifying the mechanism of interaction between nanomaterials and biomolecules on the surface of electrodes or nanofilms and using novel properties to fabricate a new generation of biosensors. Nevertheless, nanomaterial-based biosensors show great attractive prospects, which will be broadly applied in clinical diagnosis, food analysis, process control, and environmental monitoring in the near future.

ACKNOWLEDGMENTS

The work is supported by China 973 project (No.2005CB724300-G), National Natural Science Foundation of China (No.30771075 and No.30672147), 863 Project (No.2007AA022004), Shanghai Nano-project (No.0752nm024), Shanghai Pujiang Plan Project (No.06J14049), and Shanghai Foundation of Science and Technology (No. 072112006).

KEYWORDS

- **Biosensor**
- **Carbon nanotubes**
- **Gold nanoparticles**
- **Gold nanorods**
- **Quantum dots**

Chapter 4

Novel Lipid-coated Magnetic Nanoparticles for *in vivo* Imaging

Huey-Chung Huang, Po-Yuan Chang, Karen Chang, Chao-Yu Chen, Chung-Wu Lin, Jyh-Horng Chen, Chung-Yuan Mou, Zee-Fen Chang, and Fu-Hsiung Chang

INTRODUCTION

Application of superparamagnetic iron oxide nanoparticles (SPIOs) as the contrast agent has improved the quality of magnetic resonance (MR) imaging. Low efficiency of loading the commercially available iron oxide nanoparticles into cells and the cytotoxicity of previously formulated complexes limit their usage as the image probe. Here, we formulated new cationic lipid nanoparticles containing SPIOs feasible for *in vivo* imaging.

Hydrophobic SPIOs were incorporated into cationic lipid 1,2-dioleoyl-3-(trimethylammonium) propane (DOTAP) and polyethylene-glycol-2000-1,2-distearyl-3-sn-phosphatidylethanolamine (PEG-DSPE) based micelles by self-assembly procedure to form lipid-coated SPIOs (L-SPIOs). Trace amount of Rhodamine-dioleoyl-phosphatidylethanolamine (Rhodamine-DOPE) was added as a fluorescent indicator. Particle size and zeta potential of L-SPIOs were determined by dynamic light scattering (DLS) and laser doppler velocimetry (LDV), respectively. The HeLa, PC-3, and Neuro-2a cells were tested for loading efficiency and cytotoxicity of L-SPIOs using fluorescent microscopy, Prussian blue staining and flow cytometry. The L-SPIO-loaded CT-26 cells were tested for *in vivo* MR imaging.

The novel formulation generates L-SPIOs particle with the average size of 46 nm. We showed efficient cellular uptake of these L-SPIOs with cationic surface charge into HeLa, PC-3, and Neuro-2a cells. The L-SPIO-loaded cells exhibited similar growth potential as compared to unloaded cells, and could be sorted by a magnet stand over 10-day duration. Furthermore, when SPIO-loaded CT-26 tumor cells were injected into Balb/c mice, the growth status of these tumor cells could be monitored using optical and MR images.

We have developed a novel cationic lipid-based nanoparticle of SPIOs with high loading efficiency, low cytotoxicity and long-term imaging signals. The results suggested these newly formulated non-toxic lipid-coated magnetic nanoparticles as a versatile image probe for cell tracking.

Developing nanoprobe is important for diversely functional analysis in biomedical research [1]. Utilization of nanoparticles such as gold nanoparticles, quantum dots and SPIOs has been a fast growing area in molecular imaging. Nanosized gold particles are highly reactive in biological milieu, whereas quantum dots are nanocrystals of heavy metals in nature. But both of these nanoparticles are not biocompatible for human and for clinical practice. In this respect, SPIOs with polymeric coating, such as dextran, have been widely used in MR imaging [2]. With its high spatial resolution, it has been widely used for tracking cell migration [3-5] and monitoring *in vivo* status of stem cell differentiation [6-9].

Because of the low efficiency of direct loading of these commercially available iron oxide nanoparticles into cells, transfection reagents such as cationic lipids, poly-lysine, and protamine sulfate have been used for non-specific delivery of iron oxide nanoparticles into cells [10-13]. However, the formulated complexes are not stable and tend to aggregate in test tubes or even precipitate in the cell culture conditions, re-sulting in cytotoxicity [14]. Therefore, it remains a need to formulate new complexes without cytotoxicity for improving the delivery efficiencies to cells.

Polyethylene glycol (PEG)-modified lipids or polymers have been widely used to form the nanocarriers for pharmaceutical applications. Due to its small size, neutral surface charge and high hydrophilicity, PEG-modification greatly reduces the interac-tions of nanoparticles with plasma proteins and cells, ensuring their long circulation in the blood [15]. Similarly, charged phospholipids could, by themselves, form self-as-sembled film or particles in aqueous solutions. The diacyllipids, such as DOTAP with strong positive charge and low transition temperature, may facilitate the quick forma-tion of nanoparticles in physiological buffers [16, 17]. Accordingly, we developed a novel formula of nanoparticles by combining PEG-lipid and DOTAP with SPIOs, and demonstrated the cellular uptake, biocompatibility and usage in the *in vivo* imaging of these new nanoparticles. The advantages of these new formulated L-SPIOs include their small sizes, relative stability in physiological buffers, suitability for bimodal im-aging and high loading efficiency to cells, allowing their usage as a versatile image probe for *in vivo* tracking analysis.

MATERIALS AND METHODS

The DOTAP, PEG-DSPE, and Rhodamine-DOPE were obtained from Avanti Polar Lipids (Alabaster, AL, USA). These lipids were dissolved in chloroform individually, sealed in ampoules filled with argon gas and stored at –20°C before use. Dulbecco's Modified Eagle's Medium (DMEM) and RPMI-1640 medium were purchased from GIBCO™ Invitrogen Corporation (Grand Island, NY, USA). Other chemicals were purchased from Sigma-Aldrich (St. Louis, MO, USA).

Cell Cultures

The HeLa, human prostatic adenocarcinoma cell lines (PC-3), mouse neuroblastoma cells (Neuro-2a) and mouse colorectal adenocarcinoma cells (CT-26) were obtained from the American Type Culture Collections (ATCC) (MD, USA). The HeLa, PC-3, and Neuro-2a cells were cultured in DMEM supplemented with 10% heat-inactivated fetal bovine serum (10% FBS-DMEM), whereas CT-26 cells were cultured in RPMI-1640 medium plus 10% serum. Routine cell culture was carried out at 37°C with supplementation of 5% CO_2.

Synthesis of Hydrophobic SPIO Nanoparticles

The hydrophobic SPIOs with size of 10 nm were synthesized using a seed-growth method by reducing Fe(acac)3 with 1,2-dodecanediol and protected by oleic acid and oleylamine in benzyl ether. After mixing in a round bottom flask and followed by heating to reflux (300°C) under nitrogen atmosphere for 1 hr, the SPIOs were precipitated by adding ethanol, collected by centrifugation and dissolved in hexane in the presence of oleic acid and oleylamine. For lipid coating, the nanoparticles were precipitated and resuspended in chloroform before use.

Preparation of Lipid-coated SPIOs

The L-SPIOs were prepared using a combination of the standard thin-film hydration method [18] and sonication process. Briefly, DOTAP and PEG-DSPE at 3:1 molar ratio in chloroform were mixed with hydrophobic SPIOs and trace amount of Rhodamine-DOPE, and then evaporated under reduced pressure. The resulted SPIO-lipid film was hydrated in distilled water to give the final concentrations of cationic lipids at 200 μm and iron oxide at 375 μg/ml. The solution of L-SPIOs was intensively sonicated at 60°C for 20 min. They were then sorted by the magnet, filter-sterilized through a Millex 0.22-μm-diameter filter and stored at 4°C in argon gas before use.

Size and Zeta Potential Determinations of Lipid-coated SPIOs

The diameters and zeta potential of L-SPIOs were determined using a Zetasizer Nano-ZS90 (Malvern Instruments ZEN3590, Worcs, UK). The size was analyzed by DLS, and the zeta potential was calculated by LDV at 25°C with samples suspended in deionized water. Prior to use, all glass and plastic wares were pre-washed with filtered water to minimize particulate contamination.

Cellular Loading of Lipid-coated SPIOs

The HeLa, PC-3, and Neuro-2a cells were seeded individually in each well of the 24-well plates (2-5 × 10^4 cells/well) overnight before L-SPIOs loading. Trace amount of Rhodamine-DOPE in the L-SPIOs was an indicator for monitoring the translocation of nanoparticles. After incubating with L-SPIOs (1-2 nmole/well, equivalent 1.875–3.75 μg SPIO) at different time points with or without the application of magnetic field, cells were washed with PBS. The high resolution fluorescent images were recorded using a fluorescent microscope (Leica DM IRB with the halogen lamp).

Prussian Blue Staining

The HeLa cells were seeded in a 24-well plate (5 × 10^4/well) overnight and treated with L-SPIOs (1 nmole/well) for 2 hr, with or without magnetic field. These cells were then washed three times with PBS and fixed in 4% formaldehyde solution for 30 min. After fixation, the cells were stained for the presence of intracellular iron with fresh prepared potassium ferrocyanate solution (mixture of equal volume of 4% potassium ferrocyanate with 4% hydrochloric acid) for 30 min. After wash three times with distilled water, the cells were then counterstained with Nuclear Fast Red Counterstain (Sigma-Aldrich) at room temperature for 5 min.

Flow Cytometry Analysis of L-SPIO-loaded Cells

Cells loaded with L-SPIOs were washed with PBS and trypsinized. Their cell loading efficiency was evaluated by scoring the percentage of Rhodamine fluorescent-positive cells using a FACS Calibur System (Becton-Dickinson, San Jose, CA, USA). The experiments were performed in triplicate, and 10,000 cells were counted in each experiment.

Long-term Growth and Magnetic Properties of L-SPIO-loaded Cells

The HeLa and PC-3 Cells were seeded in a 24-well plate (5×10^4/well) overnight and treated with L-SPIOs (1 nmole/well) for 12 hr. These cells were then washed with PBS twice and trypsinized. To analyze the effects of L-SPIO-loading on cell growth, both L-SPIO-loaded and control cells (1.5×10^5/100-mm dish) were cultured for more than 10 days and cell numbers were counted at different time points by the hemocytometer. To determine the magnetic property of L-SPIO-loaded cells, cells were trypsinized at different time points and equal amount of cells (5×10^5/500-μl medium in an Eppendorf tube) were sorted by the magnet stands and the magnet-bound cells were counted with a hemocytometer.

Induction and Assessment of Neuron Cell Differentiation

For morphological differentiation analysis, the degree of neurite outgrowth was assessed by phase-contrast microscopy [19-20]. Briefly, Neuro-2a cells were treated with 5 mm retinoic acid in 1% FBS-containing medium for 60 hr. The neurite-like processes longer than the major cell body diameter were defined as differentiated cells. In each well, 200 cells in three random fields were counted. Data were expressed as percentage of neurite-bearing cells in total cell population.

Imaging of Tumor Growth

For MR imaging of tumor growth, CT-26 cells (1×10^6) were seeded in 100-mm dish overnight before L-SPIOs loading (20 nmole lipid/dish, equivalent to 37.5 μg SPIO per dish). After 16 hr, CT-26 cells were removed and resuspended in PBS. Control and L-SPIO-loaded CT-26 cells (3×10^6, respectively) were injected into the dermis tissue of Balb/c mice to initiate tumors. After 2 and 15 days, MR imaging were taken from the anesthetized animals. These experimental protocols were approved by National Taiwan University Institutional Animal Care and Use Committee. For magnetic resonance imaging (MRI) scanning of anesthetized animals, a 3T MR system (Biospec; Bruker, Ettlingen, Germany) was used. T2-weighted images were obtained by fast spin echo sequence, TR/TE of 3500/62 ms, sliced thickness = 1 mm, field of view = 40 $\times$ 20 mm^2, matrix size = 256 $\times$ 256, and NEX = 5. The total scanning time was 4 min. T2*-weighted images were obtained by gradient echo sequence, TR/TE = 250/7 ms, sliced thickness = 1 mm, field of view = 40 $\times$ 20 mm^2, matrix size = 256 $\times$ 256, and NEX = 10. The total scanning time was 8 min. After scanning at 15th day, the tumor nodules were excised, fixed in formalin, embedded in paraffin, and processed into 5-μm sections. They were then stained with haematoxylin and eosin (H&E) following the standard histological protocol.

RESULTS

Characterization of Lipid-coated SPIOs

In pharmaceutical research, PEG-modified lipids have been widely used to form nanoparticles or micelles for drug targeting. Combined with a cationic lipid DOTAP, it could form a stable and uniform nanocarrier to encapsulate hydrophobic SPIOs. The average size of these formulated L-SPIOs was about 46 nm in diameter as shown in Figure 1A, carrying a positive zeta potential in deionized water (49.5 ± 0.8 mV). Incorporation of trace amount (less than 0.1%) of Rhodamine-DOPE as a fluorescent indicator did not affect the particle size. These L-SPIOs remained stable in a small and uniform size in distilled water or 5% glucose solution for more than 3 months. Culture medium, with or without serum, did not influence the size of L-SPIOs after magnetic sorting. (data not shown).

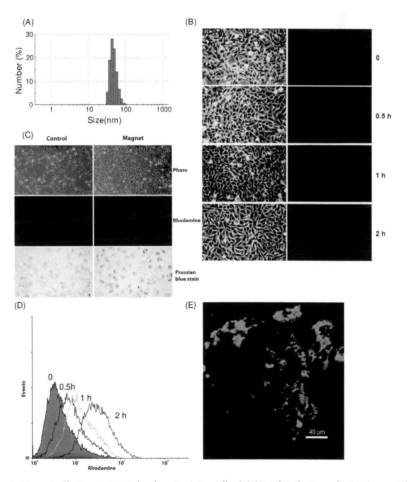

Figure 1. Magnet effect on L-SPIOs loading in HeLa cells. (A) Size distribution of L-SPIOs particles. (B) Images of HeLa cells treated with 1 nmole/well of L-SPIOs for 2 hr in 10% FBS-DMEM with

Figure 1. *(Caption Continued)*

magnet (right panel) or without magnet (left panel). Trace amount of Rhodamine-DOPE was used as an indicator for localizing the position of the nanoparticles. Images were taken after L-SPIO treatment (phase and fluorescent images: 100x magnification; Prussian blue staining: 400x magnification). (C) Time-dependent effect of magnet on cell loading. The HeLa cells were seeded overnight and treated with 1 nmole/well of L-SPIOs for 0, 0.5, 1, and 2 hr These fluorescent images were taken at different time points as indicated on the right. (D) Flowcytometric analysis of cells from (C) was shown. (E) Confocal image of L-SPIO-loaded HeLa cells. HeLa cells treated with L-SPIOs (1 nmole/well) for 16 hr. After washed with cold PBS, cells were fixed and analyzed by a confocal microscope. Red fluorescence in cells was indicated as L-SPIOs, whereas blue fluorescence was nucleus. Scale bar = 40 μm. Data represent three independent experiments with similar results.

Magnetic Effect on the Delivery of Lipid-coated SPIOs into Cultured Cells

To facilitate the delivery of L-SPIOs into cells, a magnet apparatus was used in cell culture condition. The HeLa cells (2-5×10^4/well) were seeded and treated with 1 nmole/well L-SPIOs in 10% FBS-DMEM with or without the application of magnetic field. Trace amount of Rhodamine-DOPE was used as an indicator for localizing these nanoparticles. As shown in Figure 1B, the presence of magnetic field for 2 hr remarkably improved the loading of L-SPIOs in HeLa cells without disrupting cell morphology. The magnetic field facilitated a time-dependent sedimentation and translocation of L-SPIOs into cells (Figures 1C and D). The loading efficiency with magnet for 2 hr was almost equal to that without magnet for 12 hr The loaded nanoparticles were mostly located in the cytoplasm as shown in the confocal microscopy, with only negligible L-SPIOs remaining bound to the cell membrane after 16 hr of the nanoparticle treatment (Figure 1E).

Long-term Viability and Magnetic Properties of L-SPIO-loaded Cells

After lipid coating, L-SPIOs retained the magnetic properties of SPIOs indicated on the magnet stand (Figure 2A, left). When cationic surface-charged L-SPIOs were introduced into HeLa cells with the assistance of magnetic field gradient for 12 hr, the cells demonstrated similar magnetic properties as L-SPIOs on magnet stand (Figure 2A, right). Cell proliferation assay was used to determine whether L-SPIOs affected cell viability. After seeding cells in culture plates, the cell number of control and L-SPIO-loaded cells were counted using the hemocytometer in different day. As shown in Figure 2B, there was no effect of L-SPIO loading on cell growth up to 11 days. The HeLa and PC-3 cells were loaded with L-SPIOs to further assess the duration of magnetic property of the L-SPIO-loaded cells (Figure 2C). After loading L-SPIOs at different time points, the SPIO-positive cells were separated by a magnet stand and counted with a hemocytometer. More than 80% of the SPIO-loaded cells still preserved the magnetic property when cultured for 7 days (Figure 2C).

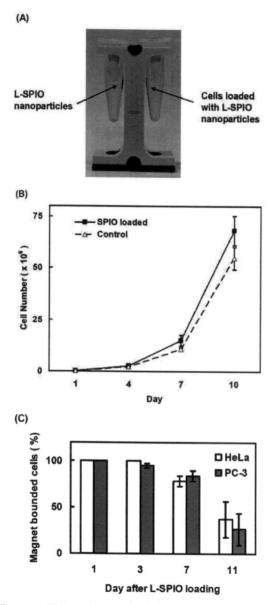

Figure 2. Magnet effect on L-SPIOs and L-SPIO-loaded cancer cells. (A) The L-SPIOs in water were placed on the magnetic stand (left); HeLa cells treated with L-SPIOs (1 nmole/ml) for 12 hr were on the right. Arrows indicate the position of L-SPIO nanoparticles or the L-SPIO-loaded cells. (B) Long-term viability of L-SPIO-loaded cells. The HeLa cells were first seeded in a 24-well plate (5 x 104/well) overnight and treated with 1 nmole/well of L-SPIOs for 12 hr in 10% FBS-DMEM. After different time periods of culture, cells were washed, trypsinized and counted with a hemocytometer. (C) Magnetic properties of L-SPIO-loaded cells. The HeLa and PC-3 cells were treated with the above L-SPIOs for 12 hr in 10% FBS-DMEM and harvested at different time points. The L-SPIO-positive cells were separated by a magnet stand and counted with a hemocytometer. All experiments were performed in triplicate. Results are mean ± SEM (n = 3).

Effects of L-SPIO-loading on Neuronal Cell Differentiation

We next evaluated the biocompatibility by testing the effect of L-SPIO-loading on neuronal cell differentiation. To this end, Neuro-2a cells, an albino mouse neuroblastoma cell, were treated with L-SPIOs (4 nmole lipid/well, equivalent to 7.5 μg SPIO per well) for 24 hr following the SPIO-loading, Neuro-2a cells were treated with 5 μm of retinoic acid for another 60 hr to induce neuronal differentiation. Figure 3 showed that neuronal differentiation was evident in both control and L-SPIO-loaded Neuro-2a cells with no significant difference in morphological differentiation between control and L-SPIO- loaded Neuro-2a cells (control: 86.5 ± 4.3%; L-SPIO-loaded: 82.7 ± 7.4%).

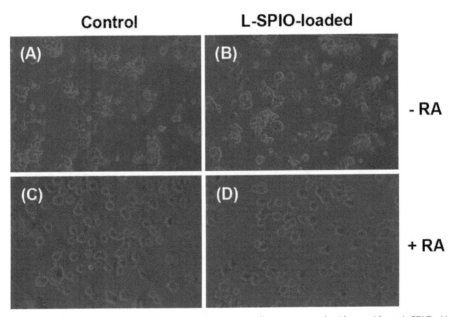

Figure 3. Differentiation status of Neuro-2a. Neuro-2a cells were treated with or without L-SPIOs (4 nmol/well) for 24 hr in 10% FBS-DMEM. After L-SPIO-loading, Neuro-2a cells were then washed and exposed to 5 μm of retinoic acid for differentiation indicated by neurites outgrowth. Phase Images of Neuro-2a cells before (A, B; 100x magnification) and after (C, D; 200x magnification) retinoic acid (RA) treatment in control or L-SPIO-loaded cells as indicated.

In Vivo Study of Tumor Cell Growth by Optical and MR Imaging

To further know whether the SPIO-loaded cells were suitable for *in vivo* study, control or SPIO-loaded CT-26 cells (3 × 10⁶) were injected subcutaneously into the back of Balb/c mice followed by MR imaging at Day 2 and Day 15 after tumor cell inoculation by a clinical 3T MR system. The T2-weighted and T2*-weighted MR images were obtained as described in Materials and Methods (Figure 4). In control cells (images A, C, and E), tumor proliferation was obvious in both T2-weighted (image A) and T2*-weighted (image C) MR images at Day 2, and tumor size increased remarkably to 5 mm in diameter at Day 15 (image E). When the Balb/c mice were injected with

L-SPIO-loaded cells, the T2-weighted (image B) and T2*-weighted (image D) images both revealed cell growth at Day 2. Tumor grew to a comparable size as compared to control cell inoculation at Day 15. In order to determine cell viability of the central region (2.5 mm in diameter) of tumor, which are positive in MRI signal as indicated by arrow in Figure 4F, histological staining and examination were employed. Tissue sections showed that the subcutaneous tumor was a poorly differentiated carcinoma, consisting of mostly viable tumor in both the center and peripheral regions of the tumor (Figure 5). The histological examination also showed no difference between tumors formed by control and L-SPIO-loaded CT-26 cancer cells (data not shown). These results demonstrated that CT-26 cells loaded with L-SPIOs retained the magnetic property for *in vivo* MR studies up to 15 days. The tumor growth was not affected by L-SPIOs and the MR images were readily distinguishable from those of control cells, thus allowing for a long-term tracking.

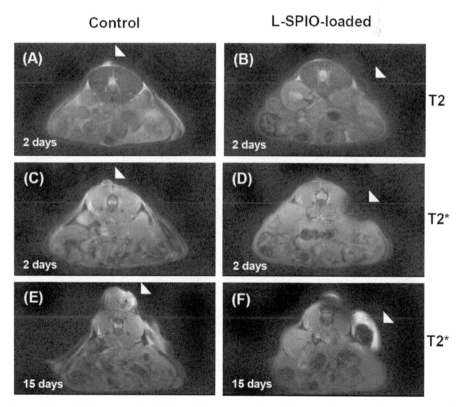

Figure 4. The MR imaging of L-SPIO-loaded cancer cells. L-SPIO loaded CT-26 cells were injected into the dermis of Balb/c mice. MRI imaging was performed in these mice bearing orthotopic colorectal tumor (triangle) 2 days (A~D) and 15 days (E, F) after subcutaneous injection of control and L-SPIO-loaded CT-26 cells. The T2-weighted images of tumor nodule from CT-26 cells without (A) and with L-SPIOs (B) were shown. T2*-weighted images of tumor nodule from CT-26 cells without (C, E) and with L-SPIOs (D, F) were shown.

(A)

(B)

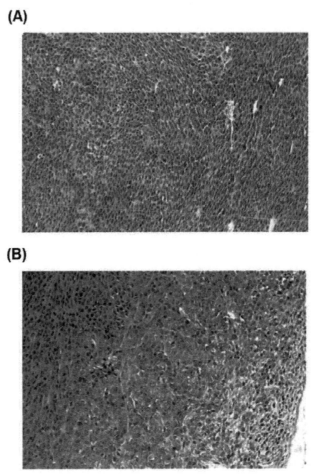

Figure 5. Histological examination of subcutaneous tumors in Balb/c mice. The L-SPIO loaded CT-26 cells were injected into the dermis of Balb/c mice to form tumors. Sections of the subcutaneous tumor growing from inoculated L-SPIO-loaded CT-26 cells are indicated. Histochemical H/E staining showed proliferation cells at central (A) and peripheral regions (B) of the tumor nodule growing from L-SPIO-loaded CT-26 cells. (200x magnification).

Nanotechnology provides size-controlled nanostructures such as quantum dots and magnetic iron oxide nanoparticles for molecular imaging [21]. For cellular MR imaging, nonspecific labeling of primary cells and stem cells with these nanoparticles has been used in biomedical research. From technical aspect, loading iron oxide nanoparticles with transfection reagents such as liposomes, polylysine, and protamine sulfate from the market is achievable [10-14]. But the sizes of these complexes in the present of serum are greater than 150 nm or even up to 1 μm in diameter, thus limiting their use in targeted delivery and live cell imaging [22-25]. In this study, we have developed a novel formulation of functionalized SPIOs by encapsulating of SPIOs into diacyl-lipid formed-micelles composing of DOTAP and PEG-DSPE. This strategy resulted in small and stable cationic nanoparticles with magnetic iron oxide in the core, allowing an efficient cellular loading with low cytotoxicity for *in vivo* imaging.

The PEG-lipid, due to its high water solubility and low toxicity, is self-assembled into nanosized micelles for drug delivery [16]. But, the long chain of PEG blocks lower the uptake of micelles by the reticuloendothelial system, thereby resulting in the long circulation times in blood [26]. In this regard, cationic lipids also increase solubility of the nanoparticles in the physiological buffers while is capable of loading nanoparticles into cells, presumably via the negatively charged biomolecules on cell membrane. With low transition temperature and low toxicity, DOTAP has been demonstrated to be an effective cationic lipid for transfection [27-29]. In our new formulation, we circumvent the disadvantage of PEG in prohibiting particle cellular uptake by including DOTAP to overcome the PEG shielding effect, thus facilitating nanoparticle translocation across cell membrane.

Another interesting feature in our formulation is that application of magnetic field in nanoparticle-loading resulted in significant increase of cellular uptake in a dose- and time-dependent manner (Figure 1). Of note, when initially loaded with high amount of magnetic nanoparticles, more than 80% of HeLa or PC-3 cells could maintain magnetic property up to 7 days, indicating the long retention of iron oxide (Figure 2). The loaded nanoparticles were predominately located in the cytoplasm rather than on the cell membrane of these cells (Figure 1D). In our previous experiments using lipid-enclosed quantum dots (LEQDs) [30], the durations of internalized LEQD could be detected by third harmonic generation (THG) microscopy in these loaded cells for more than 2 weeks. Since only lipid-coated quantum dots but not bare quantum dots could generate strong THG signal, we inferred here that the SPIOs inside cells may be still in lipid-coated form within time of these experiments. Further, it should be mentioned that magnet-assisted gene delivery has been shown to increase the efficiency of pDNA transfection but with a significant increase of cellular toxicity in human umbilical vein endothelial cell (HUVEC) [31]. In addition, long term expose of anionic iron oxide (Fe_2O_3) nanoparticles, at concentration range from 150 μm to 15 mm, caused PC12M cytotoxicity in both cell growth and neurites outgrowth [32]. Given the negligible cytotoxicity of L-SPIOs loading at high concentration both in cancer cell growth (HeLa, PC-3) and neural cell differentiation (Neuro 2a) (Figures 2B and 3), L-SPIOs thus have a potential in efficient drug delivery with low cytotoxicity and long duration.

Iron oxide nanoparticles were widely used in labeling cells due to their ability to create apparent image upon cellular internalization and particle clustering [33]. *In vitro* labeling of lymphocytes, such as T-cells and dendritic cells, and their tracking *in vivo* after injection into mice can be also visualized their pathogenesis and evolution using clinical MRI [34]. It also demonstrated that implantation of iron-oxide-labeled transplanted cells can be monitored within individual animals for a prolonged time by MR imaging [6, 35]. Therefore, it is feasible to monitor the prelabeled stem cells with iron oxide in differentiation and visualize the presence and migration of transplanted stem cells to the cerebral ischemia [36, 37]. Nevertheless, it was believed that the intracellular concentration of SPIOs decreased when cell proliferated, hence reducing the MRI signals for long-term *in vivo* tracking. However, results from our *in vivo* study of tumor growth (Figure 4) clearly demonstrated that even after 15 days the MRI signals were still detectable. Histological data showed that the tumor mass formed from injected L-SPIO-loaded tumor cells were viable both the center and peripheral regions (Figure 5), with similar growth potential as compared to non-SPIO-loaded cells (Figure 4E and F). This is consistent with a recent report by Nelson et al. showing that a long duration imaging of SPIO-loaded cells in SCID mice [38]. Thus, the biocompatibility nature of L-SPIO nanoparticles supports their use for long-term cell tracking and imaging in the *in vivo* system.

CONCLUSION

We have developed a novel cationic lipid-based nanoparticle of SPIOs with high loading efficiency, low cytotoxicity and long-term imaging signals. The potential of our lipid-coated magnetic nanoparticles is at least three folds. First, the iron oxide moiety helps the magnet-enhanced nanoparticle delivery and MR imaging, as well as for potential hyperthermia therapy. Second, targeted properties can be achieved by adding targeted ligands onto PEG linker of the lipid shell. Third, it provides size-controlled nanostructures that integrate biomolecules onto their lipid coat for efficient drug delivery. We believe that this multifunctional nanoparticle will provide an efficient way for *in-vivo* cell tracking [30], and therefore make a potential contribution to nanomedicine research in the future.

KEYWORDS

- **Dulbecco's Modified Eagle's Medium**
- **Human cervical cancer cells**
- **Magnetic resonance**
- **Polyethylene glycol**
- **Superparamagnetic iron oxide nanoparticles**

AUTHORS' CONTRIBUTIONS

Huey-Chung Huang, Chung-Yuan Mou, and Fu-Hsiung Chang conceived and designed the experiments. Huey-Chung Huang, Karen Chang, Chao-Yu Chen, Chung-Wu Lin, and Jyh-Horng Chen performed the experiments. Huey-Chung Huang, Po-Yuan Chang, Chung-Wu Lin, Jyh-Horng Chen, Zee-Fen Chang, and Fu-Hsiung Chang analyzed the data. Huey-Chung Huang, Po-Yuan Chang, Zee-Fen Chang, and Fu-Hsiung Chang wrote the chapter. All authors have read and approved the final manuscript.

ACKNOWLEDGMENTS

This study was supported by grants NSC 94-2311-B-002-029 and NSC 95-2311-B-002-023 from the National Science Council, and NTUH 95S342 from the National Taiwan University Hospital, Taipei, Taiwan. We thank Dr. Yann Hung for her helpful suggestions and comments on this manuscript.

Chapter 5

Doxorubicin Release from PHEMA Nanoparticles

Raje Chouhan and A. K. Bajpai

INTRODUCTION

Many anticancer agents have poor water solubility and therefore the development of novel delivery systems for such molecules has received significant attention. Nano-carriers show great potential in delivering therapeutic agents into the targeted organs or cells and have recently emerged as a promising approach to cancer treatments. The aim of this study was to prepare and use poly-2-hydroxyethyl methacrylate (PHEMA) nanoparticles for the controlled release of the anticancer drug doxorubicin.

The PHEMA nanoparticles have been synthesized and characterized using Fourier transform infrared spectroscopy (FTIR) and scanning electron microscopy (SEM), particle size analysis and surface charge measurements. We also studied the effects of various parameters such as percent (%) loading of drugs, chemical architecture of the nanocarriers, pH, temperature, and nature of the release media on the release profiles of the drug. The chemical stability of doxorubicin in PBS was assessed at a range of pH.

Suspension polymerization of 2-hydroxyethyl methacrylate (HEMA) results in the formation of swellable nanoparticles of defined composition. The PHEMA nanoparticles can potentially be used for the controlled release of the anticancer drug doxo-rubicin.

The number of reported cases of cancer is steadily increasing in both industrialized and developing countries. The latest world cancer statistics indicate that the number of new cancer cases will increase to more than 15 million in 2020 whereas another report issued by the World Health organization says that there are over 10 million new cases of cancer each year and over 6 million deaths annually are caused by the disease [1]. In spite of the fact that significant progress has been achieved in tumor biology, molecu-lar genetics and in the prevention, detection, and treatment of cancer over the last few years, adequate therapy remains elusive due to late diagnosis, inadequate strategies for addressing aggressive metastasis, and the lack of clinical procedures overcoming mul-tidrug resistant (MDR) cancer [2]. The integration of nanotechnology and medicine has the potential to uncover the structure and function of biosystems at the nanoscale level. Nanobiotechnology may provide a reliable and effective tool to treat diseases at a molecular scale. Nanobiotechnology offers an unprecedented opportunity to ratio-nalize delivery of drugs and genes to solid tumors following systemic administration [3]. Examples of nanotechnologies applied in pharmaceutical product development in-clude polymer-based nanoparticles, lipid-based nanoparticles (liposomes, nanoemul-sions, and solid-lipid nanoparticles), self-assembling nanostructures such as micelles and dendrimers-based nanostructures among others. In recent years, much research has gone into the characterization of nanoparticles and their biological effects and potential applications. These include bottom–up and molecular self-assembly, biological effects of naked nanoparticles and nano-safety, drug encapsulation and nanotherapeutics, and novel nanoparticles for use in microscopy, imaging, and diagnostics [4].

To be successful a cancer treatment approach needs to overcome physiological barriers such as vascular endothelial pores, heterogeneous blood supply, heterogeneous architecture to name just a few [5], and it strongly depends on the method of delivery. In the past, many anticancer drugs had only limited success and had major adverse side effects [6, 7]. Nanoparticles have attracted considerable attention worldwide because of their unique functional characters such as small particle size, high stability, lower toxicity, tuneable hydrophilic-hydrophobic balance, and the ability to bear surface features for target specific localization, and so forth. Thus, polymeric nanoparticles constitute a versatile drug delivery system [8], which can potentially overcome physiological barriers, and carry the drug to specific cells or intracellular compartments by passive or ligand-mediated targeting approaches [9]. The use of some polymers also allows, at least in principle, to achieve controlled release and the sustained drug levels for longer periods of time. Numerous biodegradable polymeric nanoparticles made of natural polymers such as proteins or polysaccharides have been tried for drug delivery and controlled drug release. More recently the focus of such studies moved onto synthetic polymers, and much progress have been achieved in this area. Recent examples include, polycationic nanoparticles for encapsulation and controlled release of amphotericin B by Vieria and Carmona-Ribeiro [10]; or encapsulation of curcumin for human cancer therapy by Maitra et al. [11].

Ideally, a successful nanoparticulate system should have a high drug loading capacity thereby reducing the quantity of matrix material for administration. The drug may be bound to the nanoparticles either (i) by polymerization in the presence of drug in most cases in the form of solution (incorporation method) or (ii) by absorbing/adsorbing the drug after the formation of nanoparticles by incubating them in the drug solution. In the present work we set to further investigate the latter method by studying swelling and controlled release of antitumor drug doxorubicin from synthetic PHEMA nanoparticles. The PHEMA attracted significant attention and is well documented in the literature. Many useful properties which make PHEMA attractive for a wide range of biomedical applications [12] include high water content, low toxicity, and tissue compatibility. The PHEMA has been used in applications such as soft contact lenses [13], drug delivery systems [14], kidney dialysis membranes [15], artificial liver support systems [16], and nerve guidance channels [17]. The presence of polar groups of hydroxyl and carboxyl on each repeat unit makes this polymer compatible with water and the hydrophobic α-methyl groups of the backbone convey hydrolytic stability to the polymer and enhance mechanical strength of the polymer matrix [18]. The drug chosen for this study was doxorubicin hydrochloride, which belongs to the family of anti-tumor drugs (Figure 1). Doxorubicin is a cytotoxic anthracycline antibiotic [19], it is widely used in the treatment of non-Hodkin's lymphoma, acute lymphoblastic leukemia, breast carcinomas, and several other types of cancer. We aimed to design a better PHEMA nanoparticulate delivery system for clinical administration of doxorubicin to achieve higher therapeutic efficacy and reduce side effects, with the overall aim to develop effective oral chemotherapy system.

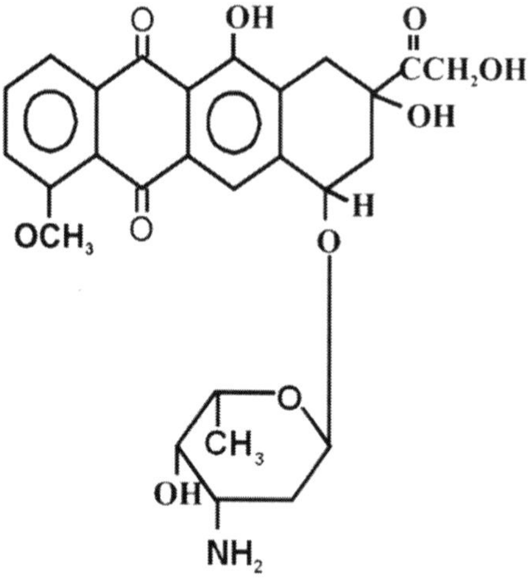

Figure 1. Structure of drug doxorubicin. Chemical structure of anticancer drug (Doxorubicin).

PREPARATION AND CHARACTERIZATION OF NANOPARTICLES

The FTIR spectra of the pure drug (doxorubicin) and loaded nanoparticles are shown in Figures 2a and b, respectively. The IR spectra (b) of loaded nanoparticles clearly indicate the presence of HEMA as evident from the observed bands at 1728 cm^{-1} (C=O stretching), 1172 cm^{-1} (O-C-C stretching), 3556 cm^{-1} (O-H stretching), 2951 cm^{-1} (asymmetric stretching of methylene group), and 1454 cm^{-1} (O-H bending) respectively. The spectra (b) also mark the presence of drug (doxorubicin) as evident from the observed bands at 10001260 cm^{-1} (C-O stretching of alcohol) and 675900 cm^{-1} (out of plane O-H bending). The resemblance of spectra shown in Figure 2a (the pure drug) and in Figure 2b (loaded nanoparticles) confirms the presence of drug in the loaded nanoparticles.

The SEM image of nanoparticles is shown in Figure 3a, which also reveals the morphology of PHEMA nanoparticles. The size of nanoparticles was estimated using SEM images. Under our experimental conditions it has been shown to vary between 100 and 300 nm. The particle size distribution curve of prepared nanoparticles is shown in Figure 3b. The small (defined) size of the nanocarriers results in the increased surface to volume ratio [20], enhanced frictional forces as well as adsorption [21]. These properties allow nanoparticles to be held in suspension and largely define their biological fate, toxicity, and targeting ability, as well as drug loading potential, their stability, and drug release properties. The interaction of nanoparticles with living systems depends on their characteristic dimensions. Previously published studies proved the ability of ultra small nanoparticles to translocate throughout the body (see [22] and references therein).

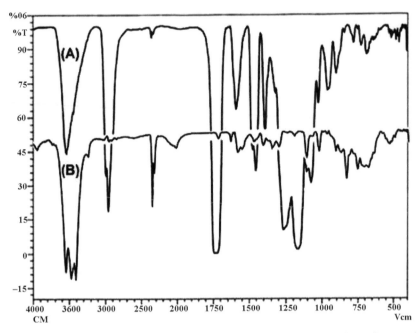

Figure 2. The FTIR spectra of nanoparticles. The FTIR spectra of (a) Pure drug doxorubicin and (b) Drug loaded PHEMA nanoparticles.

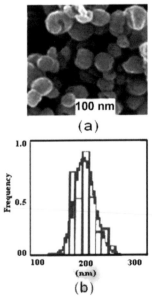

Figure 3. The SEM image and particle size distribution curve of nanoparticles. (a) The SEM images of cross-linked PHEMA nanoparticles and (b) Particle size distribution curve of cross-linked PHEMA nanoparticles.

The ξ-potential is the difference in the electrical charge developed between the dense layers of ions surrounding the particles and the charge of the bulk of the suspended fluid surrounding the particle; it gives information about the overall surface charge of the particles [23]. Thus, the measurements of ξ-potential may indicate the colloidal stability of nanoparticles. The interactions between the particles play an important role in determining colloidal stability. The use of ξ-potential measurements to predict stability attempts to quantify such interactions. Most nanoparticles have a tendency to aggregate which may lead to precipitation that could prove dangerous if those particles are injected intravenously. Since most aqueous colloidal systems are stabilized by electrostatic repulsion, the larger the repulsive forces between particles, the less likely they will come closer together and form an aggregate [24]. Therefore, it is important to know the surface charge, which directly controls aggregation behavior of the particles in the blood. The values of ξ-potential for unloaded and drug loaded nanoparticles are summarized in Table 1 which clearly indicates that loading of doxorubicin onto nanoparticles surfaces increases positive potential of the nanoparticles surface. The observed increase may be explained by the fact that drug molecules bear a positive charge and due to their loading onto the particle surface the positive charge increases on the surface, which also indicates towards the drug-surface interaction.

Table 1. Surface potential of nanoparticles.

pH	EMF(mV) Buffer solution	EMF(mV) Unloaded	EMF(mV) Loaded
1.2	276.5	260	280.1
7.4	−34.2	−34.4	−19.9
8.6	−82.8	−86.1	−45.2

Surface potential of unloaded and loaded PHEMA nanoparticles.
*EMF data have been expressed as Mean ± S.D. of at least three determinations.

MODELING OF THE RELEASE MECHANISM

The drug loaded PHEMA nanoparticles may be visualized as a three dimensional network of PHEMA macromolecules containing doxorubicin molecules, occupying the free space available between the network chains. During the drug release process, the drug diffuses through the hydrated polymer matrix into the aqueous phase. The process of hydration relaxes the polymer chains and enhances the diffusion of drug molecules. The rate of water uptake (hydration) of polymer particles increases with the hydrophilicity of polymer [25]. The doxorubicin molecules dissolve into water and release out through water permeation channels present in the macromolecular network. The diffusion of doxorubicin molecules and relaxation of PHEMA chains determine the type of release mechanism being followed by the drug molecules. According to Higuchi equation [26] when $n = 0.43$, the release is said to be diffusion controlled (Fickian), and when $n = 0.84$, the release is said to be non-Fickian (or case II). For n being in between 0.43 and 0.84, the mechanism becomes anomalous. In some cases n has been found to exceed 0.84 and the mechanism is known as super case II. The values of D and n are summarized in Table 2. The data demonstrate that the value of n lies between 0.43 and 0.84 in the majority of cases and, therefore, the release of doxorubicin may be considered as non-Fickian and swelling controlled.

Table 2. Release exponent and diffusion coefficient of nanoparticles.

S. No.	HEMA (mM)	EGDMA (mM)	BPO (mM)	pH	n*	$D \times 10^{15}$*cm^2 min^{-1}
1	12.37	1.06	0.248	7.4	0.46 ± 0.014	1.81 ± 0.054
2	16.49	1.06	0.248	7.4	0.61 ± 0.018	1.98 ± 0.059
3	20.61	1.06	0.248	7.4	0.50 ± 0.015	1.81 ± 0.054
4	24.73	1.06	0.248	7.4	0.55 ± 0.016	1.69 ± 0.051
5	12.37	0.53	0.248	7.4	0.40 ± 0.012	2.16 ± 0.065
6	12.37	1.59	0.248	7.4	0.44 ± 0.013	2.04 ± 0.061
7	12.37	2.12	0.248	7.4	0.45 ± 0.013	2.04 ± 0.061
8	12.37	1.06	0.082	7.4	0.58 ± 0.017	1.81 ± 0.054
9	12.37	1.06	0.165	7.4	0.47 ± 0.014	2.48 ± 0.074
10	12.37	1.06	0.33	7.4	0.58 ± 0.017	1.98 ± 0.059
11	12.37	1.06	0.248	1.8	0.86 ± 0.026	2.61 ± 0.078
12	12.37	1.06	0.248	8.6	0.60 ± 0.018	1.81 ± 0.054

Data showing the release exponent and diffusion coefficient under varying experimental conditions
*Data have been expressed as Mean ± S.D. of at least three determinations.

EFFECT OF PERCENT LOADING ON DRUG RELEASE

An important aspect in using nanoparticles as drug vehicle is the effect of the drug loading levels on the drug release rates. Higher drug loading may be achieved either by using highly concentrated drug solution or repeated soaking of nanoparticles in the drug solution and then drying them. In the present work, nanoparticles of defined composition were loaded with different amounts of doxorubicin by allowing the particles to swell in the drug solution of varying concentrations ranging between 1.2 and 2.4 mg/ml. The loaded particles were then allowed to release the entrapped drug into the release medium. Drug release results are shown in Figure 4. The amount of released doxorubicin increases with increasing percent loading. Similar results were reported previously by us and others for different drug release systems [27].

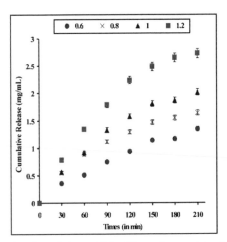

Figure 4. Effect of percent loading of doxorubicin. Effect of percent loading of doxorubicin on its release profiles from loaded nanoparticles of definite composition (HEMA) = 12.37 mm, (EGDMA) = 1.06 mm, (Bz_2O_2) = 0.248 mm, pH = 7.4, temp. = 37°C.

EFFECT OF MONOMER ON DRUG RELEASE

Doxorubicin release profiles are sensitive to chemical architecture of the carrier as well as the experimental conditions used to prepare the drug carrier. The effect of HEMA on the release of doxorubicin has been investigated by varying the monomer concentration in the range 12.3–24.7 mm. The swelling ratio and release results are shown in Figures 5a and b. Our data indicate that the swelling ratio and cumulative release of doxorubicin decreases with increasing concentration of HEMA. The results may be explained by the fact that as the content of PHEMA increases in the nanoparticles, the polymeric nanoparticles becomes largely crowded with polymer chains and this consequently reduces the free volume accessible to the penetrant water molecules. This obviously brings about a fall in the swelling ratio as well as the released amounts of drug. Another possible reason may be that with increase in PHEMA content the interaction between the polymer chains and the drug molecules increases which also results in a lower release of doxorubicin.

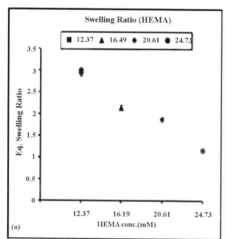
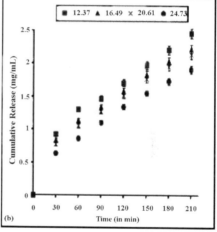

Figure 5. Effect of monomer (HEMA). Effect of monomer (HEMA) content on the (a) Swelling profile and (b) Release profile of the nanoparticles of definite composition (EGDMA) = 1.06 mm, (Bz_2O_2) = 0.248 mm, % loading = 28%, pH = 7.4, temp. = 37°C.

EFFECT OF CROSS-LINKER ON DRUG (DOXORUBICIN) RELEASE

In the cross-linked polymeric structures the swelling process may be controlled by the introduction of an appropriate amount of a second monomer with hydrophobic character. Chemically cross-linked hydrogels have been developed as carrier for drugs [28] in the last decade. Cross-linkers have pronounced effect on the swelling ratio as well as on kinetics of the drug release. We decided to use EGDMA, which is a known hydrophobic cross-linker, as a cross-linking agent in the present study. The effect of the degree of cross-linking on the swelling and drug release has been investigated by varying the concentration of EGDMA in the range of 0.53–2.12 mm in the feed mixture of the polymerization recipe. The swelling and release results are shown in

Figures 6a and b respectively. Initially the swelling ratio and drug release increase (up to 1.06 mm of EGDMA). Beyond this concentration, both the swelling ratio and drug release decrease. The observed increase is unusual and may be explained by loosening of the macromolecular chains of the nanoparticles. The latter may be due to the hydrophobic nature of EGDMA and the hydrophobic interactions occurring along EGDMA segments. The observed decrease in the released amounts of drug above 1.06 mm EGDMA could be because of the reduced free volume accessible to water molecules in more densely cross-linked polymers. Similar results have been reported for chitosan hydrogels (by Singh et al. [29]). We showed earlier that the introduction of a cross-linker increases the glass transition temperature (Tg) of the polymer, which restrains the mobility of network chains at experimental temperature and, therefore, lowers both the amount of water sorption as well as drug release [30].

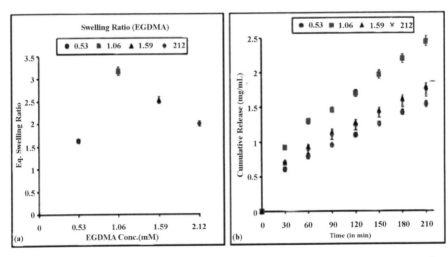

Figure 6. Effect of cross-linker (EGDMA). Effect of cross-linker (EGDMA) content on the (a) Swelling profile and (b) Release profile of the nanoparticles of definite composition, (HEMA) = 12.37 mm, (Bz$_2$O$_2$) = 0.248 mm, % loading = 28%, pH = 7.4, temp. = 37°C.

EFFECT OF INITIATOR ON DRUG RELEASE

In free radical polymerization the concentration of initiator has a direct impact on the molecular weight of the polymer [31]. We used Bz$_2$O$_2$ as a polymerization initiator and its concentration in the reaction mixture was varied in the range of 0.0820.330 mm. Swelling and release results are depicted in Figures 7a and b. An initial increase in concentration of Bz$_2$O$_2$ in the range of 0.0820.248 mm results in an increased swelling as well as drug release. The increase in the concentration of initiator may bring about an increase in the number of primary free radicals, which may eventually result in lower molecular weight of the PHEMA. Since a polymer with lower molecular weight has lower hydrodynamic volume in aqueous solution, the PHEMA chains acquire greater mobility and, therefore, show increased swelling and increased drug release. However, the higher concentration of the initiator results in shorter PHEMA chains and smaller mesh size of the polymer network and reduced drug loading and release.

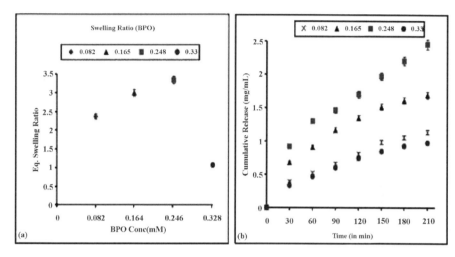

Figure 7. Effect of initiator (Bz_2O_2). Effect of initiator (Bz_2O_2) content on the (a) Swelling profile and (b) Release profile of the nanoparticles of definite composition, (HEMA) = 12.37 mm, (EGDMA) = 1.06 mm, % loading = 28%, pH = 7.4, temp. = 37°C.

EFFECT OF PH ON DRUG RELEASE

Since the pH change occurs at many specific or physiological sites in the body, it is one of the important parameters in the design of drug delivery systems. Several methods have been proposed for targeting the specific regions. Among these, utilization of pH changes within the gastrointestinal (GI) tract and exploitation of bacterial enzymes localized within the colon are of especial interest for the controlled drug delivery [31]. Differences in pH in the target site may allow a specific drug to be delivered to that target site only. The underlying principle for targeted drug delivery is the pH controlled swelling of hydrogel which normally results from the change in relaxation rate of network chains with changing pH of the medium. The pH profile of normal tissue is different from that of pathological tissues such as cancerous and infected tissues. Amiji et al. [32] reported that pH of normal tissue is higher than the pH of infected and tumor tissues. The physical properties of stimuli-responsive carriers such as swelling/deswelling, particles disruption and aggregation vary according to changing environmental conditions. These change the nanocarriers-cells interactions, and therefore the release of the drug at tumor site may be achieved. Drug loaded nanoparticles undergo rapid dissolution and release the drug content in the acidic microenvironments of a tumor [33].

In the present work, the release dynamics of the doxorubicin have been studied under varying pH conditions. The results are shown in Figures 8a and b. We found that less drug is released at physiological and alkaline pH, while the most efficient release is achieved at acidic (pH = 1.2) conditions. These results are not fully consistent with the swelling results. The unloaded PHEMA nanoparticles show maximum swelling at physiological pH solution, whilst loaded PHEMA nanoparticles show maximum release at acidic pH. The reason behind the lower swelling of unloaded nanoparticles in acidic and alkaline conditions is that the nanoparticles do not swell sufficiently in acidic and alkaline solutions. Drug loaded nanoparticles swell better in the acidic solution rather than at physiological pH or under alkaline conditions. Similar results have been reported recently for pH-sensitive liposomes [34]. These were stable at physiological pH of 7.4, but degraded to release active drug in target tissues in which the pH is less than physiologic values, such as in the acidic environment of tumor cells.

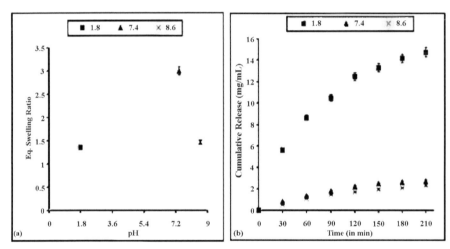

Figure 8. Effect of pH. Effect of pH on the (a) Swelling profile and (b) Release profile of the nanoparticles of definite composition (HEMA) = 12.37 mm, (EGDMA) = 1.06 mm, (Bz$_2$O$_2$) = 0.248 mm, % loading = 23%, temp. = 37°C.

EFFECT OF TEMPERATURE ON DRUG RELEASE

Temperature sensitivity is one of the most important characteristics in drug delivery technology. It has a direct influence on the swelling and release behavior of a hydrogel. Temperature affects both the segmental mobility of the hydrogel chains as well as the diffusion of penetrant molecules. In this study, the effect of temperature on the swelling ratio and drug release through PHEMA nanoparticles has been investigated by varying the temperature of the swelling medium in the range of 1237°C. The results are summarized in Figures 9a and b. Swelling increased with temperature up to 25°C, but decreased above this temperature. However, cumulative release was highest at 37°C. The increased temperature results in faster relaxations times of the polymer network due to the increased kinetic energy, which facilitates water sorption process [35].

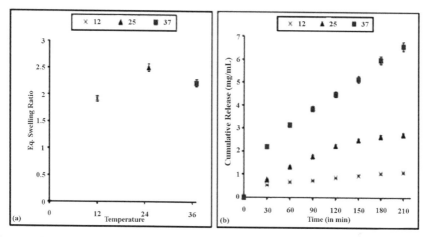

Figure 9. Effect of temperature. Effect of temperature of the release medium on the (a) Swelling profile and (b) Release profile of the nanoparticles of definite composition (HEMA) = 12.37 mm, (EGDMA) = 1.06 mm, (Bz_2O_2) = 0.248 mm, % loading = 28%, pH = 7.4.

EFFECT OF PHYSIOLOGICAL FLUIDS ON DRUG RELEASE

The influence of solutes on the swelling behavior and doxorubicin release kinetics was examined by performing swelling and release experiments in the presence of solutes such as urea (15% w/v) and D-glucose (5% w/v) and in physiological fluids such as saline water (0.9% NaCl) and synthetic urine. The results are shown in Figures 10a and b. The presence of additives reduces both the swelling ratio as well as drug release. Salts present in the release medium are likely to reduce the osmotic pressure in the system thus resulting in lower extent of swelling of loaded nanoparticles and the drug release.

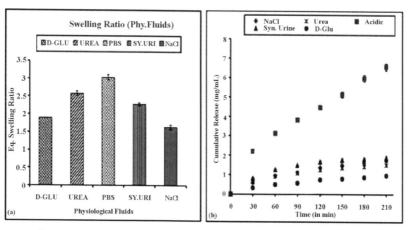

Figure 10. Effect of physiological fluids. Effect of physiological fluids on the (a) Swelling profile and (b) Release profile of the nanoparticles of definite composition (HEMA) = 12.37 mm, (EGDMA) = 1.06 mm, (Bz_2O_2) = 0.248 mm, % loading = 28%, pH = 7.4, temp. = 37°C.

CHEMICAL STABILITY OF THE ENTRAPPED DOXORUBICIN

The chemical stability of the entrapped drug was investigated by recording the UV-visible absorbance spectra of pure doxorubicin and the drug released into the release medium at different pH (Figure 11). There are no noticeable differences in the obtained absorbance spectra at all pH tested (pH1.8, pH7.4, pH8.6), suggesting no significant changes in the physical properties of the drug, and most likely of its chemical structure during nanoparticle loading and drug release.

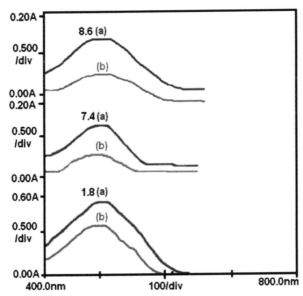

Figure 11. Chemical stability of doxorubicin. The UV spectra showing the chemical stability of doxorubicin in its pure solution (a) and released media (b) at different pH (1.8, 7.4, 8.6).

MATERIALS AND METHODS

The HEMA and ethyleneglycol dimethacrylate (EGDMA) were purchased from Sigma Aldrich Co. USA. Benzoyl peroxide (BPO) (MERCK) and polyvinyl alcohol (PVA) (Mol. Wt. 14000) (MERCK) were used as the initiator and the stabilizer, respectively. Toluene (MERCK) was used as the diluent. All chemicals were of analytical grade and doubly distilled water was used throughout the experiments.

The HEMA monomer was purified by using a previously reported method [30]. The purity of distilled HEMA was determined by high-pressure liquid chromatography (HPLC), (Backmen System (Gold 127)) equipped with a ultraviolet detector and a 25 cm × 46 mm id separation columns ODS (C_{18}) of 5 μm particle size. The UV detector was set at 217 nm. The mobile phase was methanol-water (60:40 v/v) and the flow-rate was kept at 1 ml/min. All samples were diluted with pure methanol to 1/1600. 10 μl samples were injected for each analysis. Samples of known concentrations of MAA and EGDMA were injected into the HPLC and the resultant chromatogram was used to construct a standard curve of known concentrations versus area under the curve. The chromatogram showed two distinct peaks. The first peak, at 3.614 min was identified as MAA. The next peak at 5.503 min was the major peak due to HEMA monomer. The amounts of impurities of MAA and EGDMA in the monomer samples were found to be less than 0.01 mol% MAA and 0.001 mol% EGDMA.

Preparative methods for making nanoparticles for pharmaceutical use are broadly divided into two categories, those based on physiochemical properties such as phase separation and solvent evaporation [36], and chemical reactions such as polymerization [37], and polycondensation. In the present study cross-linked PHEMA nanoparticles of defined composition were prepared by using a modified suspension polymerization technique, as previously reported by Kaparissides et al. [38]. In particular, the polymerization was carried out in an aqueous phase containing PVA, which was used as the stabilizing agent. The mixture containing the 12.37 mm HEMA (the monomer), 1.06 mm EGDMA (the cross-linker) and 0.248 mm Bz_2O_2 (the initiator) dispersed in toluene was added into 500 ml conical flask containing the suspension medium (200 ml aqueous PVA solution (0.5% W/V)). The reaction mixture was flushed by bubbling nitrogen and then sealed. The reaction mixture was then placed on magnetic stirrer and heated by vigorous stirring (600700 rpm) at 80°C for 2 hr and then at 90°C for 1 hr. The cross-linking reaction was completed within3 hr. After cooling, the polymeric particles were separated from the polymerization medium by washing thrice with toluene and twice with acetone. The collected nanoparticles were dried at room temperature to obtain the fine white powder and thereafter stored in airtight polyethylene bags.

The IR spectra of cross-linked PHEMA nanoparticles were recorded on a FTIR spectrophotometer (Perkin-Elmer, 1000 Paragon) (Shimadzu). While recording FTIR spectra KBr disc method was used for preparation of samples.

Morphological studies of cross-linked PHEMA nanoparticles were performed on scanning electron micrographs (SEMs). The SEM observations were carried out with a Philips, 515, fine coater. Drops of the polymeric nanoparticles suspension were placed on a graphite surface and freeze dried. The sample was then coated with gold by ion sputter. The coating was performed at 20 mA for 4 min, and observation was made at 10 KV. Nanoparticles were further characterized by particle size analysis for size and size distribution. The particle size analysis of prepared nanoparticles was performed on a particle size analyzer (Malvern Mastersizer, 2000).

Zeta potential studies were performed with a digital potentiometer (Model No. 118, EI Product, Mumbai, India). In a typical experiment 0.1 g nanoparticles were dispersed in 20 ml of respective pH solution and emf was recorded using a compound electrode system. A similar experiment was also repeated for drug-loaded nanoparticles.

Swelling properties of hydrogels can be used as a method to trigger drug release [39]. Swelling of nanoparticles was studied by a conventional gravimetric procedure. In a typical experiment, 0.1 g of nanoparticles were allowed to swell in a definite volume (10 ml) of PBS taken in a pre weighed sintered glass crucible (pore size 510 μm) and weighed after a definite period by removing excess PBS by vacuum filtration. The swelling process of nanoparticles was monitored continuously up to 15 min after which no weight gain of swollen nanoparticles was recorded which clearly indicates equilibrium swelling condition. The weight swelling ratio of nanoparticles was calculated from the following equation,

$$\text{Swelling Ratio} = \frac{W_t}{W_0},\tag{1}$$

where W_t is the weight of swollen nanoparticles at time t, and W_0 is the initial weight of dry nanoparticles (at time 0).

For loading of drugs onto nanoparticles, a known volume of drug doxorubicin was taken and diluted with the appropriate amount of PBS solution and shaken vigorously for mixing of drug and PBS solution. Drug-loaded nanoparticles were prepared by swelling 0.1 g of nanoparticles in freshly prepared drug solution (10 ml) until equilibrium swelling was reached. The percent loading of drug onto nanoparticles was calculated by the following equation:

$$\% \text{ Loading} = \frac{W_d - W_0}{W_0} \times 100,\tag{2}$$

where W_d and W_0 are the weights of loaded and unloaded nanoparticles, respectively.

In-vitro release of the loaded doxorubicin was carried out by placing the dried and loaded nanoparticles (0.1 g) in a test tube containing a definite volume (10 ml) of phosphate buffer saline (PBS) as the release medium (pH = 7.4) (1.2 mm KH_2PO_4, 1.15 mm Na_2HPO_4, 2.7 mm KCl, 1.38 mm NaCl). The amount of doxorubicin released from the polymeric nanoparticles was measured spectrophotometrically at 496 nm (Shimandzu 1700 Phama Spec.) and the released amount of drug was determined from the calibration plot.

To study the kinetics of the release process, drug-loaded nanoparticles were added to the release medium and the suspension was shaken for 3.5 hr. For monitoring the progress of the release process, aliquots were withdrawn at desired time intervals and the amount of drug released was estimated spectrophotometrically.

The drug release from polymeric nanoparticulate systems is actually the combination of Fickian (diffusion) and non-Fickian movements [40] of drug molecules through polymer chains. In the present study the kinetic data were analyzed with the help of the following equation, which could be helpful in determining the mechanism of the release process,

$$\frac{W_t}{W_\infty} = Kt^n,\tag{3}$$

where W_t and W_∞ are the amount of the drug release at time t and at infinity time (equilibrium amount of drug released), respectively, and K is rate constant. The constants K and n are characteristics of the drug-polymer system. An introduction to the use and the limitations of these equations was fist given by Peppas et al. [41]. For evaluating the diffusion constant of loaded drugs, the following equation can be used:

$$\frac{W_t}{W_\infty} = 4\left(\frac{Dt}{\pi L^2}\right)^{0.5},\tag{4}$$

where D is the diffusion constant of the drug and L being the diameter of the dry nanoparticles.

In order to check the chemical stability of entrapped drug in different release media, the UV spectral studies (Shimandzu 1700 Pharma Spec.) were performed as described in [42].

All experiments were done at least thrice and a fair reproducibility was observed. The data summarized in tables have been expressed as mean ± SD of at least three independent determinations. The plots were drawn taking the mean values and each curve has been shown to include error bars.

CONCLUSION

The PHEMA nanoparticles can be prepared by suspension polymerization method and characterized by techniques such as FTIR, SEM, and particle size analysis. The addition of model drug, doxorubicin, to polymeric nanoparticles results in 28% drug entrapment. Release profiles of doxorubicin can be greatly modified by varying the experimental parameters such as percent loading of doxorubicin and concentrations of HEMA, cross-linker, and initiator. Swelling of nanoparticles and the release of doxorubicin increases with the increase in percentage loading of drug. The amount of released drug decreases with increasing HEMA and EGDMA content of the nanoparticles. Increase in the concentration of the initiator, benzoyl peroxide, from 0.082 to 0.248 mm results in the increase of drug release, but this effect is reduced at higher concetration of benzoyl peroxide. The best combination of individual components for making PHEMA nanoparticles for doxorubicin delivery was 12.37 mm HEMA, 1.06 mm EGDMA, and 0.248 mm Bz_2O_2. Fast drug release was observed at acidic pH1.2 at 37°C whilst physiological and alkaline pH and lower temperature slow down the release of doxorubicin. Salts and additives affecting osmotic pressure also suppress the extent of drug release. Absorption spectra of doxorubicin do not change following its capture and release form the nanoparticles, indicating that chemical structure of the drug is likely to be unaffected by the procedure.

KEYWORDS

- **Anticancer agents**
- **Doxorubicin**
- **High-pressure liquid chromatography**
- **2-Hydroxyethyl methacrylate**
- **PHEMA nanoparticles**

AUTHORS' CONTRIBUTIONS

Raje Chouhan synthesized PHEMA nanoparticles and performed *in vitro* functional assays, A. K. Bajpai conceived the idea of nanoparticles, guided to conduct the studies, supervised data analysis, and authored the manuscript.

COMPETING INTERESTS

The authors declare that they have no competing interests.

Chapter 6

Monocyte Subset Dynamics in Human Atherosclerosis

Moritz Wildgruber, Hakho Lee, Aleksey Chudnovskiy, Tae-Jong Yoon, Martin Etzrodt, Mikael J. Pittet, Matthias Nahrendorf, Kevin Croce, Peter Libby, Ralph Weissleder, and Filip K. Swirski

INTRODUCTION

Monocytes are circulating macrophage and dendritic cell precursors that populate healthy and diseased tissue. In humans, monocytes consist of at least two subsets whose proportions in the blood fluctuate in response to coronary artery disease, sepsis, and viral infection. Animal studies have shown that specific shifts in the monocyte subset repertoire either exacerbate or attenuate disease, suggesting a role for monocyte subsets as biomarkers and therapeutic targets. Assays are therefore needed that can selectively and rapidly enumerate monocytes and their subsets. This study shows that two major human monocyte subsets express similar levels of the receptor for macrophage colony stimulating factor (MCSFR) but differ in their phagocytic capacity. We exploit these properties and custom-engineer magnetic nanoparticles for *ex vivo* sensing of monocytes and their subsets. We present a two-dimensional enumerative mathematical model that simultaneously reports number and proportion of monocyte subsets in a small volume of human blood. Using a recently described diagnostic magnetic resonance (DMR) chip with 1 μl sample size and high throughput capabilities, we then show that application of the model accurately quantifies subset fluctuations that occur in patients with atherosclerosis.

Circulating monocytes in humans fall into subsets typically identified by expression of the LPS receptor CD14 and the Fcγ receptor-III CD16. The $CD14^+CD16^{lo}$ monocytes predominate in the blood and express high levels of the CCL2 (MCP-1) receptor CCR2 while $CD14^{lo}CD16^{hi}$ monocytes are less abundant and express higher levels of the fractalkine receptor CX^3CR1. These expression patterns suggest differential tissue tropism, and indicate commitment of circulating monocytes for specific functional fates [1-4].

Atherosclerosis, a major cause of myocardial infarction, peripheral arterial disease, and stroke, is characterized by continuous accumulation of monocytes in the arterial intima [5, 6]. In the circulation, $CD16^{hi}$ monocyte counts rise in patients with coronary artery disease (CAD) [7], although the biological significance of this finding requires further investigation. Recent studies in animals have promoted the idea, however, that some monocyte subsets accentuate while others attenuate disease [8]. In this context, disease progression may be characterized by a subset imbalance that favors an inflammatory cell population. In hyperlipidemic atherosclerotic mice for example, one monocyte subset (Ly-6C^{hi}, likely corresponds to CD16^{lo} in human [1]) preferentially accumulates in growing atheromata and differentiates into macrophages while another subset (Ly-6C^{lo}, CD16^{hi} in human) accumulates less and likely differentiates to dendritic cells [9-10]. The possibility that monocytes participate divergently in lesion growth necessitates evaluation of how findings obtained in animals translate to humans, and whether monocyte subsets represent therapeutic targets and prognostic biomarkers.

High-throughput profiling of peripheral monocytes in humans requires tools of exceptional selectivity and sensitivity. Clinically, automated blood differentials quantify circulating monocytes but do not inform on subsets [11-12], and no other diagnostic tools are routinely available that can delineate how specific subsets fluctuate in population samples. In this study, we profiled phenotypic and functional characteristics of human monocyte subsets and focused on two distinguishing features: shared expression of MCSFR among subsets and differential capacity of monocyte subsets for phagocytosis. We custom-engineered novel magnetic nanoparticles and, using a recently developed DMR chip technology [13], simultaneously profiled monocyte and monocyte-subset changes for use in patients with atherosclerosis.

DIVERGENT PHENOTYPIC AND FUNCTIONAL PROPERTIES OF HUMAN MONOCYTE SUBSETS FURNISH PROSPECTIVE LABELING TARGETS

Monocyte heterogeneity may represent an as yet unexplored target for imaging and treatment [8, 14], but currently no clinical assays can simultaneously discriminate between monocyte subsets. Thus, we first focused on the biology of human monocyte subsets with the aim of uncovering a phenotype or function readily exploitable for specific and selective targeting. Mononuclear leukocytes were obtained from peripheral blood of healthy volunteers. The procedure uses density-gradient centrifugation, and thus enriches for mononuclear cells by removing neutrophils and other granulocytes. Monocytes are 10–30 μm in diameter, are therefore larger than lymphocytes, and occupy a distinct gate on a forward scatter/side scatter (FSC/SSC) flow cytometric dot plot (Figure 1A). Labeling with antibodies against CD14 and CD16 allows identification of two monocyte subsets: a dominant CD14^{+}CD16^{lo} (thereafter referred to as CD16^{lo}) population and a minor CD14^{lo}CD16^{hi} (CD16^{hi}) population (Figure 1A), as previously reported [15, 16]. Both subsets bear MCSFR (also known as CD115) (Figure 1B) but fall into two subsets identified by distinct expression profiles of trafficking (CCR2, CX$_3$CR1) and myeloid function/differentiation (CD11b, MPO, CD68, HLA-DR) markers (Figure 1C). Importantly, neutrophils, which were absent in the preparations, also express markers such as CD14, CD16, CD11b, and MPO, but not MCSFR. When cultured for 6 days with LPS and IFNγ, mediators that promote the acquisition of the M1-macrophage phenotype, both subsets, but not other cells, acquire morphologic characteristics of mature macrophages, and both subsets display increased levels of the macrophage marker CD68 (Figure 1D). Similar macrophage morphology and expression of CD68 occur in cells cultured with the M2-phenotype-promoting mediators IL-4/IL-13 (data not shown). However, freshly-isolated monocyte subsets phagocytose fluorescently-labeled latex beads differently: both subsets are positive for bead uptake, but the cellular bead concentration, as assessed by the beads' mean fluorescent intensity (MFI), is significantly higher in the CD16^{lo} population, indicating higher phagocytosis by this subset compared to its CD16^{hi} counterpart (Figure 1E). Altogether, these data identify at least two promising targets to label and track human monocytes and their subsets. The MCSFR expression is a potential candidate to selectively target monocytes because monocytes express MCSFR uniquely and at similar intensities. Phagocytosis lends itself to discrimination between monocyte subsets because targeting phagocytosis is simpler than targeting differential expression of membrane-bound proteins.

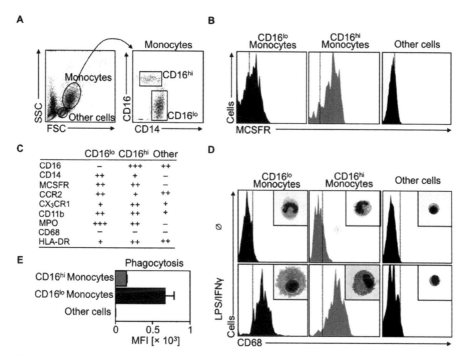

Figure 1. Human monocyte subsets differ phenotypically and functionally. A. Flow cytometry dot plots show forward scatter (FSC) versus side scatter (SSC) of mononuclear cells obtained from fresh blood. A monocyte gate is drawn and monocyte subsets are identified according to their CD14 and CD16 expression profile. B. Histograms depict MCSFR expression of CD16^{lo} monocytes, CD16^{hi} monocytes and other cells (mostly lymphocytes). C. Table summarizes relative expression profiles of selected markers for CD16^{lo}, CD16^{hi} monocytes and other cells. D. Representative histograms and H&E cytospin preparations show CD68 expression and morphology of CD16^{lo}, CD16^{hi} monocytes and other cells freshly isolated or after *in vitro* culture for 6 days with LPS/IFNγ. E. Bar graph depicts *ex vivo* phagocytosis of fluorescently labeled latex beads in CD16^{lo}, CD16^{hi} monocytes, and other cells (n = 4).

TARGETED FLUORESCENT IRON-OXIDE NANOPARTICLES ALLOW DISCRIMINATION OF MONOCYTE SUBSETS OPTICALLY

Having identified suitable targets to phenotype human monocytes, we next considered whether functionalized nanoparticles report on these targets with sufficient sensitivity and selectivity. We engineered a putative monocyte-targeted nanoparticle by covalently attaching antibodies against MCSFR to cross-linked iron oxide (CLIO) which is a dextran-coated, superparamagnetic nanoparticle [17]. The advantages of a nanoparticle-based strategy compared to antibody alone include the ability to conduct magnetic resonance sensing in optically turbid media such as blood, and the likely improvement in the stoichiometry of MCSFR targeting because, on average, a single CLIO molecule can bear 2–3 MCSFR antibodies. Further covalent attachment of the NIR fluorochrome VT680 allows for an independent, optical read-out. Incubation of mononuclear cells for 10 min at room temperature (RT) with increasing doses of fluorescently-tagged CLIO-MCSFR leads to similar labeling of both CD16^{hi} and CD16^{lo} monocyte subsets across all concentrations (Figure 2A), with no detectable labeling of

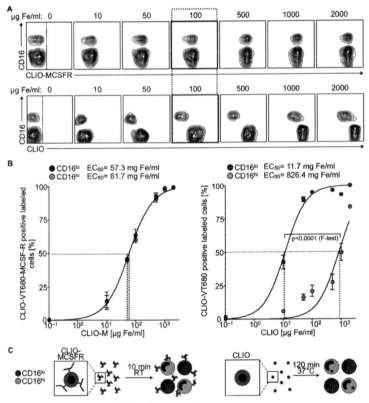

Figure 2. Nano-sensors discriminate between monocyte subsets optically. A. Representative flow cytometry contour plots of human monocytes labeled with increasing concentrations of fluorescent superparamagnetic nano-particles CLIO-MCSFR (top row) and CLIO (bottom row). To discriminate between subsets the mean fluorescent intensity of the particle (x-axis) is plotted against expression of CD16 (y-axis). B. The Fe concentrations for both nano-particles (CLIO-MCSFR and CLIO) are log-transformed (x-axes) and plotted against % of positive labeled cells (y-axes). A sigmoidal dose-response curve is generated to calculate the corresponding EC50 (nano-particle concentration at which 50% of each monocyte subset is labeled). N = 4. C. Principle of the assay. The principle postulates that equal binding of subsets with CLIO-MCSFR will occur after 10 min at RT while incubation of subsets with CLIO at 120 min at 37°C will result in preferential uptake of the particle by CD16^{lo} monocytes.

other cells (data not shown) and no toxicity at doses up to 1000 µg Fe/ml as detected by Trypan blue and Annexin V staining, in accord with the literature [18-19]. The labeling is antibody-specific because incubation of non-derivatized fluorescent CLIO for the same duration and temperature does not result in significant particle uptake (data not shown). The high MFI, especially at doses of 500–2000 µg Fe/ml suggests enhanced binding by the MCSFR antibody. Notably, the dose of 100 µg Fe/ml approximates Fe concentrations in human plasma detected shortly after intravenous injection of clinically approved doses of nanoparticles such as MION [20].

To test for phagocytosis, we hypothesized that non-derivatized CLIO, when incubated with cells at physiological conditions for a longer period of time, will report on differences in phagocytic capacity between subsets. In contrast to the similar labeling achieved with fluorescent CLIO-MCSFR, incubation of mononuclear cells for 120 min at 37°C with increasing doses of fluorescent CLIO leads to preferential uptake of the agent by CD16^{lo} monocytes at all concentrations tested (Figure 2A), reflecting the heightened capacity of these cells for phagocytosis. Leukocytes other than monocytes do not accumulate the particle significantly (data not shown).

Calculation of the concentration of the nanoparticles at which 50% of the cells are labeled (EC_{50}) used an equation derived from fitting of a sigmoidal concentration-dependence relationship (Figure 2B). The CLIO-MCSFR yielded a similar calculated EC_{50} between subsets: 57.3 mg Fe/ml for CD16^{lo} and 61.6 mg/Fe/ml for CD16^{hi} monocytes (p = 0.3845). In contrast, unconjugated CLIO yielded a significantly different calculated EC_{50} between the subsets: 11.7 mg Fe/ml for CD16^{lo} and 826.4 mg Fe/ml for CD16^{hi} subsets (p < 0.0001). Thus, the equal labeling of monocytes with fluorescent CLIO-MCSFR permits identification of monocytes while differential uptake of fluorescent CLIO effectively discriminates between monocyte subsets (Figure 2C).

MONOCYTE SUBSET FLUCTUATIONS CAN BE RESOLVED WITH A DIAGNOSTIC MAGNETIC RESONANCE (DMR)-CHIP AND MODELED MATHEMATICALLY FOR ENUMERATION STUDIES

An emerging application of nanotechnology for clinical high-throughput screening and diagnosis utilizes a chip-based DMR system [13]. The sensitivity of DMR technology permits analysis of rare targets in sample volumes ≤1 µl with few or no sample purification. Before testing with the DMR chip, we first sought to determine with conventional approaches the feasibility of discriminating between subsets by magnetic resonance. Equal numbers of sorted monocyte subsets were labeled at different Fe concentrations with either CLIO-MCSFR or CLIO and were investigated by Magnetic Resonance Imaging (MRI) with a multi-slice multi-echo sequence at 7 T (Figure 3A) and with a conventional bench top relaxometer at 0.5 T (Figure 3B). Leukocytes other than monocytes were used as controls. Both CD16^{lo} and CD16^{hi} monocyte subsets experienced a similar decrease in spin–spin relaxation time (T_2) when labeled with CLIO-MCSFR. The decrease was concentration-dependent, significantly less pronounced than in other cells, and evident with both methods. In contrast, CD-16lo monocytes showed an accentuated decrease in T_2 values compared to CD16^{hi} monocytes when labeled with CLIO alone. The T_2 decrease for CD16^{hi} monocytes was intermediate when compared to CD16^{lo} and control leukocytes, and was likewise concentration-dependent and method-independent. These data indicate the feasibility of discriminating between subsets with targeted superparamagnetic nanoparticles, and suggest that the DMR chip, which is expectedly more sensitive and requires smaller samples volumes, provides a promising approach.

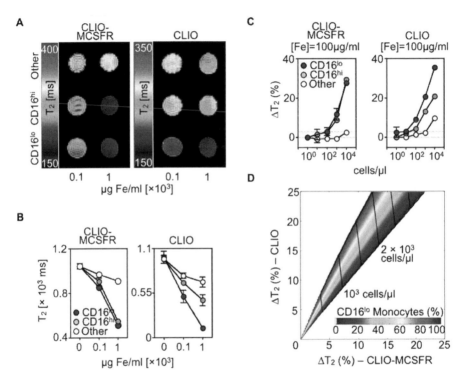

Figure 3. *Ex-vivo* nuclear magnetic resonance generates an enumerative mathematical model for monocyte subsets. A. Representative NIH-color coded map generated from T$_2$-weighted magnetic resonance imaging. Data show equal number of CD16^{lo}, CD16^{hi} monocytes, and other leukocytes labeled with two CLIO-MCSFR (left panel) and CLIO (right panel) concentrations. B. The T$_2$ measurements detected with a conventional benchtop-relaxometer. Data show equal number of CD16^{lo}, CD16^{hi} monocytes, and other leukocytes labeled with two CLIO-MCSFR (left panel) and CLIO (right panel) concentrations. N = 3–5. Mean ± SEM. C. The T$_2$ changes detected with a diagnostic magnetic resonance (DMR) chip. Data show increasing number of CD16^{lo}, CD16^{hi} monocytes, and other leukocytes labeled with one CLIO-MCSFR (left panel) and CLIO (right panel) concentration. N = 3. Mean ± SEM. D. Two-dimensional T$_2$ map derived from data in C to simultaneously enumerate total monocyte numbers and subset proportions. Model combines T$_2$ changes for CLIO-MCSFR (x-axis) and CLIO (y-axis). Changes in predicted monocyte number are demarcated with vertical lines while the rainbow region defines monocyte subset fluctuations.

To test the DMR chip for monocyte sensing we next incubated a priori known numbers of monocyte subsets with 100 µg Fe/ml of either CLIO-MCSFR or CLIO and measured T$_2$ changes (ΔT$_2$) at different cell concentrations. The DMR chip detects as few as 100 monocytes (105 cells/ml) resuspended in human serum, with robust read-outs at higher monocyte concentrations (Figure 3C). This sensitivity suffices to detect monocytes in healthy controls in volumes of blood as low as 1 μl. Both CD16^{hi} and CD16^{lo} monocyte subsets show similar ΔT_2 when incubated with CLIO-MCS-FR. However, $\Delta T2$ are higher in CD16^{lo} monocytes when the phagocytic capacity is probed with CLIO. Labeling of leukocytes other than monocytes is minimal with either particle.

On the basis of these data, we formulated an enumerative mathematical model for monocyte populations and their subset proportions. First, we determined the cellular relaxivity (R_{ij}) for each cell type (i: CD16^{hi}, CD16^{lo}, and others) and nanoparticle (j: CLIO-MCSFR and CLIO) combination by fitting the titration curve (Figure 3C) into $\Delta(1/T_2)_j = R_{ij} \cdot N_i$, where N_i is the concentration of a given cell type (i). When a sample with heterogeneous cell composition is probed with a nanoparticle (j), the total $\Delta(1/T_2)_j$ is approximated as the sum of the individual contributions by each cell type (i): $\Delta(1/T_2)_j = \sum R_{ij} \cdot N_i$. We thus obtained two equations of $\Delta(1/T_2)_j$ for each nanoparticle (j: CLIO-MCSFR or CLIO) which can be solved to determine N_i for each monocytes subset. Note that the contribution from other cells can be exactly compensated provided that N_{others} is known. Otherwise, we could use the highest $\Delta(1/T_2)$ for other cells (Figure 3C) to obtain a conservative estimation on monocyte populations. Applying the method, we could then construct a 2-dimensional ΔT_2 map (Figure 3D) that can be used to simultaneously determine the total monocyte population and the subset proportions from observed T_2 changes.

PATIENTS WITH ATHEROSCLEROSIS HAVE AN ALTERED MONOCYTE SUBSET PROFILE DETECTABLE BY A TWO-DIMENSIONAL DMR-CHIP-BASED ASSAY

A high-throughput assay that enumerates cells and discriminates between subsets requires that the cells in question fluctuate within the assay's dynamic detection range. As a proof of principle, we investigated monocyte numbers and proportions in two cohorts: healthy volunteers and patients with CAD undergoing cardiac catheterization at Brigham and Women's Hospital. Blood from 12 healthy volunteers contains, on average, $86.5 \pm 1.0\%$ of CD16^{lo} and $11.5 \pm 0.7\%$ CD16^{hi} monocytes, whereas blood from 18 patients with CAD contains 78.4 ± 1.7 CD16^{lo} and $19.7 \pm 1.7\%$ CD16^{hi} monocytes (Figure 4A). Comparison of the relative and absolute changes between healthy volunteers and patients with CAD reveals a greater range in patients with CAD both in terms of proportion and monocyte cell number (Figure 4B). Elevation of CD16^{hi} monocytes is significant proportionally ($p = 0.0016$) and in absolute numbers ($p = 0.0192$). While larger and better-controlled clinical studies are needed to determine this finding's significance for example, whether altered monocyte subset numbers portend prognosis in patients with atherosclerosis this preliminary enumeration reveals a sufficient sensitivity of a DMR-chip assay to detect values observed in the clinic.

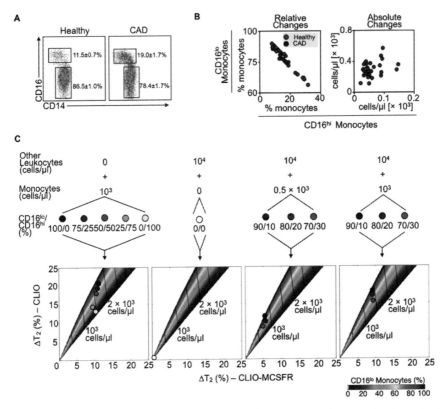

Figure 4. Magnetic nano-sensors enumerate monocyte subset variations that occur in atherosclerotic patients. A. Representative flow cytometry dot plots of monocyte subsets from healthy volunteers and patients with documented coronary artery disease (CAD). Numbers depict percentage of subsets in both groups. Mean ± SEM. B. Plots depict percentage (left plot) and absolute numbers (right plot) of CD16^{lo} and CD16^{hi} monocytes from healthy volunteers and patients with CAD. C. Validation of the enumerative mathematical model with varying number and percentage of monocyte subsets. Data show goodness of fit of defined numbers of monocytes alone (left panel), other leukocytes (middle-left panel) and combinations of leukocytes and monocytes (right two panels). Different proportions of monocyte subsets (CD16^{lo}/CD16^{hi} (%)) are color-coded and their fit is depicted on the two-dimensional T$_2$ maps.

We next evaluated the accuracy of the DMR assay to profile monocyte populations in complex and physiological cell compositions. This is important, as it would reveal whether the method has clinical, high throughput potential. Samples were incubated with CLIO-MCSFR or CLIO, and observed T$_2$ changes were plotted on the two dimensional ΔT$_2$ map already formulated and shown in Figure 3D. First, we tested the accuracy of the model to quantify purified monocytes at various subset proportions (Figure 4C, left panel). T$_2$ changes show strong agreement with the model over the proportional range (%CD16^{lo}/%CD16^{hi}: 100/0, 75/25, 50/50, 25/75, 0/100). As a negative control, leukocytes other than monocytes were incubated and, as expected, the observed ΔT$_2$ values are negligible (Figure 4C, center left panel). However, monocytes are outnumbered by other leukocytes in blood and subsets fluctuate across a range of ~30% (Figure 4B). We therefore evaluated samples in which physiological monocyte concentrations (0. 5 × 10^3 and 10^3 cells/μl) at various physiological subset proportions (%CD16lo/%CD16hi: 90/10, 80/20, 70/30) were mixed with other leukocytes, also at physiological concentrations (104 cells/μl). The T_2 changes for CLIO-MCSFR fit into the model's prediction, although the model slightly overestimates total monocyte counts, possibly due to the increased sample viscosity and the attendant T_2 shortening. For a given cell population, the T_2 changes for CLIO correctly reports the subset ratios. These experiments demonstrate that the DMR assay enumerates monocyte subsets in human samples and measures ≤10% changes of monocyte subsets even when the monocytes represent only a rare leukocyte population. Importantly, the method detects cells in volumes much smaller than is required for flow cytometry.

DISCUSSION

This study reports the development and evaluation of an assay that enumerates circulating monocyte subsets in small sample volumes (1 μl). The assay detects small changes of rare leukocyte subset populations with sensitivies not possible with conventional clinical tools. The simultaneous enumeration of total monocytes and their relative proportion is feasible because of magnetic nanoparticles that target different cellular features and properties. The method described here is therefore customizable to profile various cellular phenotypes for comprehensive disease screening.

The recognition of the involvement of inflammation in atherosclerosis promises to lead to improved detection and treatment modalities for patients at risk for atherothrombotic events. This study focused on analysis of monocyte subset fluctuations that occur in patients with atherosclerosis as these cells participate critically in disease progression and can be obtained simply with blood withdrawals [3, 21]. Moreover, the observation that monocyte subsets differentially participate in experimental atherogenesis [9-10, 22] underscores the need to assess the consequence of their fluctuations in humans. Some clues are already available: human coronary artery lesions contain macrophage subpopulations with different gene expression patterns [23], while CD16[lo] monocytes accumulate lipids preferentially in vitro [24]. It remains unknown whether functional monocyte heterogeneity in atheromata arises from divergent stimuli encountered by infiltrating but uncommitted macrophage precursors, or whether functionally committed monocyte subsets accumulate in specific lesional niches, or at defined moments of plaque evolution, to influence plaque growth or stability. An altered monocyte subset profile in patients with atherosclerosis argues for the latter possibility, even if it does not preclude the former, and suggests that monocyte subsets may serve as biomarkers and targets for discriminate monocyte-based therapeutic intervention.

Future large scale studies that determine if and how blood monocyte profiles predict complications of atherosclerosis will benefit from sensitive and efficient diagnostic tools such as the nanoparticle-based DMR chip assay described here, particularly given the assay's relatively small sample needs, low cost, and high-throughput capability. Because the assay depends primarily on R2 relaxivities of the magnetic nanoparticles and the volume of the sample [13, 25], future efforts will include de novo synthesis of magnetic nanoparticles with higher R2 compared to conventional particles, and further miniaturization and improvement of the Nuclear magnetic resonance (NMR)-microfluidic chips. The next generation of DMR systems may detect single cells in a small volume of turbid and diverse media. Application of this technology may further identify that numerous pathologies other than atherosclerosis, such as cancer, HIV infection, sepsis, or kidney failure, carry a specific "monocyte-subset signature" [11, 26-32], and it will remain to be determined how such observations associate with disease prognosis and severity. Finally, the method should have even more widespread utility for detection and profiling of other rare cell populations.

MATERIALS AND METHODS

Isolation of Human Monocytes

Ethics Statement

The protocol was approved by the Institutional Review Board at Brigham and Women's Hospital, Boston. Whole blood was obtained from healthy volunteers. CAD samples were obtained from patients undergoing cardiac catheterization at Brigham and Women's Hospital. Patients included in the protocol were diagnosed with CAD, defined by >70% stenosis in one or more epicardial coronary artery determined by coronary angiography. All donors gave written and informed consent. Fresh whole blood was drawn into heparinized collection tubes. To obtain leukocyte suspensions, whole blood was diluted 1:1 with DPBS and 20 ml diluted blood was overlaid on a 15 ml density gradient (Ficoll-Paque Plus, density 1.077 g/ml, GE Healthcare, NJ) and centrifuged (20 min, 1600 rpm, 18°C). The mononuclear cell interphase was carefully isolated and washed three times with DPBS. Resuspended cell suspensions were counted using Trypan blue (Cellgro, Mediatech Inc., Manassas, VA).

Cells

Cell suspensions were stained with the following antibodies (all from BD Bioscience, unless otherwise stated) at a final concentration of 1:100: CD11b-APC-Cy7/ICRF44, CD14-PE/M5E2, CD16 PE-Cy7/3G8, CCR2-Alexa-647/48607, CX$_3$CR1-FITC/2A91 (MBL International, Woburn, MA), MPO-FITC/2C7 (AbD Serotec, Raleigh, NC), HLA-DR-APC/L243, CD68-PE/Y1-82A, MCSFR-FITC/61708 (R&D Systems, Minneapolis, MN). Intracellular MPO staining was performed after fixation and permeabilization (BD Cytofix/Cytoperm, BD Bioscience, San Jose, CA). Cell phenotyping was performed using a LSRII Flow Cytometer (BD Bioscience, San Jose, CA) after appropriate compensations. For cell sorting, cells were labeled with CD14/CD16 and flow-sorted with a FACSAria (BD Bioscience, San Jose, CA). Purity of each monocyte subset population was >95% as determined by post-FACS flow-cytometric assessment. Non-monocyte control cells were sorted according to their Forward- and Side-Scatter profiles. These cells were ~90% lymphocytes on post-FACS analysis and were therefore considered "other cells"; they do not include neutrophils or granulocyte that were depleted with density gradient centrifugation. For *in-vitro* differentiation into macrophages, FACS sorted monocyte subsets were cultured in 200 µl cultures for 6 days in 1640 RPMI (containing glutamine, supplemented with 10% heat-inactivated FCS, 1% Penicillin/Streptomycin, Cellegro, Washington, DC) with additional cytokine stimulation: with LPS at 100ng/ml (Sigma, Saint Louis, MO) and IFNγ at 1000 U/ml as well as with IL-4 and IL-13, both at 20 ng/ml (from R&D Systems, Minneapolis, MN). Cytokine-enriched medium was changed every 2 days. Viability of cell culture was assessed by trypan blue. Flow cytometric data were analyzed using FlowJo v.8.5.2 (Tree Star, Inc., Ashland, OR). For morphologic characterizations, sorted cells were prepared on slides by cytocentrifugation (Shandon, Inc., Pittsburgh, PA) at $10 \times g$ for 5 min, and stained with HEMA-3 (Fischer Scientific, Pittsburgh, PA).

Phagocytosis Assay

For the determination of differences in phagocytosis between both monocyte subsets, yellow-green labeled latex beads were used (Bead size 2.0 μm, Sigma, Saint Louis, MO). The FACS-sorted monocyte subsets where incubated at a cell/bead ratio of 1/10 for 4 hr at 37°C in RPMI 1640, supplemented with 1% Penicillin/Streptomycin and 10% heat-inactivated fetal calf serum (FCS). After incubation, free beads were washed from the cell suspension three times and cells were analyzed by flow cytometry.

Magnetic Nanoparticles

For optical assessment a Near-infrared fluorescent (NIRF) fluorochrome (VT680, Excitation 670 ± 5 nm, Emission 688 ± 5 nm, Visen Medical, Woburn, MA) was coupled to CLIO nanopaticles as previously described [33]. Fluorescent CLIO had the following properties: Size ~ 30 nm, R1 = 28.8 s^{-1} mm^{-1} [Fe], R2 = 74.3 s^{-1} mm^{-1} [Fe]. For further experiments monoclonal MCSFR antibody (Clone 61708, R&D Systems, Minneapolis, MN) was coupled covalently to fluorescent CLIO. An average of 2.5 antibodies were immobilized per nanoparticle as determined with bicinchoninic acid assay. The resulting CLIO-MCSFR particles were protected from light and stored at 4°C. For uptake experiments non-derivatized CLIO was used (Center for Molecular Imaging Research, Massachusetts General Hospital, Charlestown, MA). For the detection of VT680, an LSR II Flow Cytometer was equipped with a 685/LP and 695/40 BP filter.

Exposure of Cell Suspensions to Magnetic Nanoparticles

The FACS sorted monocytes or unsorted leukocytes were plated at 100 K/200 μl in a 96 well plate. Cell labeling with iron-oxide nanoparticles was performed at different concentrations from 0 to 2000 μg Fe/ml in 1640 RPMI. Cell suspensions were incubated at the following conditions: CLIO-MCSFR (10 min, RT), CLIO (2 hr, 37°C, humidified CO_2 atmosphere). After the incubation period, cell suspensions were washed three times to separate labeled cells from unbound particles. For optical assessment by flow cytometry, cells were additionally stained with CD14 and CD16. After labeling, cells were counted again with trypan blue.

Magnetic Resonance Sensing

FACS-sorted monocyte subsets and control cells were incubated with CLIO-MCSFR and CLIO at 100 μg Fe/ml. 10^5 labeled cells were resuspended in 300 μl sucrose gradient solution (Ficoll-Paque Plus, density 1.077 g/ml, GE Healthcare, NJ) to prevent sedimentation of cells during sensing [34]. The Ficoll-cell suspension was subsequently embedded in an agarose-gel phantom, which minimizes susceptibility artifacts caused by interfaces with air or plastic. The MRI was performed using a 7-T horizontal-bore scanner (Pharmascan, Bruker, Billerica, MA) and a volume coil in birdcage design (Rapid Biomedical, Wuerzburg, Germany). A T_2 weighted multi-slice multi-echo sequence was used with the following parameters: TE = 8.8 ms, TR = 2330 ms, flip angle = 90 degrees excitation, and 180 degrees refocusing, slice thickness = 1 mm, matrix 128 × 128, FOV 4 × 4 cm. The NIH-color coded T_2 maps were calculated using OsiriX (Geneva, Switzerland).

Benchtop NMR Relaxometer

For T2 measurements with a conventional Benchtop Relaxometer (Minispec mq20, Bruker BioSpin, Billerica, MA), FACS-sorted monocyte subsets and control cells were incubated with CLIO-MCSFR and CLIO at 100 µg Fe/ml. 100K of labeled cells were resuspended in 300 µl 1% Triton ×100 solution and T_2 assessment was performed at 0.47 T (20 MHz).

DMR Assay

For rapid detection of monocyte heterogeneity in small sample volumes, we custom-designed a DMR chip, as previously reported [13]. The DMR system consists of a solenoidal microcoil for NMR detection, a microfluidic channel for sample handling, a single-board NMR electronics and a small permanent magnet (0.5 T). The microcoil was embedded along with the microfluidic channel to achieve high filling factor (≈1) and thereby larger NMR signal. For DMR assay, FACS sorted monocyte subsets were incubated with CLIO-MCSFR and CLIO (100 µg Fe/ml) as described above. Labeled cells were resuspended in human serum and DMR measurements were performed with 1 µl sample volumes. The T_2 were measured using Carr-Purcell-Meiboom-Gill pulse sequences with the following parameters: TE = 4 ms, TR = 6 s; the number of 180° pulses per scan, 500; the number of scans, eight. For ΔT_2 calculation, T_2 differences were calculated between magnetically labeled and cell-number matched control samples. All measurements were performed in triplicate.

Statistics

Data are expresses as Mean±SEM. For group comparisons Student's t test was used. For multiple comparisons ANOVA was used. For assessment of the EC50 a sigmoidal dose-response equation was chosen (Y = Bottom+(Top-Bottom)/1+10ˆ((Log EC50-X)*Hill Slope))), for comparison of the EC_{50} of the two different monocyte subsets F-test was used. When ΔT_2 values are presented the T_2 relaxation time of a sample with non-labeled cells, resuspended in human serum, is used as reference. Each measurement was carried out at least in triplicates, unless stated otherwise. P < 0.05 was considered statistically significant. Statistical analysis was performed with GraphPad Prism 4.0c for Mac (GraphPad Software, Inc, San Diego, CA).

KEYWORDS

- **Atherosclerosis**
- **Diagnostic magnetic resonance**
- **Fluorescent intensity**
- **Macrophage colony stimulating factor**
- **Monocytes**
- **Phagocytosis**

ACKNOWLEDGMENTS

The authors thank Nikolay Sergeyev with nanoparticle synthesis, Mike Waring and Andrew Cosgrove with Cell Sorting, Anne Yu and Ren Zhang with MRI and Timur Shtatland, with Bioinformatics (MGH and Harvard Medical School).

AUTHORS' CONTRIBUTIONS

Conceived and designed the experiments: Moritz Wildgruber, Mikael J. Pittet, Matthias Nahrendorf, Ralph Weissleder, and Filip K. Swirski. Performed the experiments: Moritz Wildgruber, Hakho Lee, Aleksey Chudnovskiy, Tae-Jong Yoon, and Martin Etzrodt. Analyzed the data: Moritz Wildgruber, Hakho Lee, Tae-Jong Yoon, and Filip K. Swirski. Contributed reagents/materials/analysis tools: Hakho Lee, Kevin Croce, Peter Libby, and Ralph Weissleder. Wrote the chapter: Moritz Wildgruber and Filip K. Swirski.

Chapter 7

Singlet-fission Sensitizers for Ultra-high Efficiency Excitonic Solar Cells

J. Michl

INTRODUCTION

Sensitizer dyes capable of producing two triplet excited states from a singlet excited state produced by the absorption of a single photon would allow an increase of the efficiency of photovoltaic cells by up to a factor of 1.5, provided that each triplet injects an electron into a semiconductor such as TiO_2. Although singlet fission (SF) in certain crystals and polymers was reported long ago, little is known about its efficiency in dyes suitable for use as sensitizers of photo-induced charge separation on semiconductors surfaces. In the present project, we have accomplished the following, in collaboration with Prof. A. J. Nozik at National Renewable Energy Laboratory (NREL) and with a subcontractor, Prof. M. A. Ratner at Northwestern University:

1. A theoretical analysis and a series of computations established that biradicaloids and alternant hydrocarbons are likely parent structures for meeting the exothermicity requirement $E(T_2)$, $E(S_1) > 2E(T_1)$ for the excitation energies of the lowest excited singlet (S_1) and the two triplet (T_1, T_2) states.

2. 1,3-Diphenylisobenzofuran (1) has been chosen as a model compound of the biradicaloid type, and a complete spectroscopic and photophysical characterization has been obtained. In the neat solid state, 1 forms triplets by SF in a yield of at least 10% and possibly as high as 50%. This appears to be the first compound displaying SF by design.

3. We have performed calculations of the degree of coupling to be anticipated in a large number of possible covalent dimers and pointed out the existence of contradictory requirements: the two halves of the dimer need to be coupled strongly enough for fast SF kinetics and weakly enough for favorable SF exothermicity (the coupling should not stabilize the excited singlet excessively to keep its excitation energy above twice the triplet excitation energy, as they are in the monomer).

4. We have begun to explore ways in which the two conditions can be met simultaneously. We have synthesized dimers of 1 in which these chromophores were attached covalently to each other in three different ways differing in the nature of coupling of the two halves.

5. Upon examining their photophysics we found that in non-polar solvents the two more weakly coupled dimers 2 and 3 have singlet excitation energies very similar to those of the monomer 1 but produce no triplets and only fluoresce.

The rate of SF is clearly too slow to be competitive, and we conclude that the coupling of the two halves is too weak. In polar solvents both form triplets in yields of up to 9% by converting into a intramolecular charge-transfer intermediate, which then undergoes intersystem crossing (ISC), but this process is not useful for our purposes.

6. The third dimer 4 is much more strongly coupled and its excited singlet is clearly stabilized significantly. This dimer provides an illustration of the above mentioned contradiction: the coupling is strong enough to secure fast SF but also strong enough to make SF endoergic. The SF is observed, but it does not proceed from the relaxed excited singlet state and proceeds only upon excitation to a higher state. Then, it proceeds fast enough to be somewhat competitive with internal conversion and vibrational deactivation, but because of this competition it affords a triplet yield of only ~3%. Details of the process, which appears to yield the first observed quintet state of an aromatic molecule, are currently under investigation funded by a different contract.

7. In summary, we have produced the first designed crystalline material for SF and we have made the first determinations of the coupling strength in a covalent dimer that is required for efficient SF We have suggested paths for overcoming a problem that has been identified and have already started to develop them in a follow-up project.

This project has been done with a subcontractor, Prof. Mark A. Ratner of the Northwestern University. The work has been performed in close collaboration with the research group of Prof. Arthur J. Nozik at NREL. In particular, time-resolved spectroscopy and steady-illumination triplet spectral measurements on the samples prepared in our laboratory were done at NREL.

Single-stage photovoltaic cells can be inexpensive but suffer from low efficiency, in part because they only utilize that part of the energy of an absorbed photon that corresponds to a semiconductor band gap or to the lowest energy transition of a dye sensitizer, and convert the rest to heat. The SF is being considered as one possibility for improving the situation [1]. In this process, a singlet excitation S_1 of a molecular chromophore is converted into triplet excitations T_1 on two molecular chromophores, both of which can then in principle generate electron-hole pairs. The SF process is the inverse of the long known and much studied triplet–triplet annihilation. It has been recognized for decades from studies of more or less randomly chosen organic molecular crystals, and also of certain polymers and oligomers, but it has not been widely studied or put to practical use. The literature on SF has been recently collected.

A quantitative analysis of the possible contribution of SF to excitonic solar cell efficiency [2] has shown that an improvement by as much as a factor of 1.5 is possible theoretically in a cell in which light first passes through a layer of semiconductor containing an adsorbed SF sensitizer with an absorption edge at ~2 eV and then through a layer of the same semiconductor carrying an ordinary sensitizer with an absorption edge at ~1 eV. It is assumed that the photons absorbed in the SF layer each generate two electron-hole pairs, and those absorbed in the second layer each generate a single electron-hole pair. For suitable sensitizer choices, all of the electrons generated are at the same potential and can be injected into the conduction band of the semiconductor and then transported to an electrode. All of the holes generated are at the same potential and can be transported to the other electrode. For efficient operation of the SF sensitizer one has to make sure that electron injection from the originally excited singlet state S_1 is slower than SF, while electron injection from the triplet state T_1 is faster than the decay of the T_1 into the ground state S_0. This appears feasible, given the huge difference between the usual lifetimes of the S_1 and T_1 states. At the same time, it needs to be recognized that the long T_1 lifetime provides opportunities for triplet-triplet recombination, which would defeat the purpose of the whole exercise.

GENERAL CONSIDERATIONS FOR A SINGLET FISSION SENSITIZER

The adsorbed sensitizer could be in the form of nanocrystals or another type of non-covalent aggregate, or it could be in the form of a covalent polymer, oligomer, or dimer. There are advantages and disadvantages to all these choices. Most of the instances of rapid and presumably efficient SF that have been reported occurred in crystals, and they might be a natural choice. The crystallites could be prepared separately and then applied to the semiconductor surface, as long as they are small enough to penetrate into the pores in the semiconductor particles. They could also be grown directly on the surface from solution. The lifetime of the triplet exciton is limited and if the time required for it to diffuse to the crystal wall and to be injected into the semiconductor is excessive, the injection efficiency will be low. Similarly, the hole that results from the injection needs to diffuse to the opposite surface of the crystal, where it is to be transferred to a shuttle such as the iodide anion, or to a hole-conducting polymer. This diffusion also needs to be fast in order to minimize electron-hole recombination. Control of crystal or aggregate size therefore becomes critical.

Covalent polymers have been observed to undergo SF, and their structure and size might be easier to control. Also the adsorption process, which would occur from solution, seems simpler. The observation of SF in carotenoids, which can be viewed as covalent oligomers of ethylene, capable of decoupling into two smaller oligomers by twisting about an internal bond, suggests that other covalent oligomers might be suitable as well. In the extreme of smallness, one could use a covalent dimer. This would appear to be the simplest approach, but the one case reported in the literature is discouraging. Although crystalline tetracene seems to undergo SF quite efficiently, a covalent dimer of tetracene exhibited only an extremely low yield of SF [3].

The next issue to address is the nature of the coupling between the individual chromophores that is optimal for the SF process to proceed fast. Very little if anything was known about this from theory. We thought at first that the coupling apparently does not have to be very strong, since SF proceeds well in crystals, where the chromophores merely touch, and our initial guess was that almost any degree of coupling will be sufficient. As we shall see below, this does not appear to be the case, and we now believe that the coupling actually needs to be quite strong when the initial excitation is localized. This complicates matters, since strong coupling will also reduce the energy of the relaxed singlet state and this will need to be taken into account in the design of the optimal chromophore.

In the simplest case only one potential energy surface describing the lowest excited singlet state of the chromophore pair needs to be invoked, although other states necessarily lie nearby. In a vibrationally relaxed system, SF is represented by a transition from an initially populated excimer-like (S_0S_1, S_1S_0) minimum in the lowest singlet surface to a minimum best described as doubly excited in the same surface (T_1,T_1; two triplets coupled into an overall singlet), usually over a barrier separating the two minima. This final minimum is located at a geometry in which both halves of the system are at equilibrium T_1 geometries, and their interaction is minimized. For efficient SF, it needs to lie below or at most only a little above the starting "excimer-like" minimum. It is likely that an energy barrier separating the two minima, or two barriers if a third intermediate minimum intervenes as discussed below, will be minimized when the difference in the S_1 and T_1 equilibrium geometries is small, and this may be one of the criteria for the selection of SF chromophores.

The states and electron configurations involved in the description of SF in a dimer are already familiar from earlier analyses of the photophysics of excimers and of photochemical dimerization, since the "double triplet" state correlates with the ground state of the photodimer [4], but they may be worth a closer description. In addition to providing the starting minimum in the lowest excited singlet surface, the S_0S_1, S_1S_0 combination also gives rise to another excimer-like state, expected to lie only a little above the lowest excited singlet surface at the initial geometry, and both excimer-like states may have smaller or larger contributions from charge-transfer configurations in which an electron is transferred from one chromophore to the other (I^+I^-, I^-I^+). We first ignore these complications and only consider the lowest singlet surface.

At the geometry of the final "double triplet" singlet state, composed of two only weakly interacting triplet excited halves, the minimum in the lowest singlet surface is nearly degenerate with two additional states. One is a triplet and the other a quintet, and they result from the other possible couplings of local triplet excitations. After transition from the "excimer-like" into the "double triplet" minimum in the excited singlet state, the system can undergo rapid dephasing of the two only weakly coupled local triplet excitations from an initial overall singlet into an overall triplet or quintet, induced by minor perturbations such as the magnetic fields of protons or other magnetic nuclei present in the system.

Two cases of initial singlet excitation need to be distinguished. If this excitation is localized, as would usually be the case in covalent molecular dimers, SF starts from one of two isoenergetic minima that correspond to a combination of one chromophore in its S_1 state and the other chromophore in its S_0 state, both at their relaxed equilibrium geometries. These minima are usually separated by a very low energy barrier and both are normally populated in a rapid equilibrium, which corresponds to an electronic excitation transfer between the two halves of the molecule, accompanied by appropriate adjustment in nuclear positions. If the initial singlet excitation is delocalized, as would often be the case in molecular crystals where the singlet exciton can extend over many molecules (but not if it is trapped on a pair of molecules), there is no barrier between the S_0S_1 and S_1S_0 excitations and the lowest excited singlet surface contains only one "excimer-like" minimum. Which of the two situations prevails depends on the strength of the coupling between the two halves of the chromophore pair and on the site distortion energy, related to the difference in the equilibrium geometries in the S_0 and the S_1 states.

The difference between a localized and a delocalized starting singlet excitation may appear unimportant, as after all, the latter is just a limiting case of the former. However, this difference may actually have important consequences for SF. If we assume that the primary effect of the geometry change upon travel from the initial minimum to the final minimum on the lowest singlet surface is to act as a one-electron perturbation, and consider that the initial and the final electronic configurations differ in the occupancies of two spinorbitals as illustrated schematically in Figure 1, it is apparent that the perturbation operator needs to be applied twice and that there is a virtual or real intermediate state, represented by a configuration in which only one electron has been moved (only one choice is shown in Figure 1). Now, if the excitation is delocalized, the molecular orbitals (MOs) whose energies and occupancies are shown in Figure 1 are delocalized equally over both halves of the system, and the creation of the intermediate configuration involves no charge separation. In contrast, if the excitation is localized, the MOs whose energies and occupancies are shown in Figure 1 are localized on one or the other half of the system, and the intermediate configuration involves an electron transfer from one to the other half (I^+I^-, I^-I^+). It is strongly dipolar and its energy will be greatly affected by the polarity of the medium. One might therefore suspect that SF in delocalized systems will proceed well in non-polar environments, but that SF in localized systems will be promoted by polar environments, which will lower the energy of the virtual or real intermediate charge-transfer state. If the intermediate state is stabilized sufficiently to become real and represent a third minimum on the lowest excited singlet state surface, located between the initially considered starting ("excimer-like") and final ("double triplet") minima, SF becomes a two-step process. The intervention of a real intermediate is not likely to be a welcome phenomenon, since it will most likely offer opportunities for various deactivation channels, such as ISC from the singlet to the triplet state, or back electron transfer to form the S_1 or S_0 state. These processes would result in a decreased triplet yield.

THEORETICAL REQUIREMENTS FOR THE MOLECULAR STRUCTURE OF A SINGLET FISSION SENSITIZER

We consider next the molecular structure of the SF chromophore that underlies the sensitizer, whether it is used in the form of a dimer, oligomer, polymer, aggregate, or crystal, and the results have been published. Because of the requirement of large absorption cross-sections, π-electron systems are most likely candidates and we shall focus on them. The compounds that have been observed to produce SF so far have been extremely limited structurally and have essentially all been alternant hydrocarbons, that is π-electron systems without odd-membered rings. In almost all of them, the excitation energy from S_0 into S_1 is less than twice the excitation energy from S_0 into T_1, and as a result SF is endothermic and only possible when the missing amount of energy is delivered by the thermal bath, or by an initial excitation into a higher vibrational level of S_1 or into a higher excited singlet state S_n. This is a clear disadvantage.

The need for thermal excitation will slow down SF, and vibrational deactivation will compete with SF from an initially vibrationally hot sensitizer. Also, the undesirable T_1–T_1 annihilation to produce S_1 and S_0 will be exothermic and will occur at a rate that is close to diffusion-controlled.

A search for suitable SF chromophores should therefore probably impose the level energy condition $E(S_1) > 2 E(T_1)$ as one of the requirements. This highly unusual ordering immediately excludes most standard dyes. To make matters worse, it does not in itself guarantee that triplettriplet annihilation will be slow. Although the T_1 + $T_1 \rightarrow S_1 + S_0$ channel will be suppressed, the $T_1 + T_1 \rightarrow T_2 + S_0$ channel may still be open, and in order to close it, the condition $E(T_2) > 2 E(T_1)$ should be imposed as well. We probably do not need to worry about the third obvious channel, $T_1 + T_1 \rightarrow Q_1 + S_0$, since the energy of the lowest quintet state Q_1 is likely to be always too high. The channels $T_1 + T_1 \rightarrow S_0 + S_0$ and $T_1 + T_1 \rightarrow T_1 + S_0$ are likely to be too exothermic to compete and will not cause trouble.

There are two obvious classes of candidate structures, which are not mutually exclusive. One of these are systems derived from molecules with an ordinary closed shell ground state, and the other, systems derived from molecules with an open shell ground state (biradicals). Once a suitable fundamental structure is identified, it will undoubtedly be necessary to modify it by the attachment of heteroatoms or substitutents that will fine-tune the absorption wavelength and assure other desired properties.

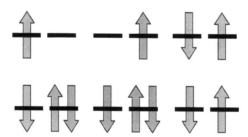

Figure 1. A symbolic representation of MO occupancies in the initial excited singlet state (left) and the final singlet state (right) in SF. Occupancy in one of the possible intermediate singlets is shown in the center.

Among π-electron parent systems with closed shell ground states, alternant hydrocarbons are the most likely to have a large S_1–T_1 splitting, and therefore to satisfy the condition $E(S_1) > 2 E(T_1)$. This follows from the alternant pairing theorem long known from simple MO theory of π-electron systems,5,6 according to which the highest occupied MO (HOMO) and the lowest unoccupied MO (LUMO) have large absolute amplitudes on the same set of atoms. As a result, the exchange integral $K_{HOMO,LUMO}$ between these two MOs is large, and so is the S_1–T_1 splitting, which is approximately equal to $2K_{HOMO,LUMO}$.

In order to actually satisfy the condition $E(S_1) > 2 E(T_1)$, the alternant hydrocarbon has to have a small S_0 to S_1 excitation energy. Since a singlet excitation energy of only about 2 eV is desired in any event, this requirement is not bothersome, but it means that the hydrocarbon has to be quite large. For instance, in the polyacene series, which has been investigated most thoroughly, the first hydrocarbon that meets the requirement is pentacene. Tetracene is not quite large enough, and the energy of its S_1 lies a little above twice the energy of T_1. Fortunately, the misfit is small and the difference can be made up easily by thermal excitation at room temperature. Nonalternant systems, such as azulene, can have similarly low S_0–S_1 excitation energies already at much smaller molecular size. However, their HOMO and LUMO are typically predominantly located on different sets of atoms. This results in a small $K_{HOMO,LUMO}$ value and a small S_1–T_1 gap, which does not permit them to meet the condition $E(S_1) > 2 E(T_1)$.

There seems to be no particular reason to expect T_2 to lie at or above S_1 in large alternant hydrocarbons, but in pentacene it does. This chromophore therefore looks promising for a possible use in SF, and it is unfortunate that it is highly reactive and difficult to handle. In this regard, the rylenes, the higher analogues of perylene, may offer advantages.

The other class of likely π-electron hydrocarbon parent structures are those with an open-shell ground state, biradicals. In these, T_1 may actually be the ground state, and in any event, it is not far in energy from the lowest singlet S_0. Two additional singlet states result from intra-shell excitation. The upper one, S_2, is usually significantly above S_0 and T_1 in energy. The S_1 state is degenerate with S_0 in axial biradicals, such as charged 4N-electron annulenes, and nearly degenerate with S_2 in pair biradicals, such as uncharged 4N-electron annulenes. Inter-shell excitation is needed to produce additional states, including T_2, and usually requires higher energies. As a result, in the parent π-electron biradicals, both conditions $E(S_1)$, $E(T_2) > 2 E(T_1)$ are generally satisfied effortlessly. However, the S_0 to T_1 excitation energy is usually far too low, and possibly even negative, and the compounds are much too reactive to be useful. Both of these detrimental characteristics are removed simultaneously when the parent biradical is modified by a polar or a covalent perturbation to produce a biradicaloid. Such perturbations stabilize the S_0 state relative to the T_1 state and give the former a more or less ordinary closed shell structure, often endowed with fairly normal chemical stability. For instance, the parent antiaromatic dication biradical $C_6H_6^{2+}$ can be formally converted into quinone, $C_6H_4O_2$, by the attachment of two phenolate oxygen substituents O. One can hope that when the extent of the perturbation is just right, the S_0T_1 energy gap will be only a little smaller than the T_1S_1 energy gap, T_2 will lie above S_1, and the compound will be perfectly stable.

In addition to the conditions $E(S_1)$, $E(T_2) > 2 E(T_1)$, the requirement that ISC from S_1 to T_1 be negligibly slow, and possibly the requirement that the equilibrium S_1 and T_1 geometries be as close as possible, the SF chromophore needs to meet the same series of criteria as ordinary sensitizers. The redox properties have to be matched to the semiconductor and the shuttle used, the electron injection kinetics from T_1 have to be fast and back electron transfer insignificant, the material needs to adhere well to the semiconductor, its photostability needs to be outstanding, etc. We shall be concerned with these matters only if it turns out that an SF sensitizer with a T_1 quantum yield close to 200% can actually be produced.

We then performed the next logical step in the search for a suitable SF sensitizer, a series of calculations for a many candidate chromophore structures, and found which ones satisfy the conditions $E(S_1)$, $E(T_2) > 2 E(T_1)$. We used a semiempirical method that ordinarily provides an accuracy of about 0.20.3 eV. We have selected a few chromophores that appeared promising for proof of principle studies and verified that their energy levels behave as calculated.

RESULTS FOR A MODEL CHROMOPHORE 1

We have selected one of these chromophores, 1,3-diphenylisobenzofuran (structure 1 in Figure 2), for a detailed study [9]. The chromophore 1 can be formally derived from a biradical in which two CH_2 groups are attached in the ortho positions of a benzene ring, and twisted by 90ª out of conjugation with its π system. This biradical is then converted into a biradicaloid by twisting the CH_2 groups into conjugation to make a planar π-electron system. The resulting molecule, O-quinodimethane, [10] is still extremely reactive. It is stabilized further by formation of a five-membered heterocyclic ring with oxygen and addition of two phenyl substituents to yield 1, a stable commercially available compound, albeit sensitive to oxygen under irradiation and therefore ultimately impractical.

We have obtained the single-crystal X-ray structure of 1 (Figure 3) and found that bond lengths and angles agree well with calculations. A comparison of results calculated for various electronic states of the monomer 1 as well as the various dimers 24, using DFT procedures, is given in Table 1. There is nothing surprising about the molecular structure of the ground state of 1. The molecules are packed in parallel stacks that appear quite favorable for providing a weak coupling between neighbors. In accordance with this notion, the absorption spectrum of the polycrystalline layer is ~20 nm red shifted relative to the solution spectrum. The two computed conformers have virtually identical energies and differ by the sense of the angle of rotation of the phenyl substituents out of the heterocyclic plane; one has C_2 and the other has C_s symmetry. Only the former is observed in the crystal, but both are undoubtedly present in solutions. Their spectral properties are calculated to be nearly identical.

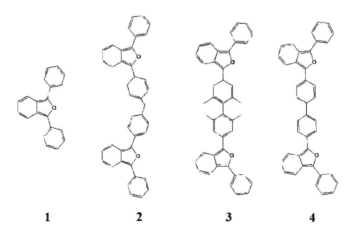

| 1 | 2 | 3 | 4 |

Figure 2. Molecular structures 14.

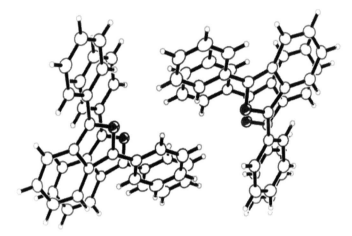

Figure 3. The crystal structure of 1. Oxygen is shown in gray.

In the singlet and triplet excited states, the phenyl substituents are essentially co-planar with the heterocyclic ring, and bond lengths in the heterocycle alternate much less. The results given in Table 1 for S_1 are only preliminary and unreliable because of convergence difficulties.

Table 1. Calculated geometries of 14 in various electronic states (TD-DFT, B3LYP/6-31G*, bond lengths, E; angles; degrees).

	$1(S_0)$	$1(S_1)$	$1(T_1)$	$2(S_0)$	$2(S_1)^y$	$2(T_1)$	$2(Q_1)$	$3(S_0)$	$3(S_1)^y$	$3(T_1)$	$3(Q_1)$	$4(S_0)$	$4(S_1)$	$4(T_1)$	$4(Q_1)$
$\Delta E^{w,x}$	0.00	61.79	33.07	0.00		32.82	65.67	0.00		32.77	65.48	0.00	68.35	31.68	66.22
a^y	1.456	1.429	1.423	1.456		1.423	1.423	1.456		1.456	1.423	1.456	1.442	1.455	1.423
b	1.399	1.423	1.444	1.399		1.444	1.444	1.399		1.399	1.444	1.399	1.411	1.400	1.444
c	1.428	1.412	1.399	1.428		1.399	1.399	1.428		1.428	1.399	1.428	1.421	1.427	1.399
d	1.374	1.411	1.411	1.374		1.411	1.411	1.374		1.374	1.411	1.374	1.385	1.375	1.412
e	1.432	1.396	1.390	1.432		1.391	1.391	1.432		1.432	1.391	1.432	1.417	1.431	1.391
f	1.374	1.411	1.411	1.374		1.411	1.411	1.374		1.374	1.411	1.375	1.389	1.375	1.411
g	1.428	1.412	1.399	1.428		1.399	1.399	1.428		1.428	1.399	1.427	1.415	1.426	1.399
h	1.399	1.424	1.444	1.399		1.444	1.444	1.399		1.399	1.444	1.401	1.421	1.403	1.442
i	1.358	1.380	1.386	1.358		1.387	1.387	1.358		1.358	1.387	1.359	1.369	1.360	1.385
j	1.358	1.380	1.386	1.359		1.386	1.386	1.359		1.359	1.385	1.385	1.365	1.358	1.387
k	1.452	1.443	1.436	1.452		1.436	1.436	1.452		1.452	1.436	1.451	1.446	1.451	1.437
l	1.456	1.429	1.423	1.455		1.421	1.421	1.456		1.456	1.422	1.453	1.426	1.449	1.423
a′				1.456		1.456	1.423	1.456		1.423	1.423	1.456	1.442	1.425	1.423
b′				1.399		1.399	1.444	1.399		1.444	1.444	1.399	1.411	1.442	1.444
c′				1.428		1.428	1.399	1.428		1.399	1.399	1.428	1.421	1.401	1.399
d′				1.374		1.374	1.411	1.374		1.411	1.411	1.374	1.385	1.408	1.412
e′				1.432		1.432	1.391	1.432		1.391	1.391	1.432	1.417	1.393	1.391
f′				1.374		1.374	1.411	1.374		1.411	1.411	1.375	1.388	1.409	1.411
g′				1.428		1.428	1.399	1.428		1.399	1.399	1.427	1.415	1.398	1.399
h′				1.399		1.399	1.444	1.399		1.444	1.444	1.400	1.421	1.446	1.442

Table 1. (Continued)

	1(S₀)	1(S₁)	1(T₁)	2(S₀)	2(S₁)ʸ	2(T₁)	2(Q₁)	3(S₀)	3(S₁)ʸ	3(T₁)	3(Q₁)	4(S₀)	4(S₁)	4(T₁)	4(Q₁)
i′			1.358			1.358	1.387	1.358		1.387	1.387	1.359	1.369	1.385	1.385
j′			1.358			1.359	1.386	1.359		1.385	1.385	1.358	1.365	1.384	1.387
k′			1.452			1.452	1.436	1.452		1.436	1.436	1.451	1.446	1.435	1.437
l′			1.455			1.455	1.421	1.456		1.421	1.422	1.453	1.427	1.412	1.423
Aᶻ	18.0	5.1	2.1	-17.9		-3.7	-2.7	-17.5		-18.4	-3.9	-18.6	-9.8	-18.1	-3.9
B	18.0	5.1	2.1	-19.2		-4.9	-1.9	17.8		18.9	5.1	-16.4	-4.6	-14.3	-6.0
C				-56.2		-54.6	-54.9	-89.3		-88.7	-88.9	31.1	14.9	23.6	30.3
D				-19.2		-19.2	-2.9	17.8		5.1	5.1	-16.4	-5.1	-4.2	-5.9
E				-17.9		-17.8	-0.6	-17.5		-4.6	-3.9	-18.6	-10.1	-5.2	-3.9

Calculations are still in course. "The absolute energies for the ground state of **1, 2, 3**, and **4** are -844.653483, -1727.374540, -1845.151430, and -16.88.123380 respectively. ˣRelative energies are kcal/mol. ʸBond lengths are in Å. ᶻDihedral angles are in degrees.

Cmpd.	ΔE (kcal/mol)	A	B	C	D	E
1(S₀)ᵃ	0.00	18.0				
1(S₁)	61.79	5.1	5.1			
1(T₁)	33.07	2.1	2.1			
2(S₀)ᵃ	0.00	-17.9	-19.2	-56.2	-19.2	-17.9
2(S₁)ᵇ						
2(T₁)	32.82	-3.7	-4.9	-54.6	-19.2	-17.8
2(Q₁)	65.67	-2.7	-1.9	-54.9	-2.9	-0.6
3(S₀)ᵃ	0.00	-17.5	17.8	-89.3	17.8	-17.5
3(S₁)ᵇ						
3(T₁)	32.77	-18.4	18.9	-88.7	5.1	-4.6
3(Q₁)	65.48	-3.9	5.1	-88.9	5.1	-3.9
4(S₀)ᵃ	0.00	-18.6	-16.4	31.1	-16.4	-18.6
4(S₁)ᵇ						
4(T₁)	31.68	-18.1	-14.3	23.6	-4.2	-5.2
4(S₀)	66.22	-3.9	-6.0	30.3	-5.9	-3.9

ᵃ The total ground state energies of **1, 2, 3**, and **4** are -844.653483, -1727.374540, -1845.151430, and -1688.123380 a. u., respectively.
ᵇ Calculations are still in course.

We have performed a thorough experimental characterization of the electronic and vibrational excited states of 1 and accompanied it with semiempirical, DFT, and *ab initio* calculations. The following spectra were measured (mostly at room temperature and 77 K): absorption, fluorescence, fluorescence anisotropy, fluorescence excitation, fluorescence excitation anisotropy, magnetic circular dichroism, and electron energy loss (EELS) in the ground state. We have also measured the linear dichroism in stretched polyethylene and deduced the purely polarized spectra from it. We have further obtained the absorption spectra of the first excited singlet and triplet states, including absolute intensities. The $T_1 \rightarrow T_n$ absorption spectrum was obtained both from laser flash photolysis and from a steady-state absorption experiment. In these measurements, a mixture of 1 and anthracene sensitizer in cyclohexane solution was irradiated at a wavelength absorbed nearly exclusively by anthracene, and absorbance in the visible was measured. In the absence of the anthracene sensitizer, the absorption was not detectable. We concluded that ISC in 1 is negligible, at least in solution. The absorption spectra of the radical cation and the radical anion of 1 were measured for us by Dr. John Miller at Brookhaven National Laboratory. A selection of these results is displayed in Figures 4.

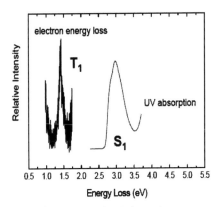

Figure 4. The T$_1$ and S$_1$ states of 1 from EELS and UV absorption spectroscopy.

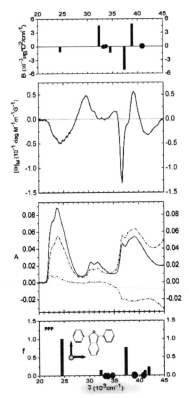

Figure 5. Magnetic circular dichroism of 1 (top) and its polarized absorption in stretched polyethylene (bottom) along with results of PPP calculations.

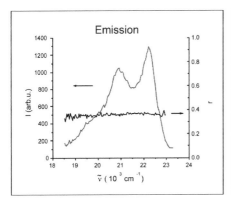 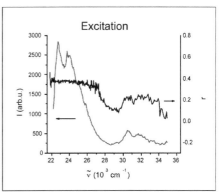

Figure 6. Anisotropy of fluorescence and fluorescence excitation of 1 (3-methylpentane glass at 77 K).

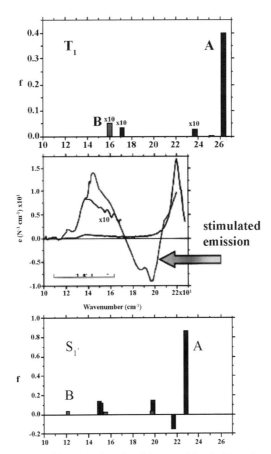

Figure 7. Absorption spectra of the S_1 (red) and T_1 (blue, sensitized with anthracene) states of 1 and results of CASPT2 calculations.

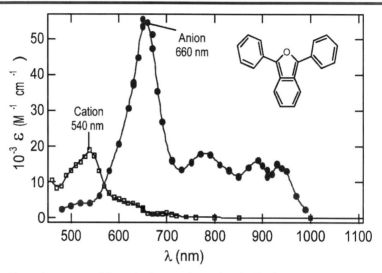

Figure 8. Absorption spectra of the radical cation (square) and radical anion (circle) of 1 from pulse radiolysis (courtesy of J. Miller, Brookhaven National Laboratory).

We have also measured the IR spectra of 1 in stretched polyethylene at room and low temperature and calculated the frequencies and polarizations of the transitions. These results were used to analyze the dichroism observed in the UV region. The detailed interpretation of all the spectra is being published. Here we merely note the presence of an unusual discrepancy between the conclusion that seems obvious from the fluorescence polarization and magnetic circular dichroism spectra, as well as the PPP, TD-DFT, and CASPT2 computational results, namely that the first excited singlet state is quite isolated, and the clear conclusion from the linear dichroism in stretched polyethylene, supported by the results of SORCI computations, namely that two oppositely polarized nearly degenerate states are present in the region of the lowest singlet excitation. Our best interpretation is that the first singlet excitation is indeed isolated and that the linear dichroism results are distorted by the simultaneous presence of the C_2 and C_s conformers, which happen to have mutually perpendicular principal orientation axes in stretched polyethylene.

The calculated and experimental excitation energies of 1 are listed in Table 2. The S_1 excitation energy was determined from absorption and fluorescence spectra and the T_1 excitation energy was known from a study [11] of solution sensitization with donors of different triplet energies. We have added a value for the polycrystalline solid using EELS, and are presently attempting to use the same method to ascertain the energy of the T_2 state as well. Within the experimental uncertainty of about 0.1 eV the singlet excitation energy is twice the triplet excitation energy, both in an isolated molecule in solution and in a polycrystalline solid, and SF should be roughly thermoneutral as long as the chromophore coupling is small. The same should be true in covalent dimers of 1, again provided that chromophore coupling does not change the state energies much.

The photophysical data for 1 are collected in Table 3 and it is seen that they are also highly favorable for a hoped-for SF chromophore. In particular, the fluorescence yield is very high, suggesting that there are no active channels likely to compete with SF in crystals or dimers, such as ISC, internal conversion, or photoproduct formation. The fluorescence lifetime is quite long, giving SF a good chance to compete successfully with emission. The results are in good agreement with the absence of any detectable triplet–triplet absorption upon excitation of solutions of 1, in which SF cannot take place.

Table 2. Excited state energies for 1,3-Diphenylisobenzofuran (1) in eV.

	Calcd. Obsd. (solution)	**Obsd. (solid)**	
S_1	3.0	~2.8	~2.7
T_1	1.7	~1.5[a]	~1.4[b]
T_2	3.3		

[a] From bracketing with sensitizers. [b] From electron energy loss.

Table 3. Photophysical characteristics of 1 in solution.

Solvent	$\tilde{V}_{abs}$ $(10^3 \times cm^{-1})$	$\tilde{V}_{F}$ $(10^3 \times cm^{-1})$	$\varnothing_F$	$K_F (ns^{-1})$	$\tilde{V}_{T-T}$ $(10^3 \times cm^{-1})$	$k_T (ms^{-1})$
CH	24.33 ± 0.03	22.32 ± 0.03	0.95 ± 0.03	0.154 ± 0.008	22.12 ± 0.06	6.8 ± 0.5
THF	24.21	22.17	0.99	0.150	22.12	4.9
AN	24.36	22.12	0.98	0.132	22.08	3.9
DMSO	24.04	21.83	0.97	0.175	21.98	5.4
TOL	24.15	22.03	0.99	0.156	22.12	5.0
DMF	24.15	22.03	0.96	0.147	22.03	4.2

[a] The position of the first absorption and fluorescence peaks, the fluorescence quantum yield, the fluorescence rate constant, the position of the intense triplet-triplet-triplet absorption peak, and the triplet decay rate constant, respectively. CH: cyclohexane, THF: tetrahydrofuran, AN: acetonitrile, DMSO: dimethylsulfoxide, TOL: toluene, DMF: dimethylformamide.

Since most of the reported successful observations of SF were performed on crystalline material, we have attempted the same type of experiment on a polycrystalline film of 1. We found that unlike an unsensitized solution, the polycrystalline sample shows an intense triplet–triplet absorption spectrum upon UV irradiation. It turned out to be relatively difficult to obtain an accurate value of the triplet yield but the data we have obtained so far leave little doubt that it is at least 10% and possibly as high as 50%. The triplet formation action spectrum (efficiency of triplet formation as a function of energy of the exciting photon) follows accurately the absorption spectrum. This is exactly the behavior anticipated for exothermic or thermoneutral SF which we attempted to achieve by choosing 1 as the chromophore. We cannot strictly rule out the possibility that ISC from S_1 to T_1, which is negligible in the isolated molecule, is for some unanticipated reason efficient in the crystalline material and generates the triplet state that we observe, but it appears highly improbable. Definitive evidence will be obtained if the triplet quantum yield is found to exceed 100%, or from future measurements in magnetic field. For the moment, we consider it virtually certain that 1 is the first crystalline material in which SF has been observed by design. The next question is, will it also occur in covalent dimers of 1?

CALCULATIONS OF COUPLING IN COVALENT DIMERS

The next task was the covalent coupling of two molecules of the chromophore 1 into a dimer. Nothing was known about the strength of coupling that is required for SF to be fast relative to fluorescence, which has a natural lifetime of about 5 ns in the monomer and could be somewhat faster in the dimer. A subcontractor on the project, Prof. M. Ratner, and his student have performed a series of calculations for a variety of dimers and evaluated the coupling constant. Selected results are shown in Table 4. In themselves, these numbers have no absolute significance, but they do provide a feeling for the relative strength of coupling between the two chromophores present in the molecules. They need to be calibrated by measurement of absolute SF rates in several dimers (see Figure 9).

Table 4. Electronic matrix elements and free energies for various covalent dimers of 1a.

DPIBF	t_h (eV)	t_l (eV)	S_1-S_0 (eV)	ΔG_f (eV)
monomer	N/A	N/A	2.917	−0.056
D1	0.027	0.014	2.835	0.026
D2	0.027	0.027	2.770	0.092
D3	0.203	0.133	2.356	0.505
D4	0.190	0.200	2.152	0.710
D5	0.122	0.136	2.441	0.420
D6	0.122	0.122	2.525	0.337
D7	0.041	0.054	2.715	0.147
D8	0.027	0.027	2.772	0.089
D9	0.054	0.068	2.693	0.168
D10	0.004	0.009	2.798	0.064
D10@70°	0.068	0.068	2.659	0.203
D11	0.002	0.006	2.805	0.056
D12	0.014	0.014	2.789	0.073
D13	0.014	0.014	2.797	0.064
D14	0.019	0.006	2.822	0.039
D15	0.014	0.014	2.822	0.040
D16	0.176	0.216	2.269	0.593
D17	0.036	0.054	2.713	0.149
D18	0.035	0.025	2.766	0.095
D19	0.079	0.105	2.629	0.232
D20	0.118	0.155	2.523	0.339
D21	0.085	0.082	2.611	0.251
D22	0.027	0.013	2.733	0.129
D23	0.064	0.073	2.583	0.278
D24	0.094	0.082	2.448	0.414

[a] Matrix element for HOMO-HOMO (t_h) and for LUMO-LUMO (T_l) interaction, free energy of S_0 to S_1 excitation, and of singlet fission. Structural formulas of the dimers are shown in Figure 9.

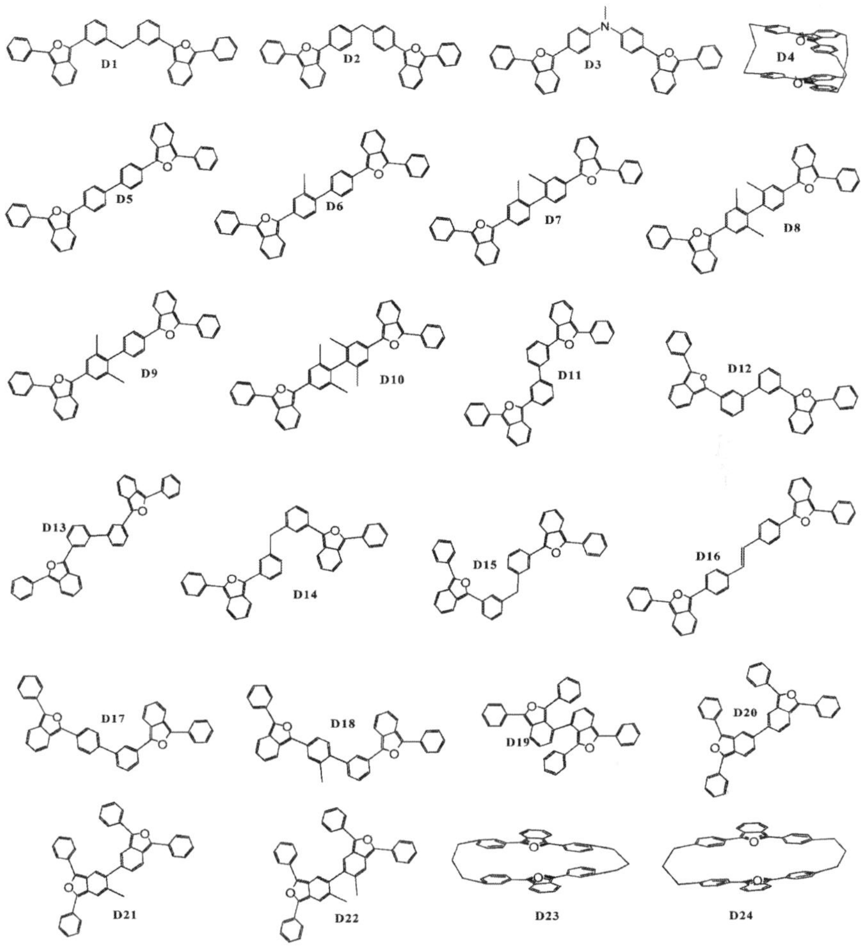

Figure 9. Structures of dimers of 1 for which DFT computations have been performed.

SYNTHESIS OF COVALENT DIMERS

Figure 2 presents the chemical structures of the dimers 24 that were synthesized for the purpose. The synthetic paths to 24 are shown in Figures 10, 11, and 12, respectively. The routes shown for 3 and 4 are particularly simple and efficient, and the same procedure was subsequently also employed for 2. We will not discuss the details of the synthetic procedures and their pros and cons here, and merely note that the purification of the products is quite demanding, especially due to their sensitivity to the simultaneous action of light and atmospheric oxygen. It was done very carefully, and we believe that the disagreement between the absorption and fluorescence excitation spectra of 4 that will be noted below is genuine and not due to a minor impurity.

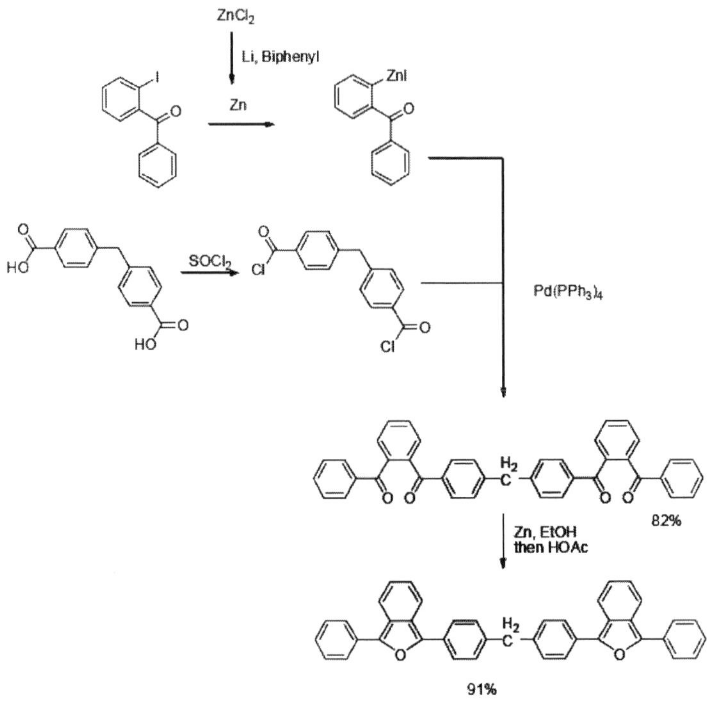

Figure 10. Synthetic route to 2.

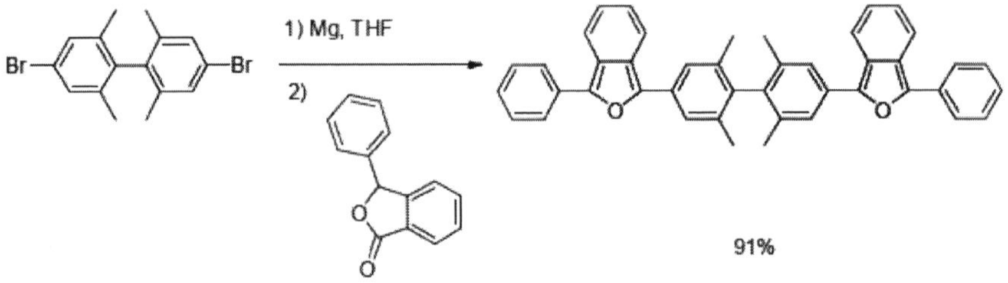

Figure 11. Synthetic route to 3.

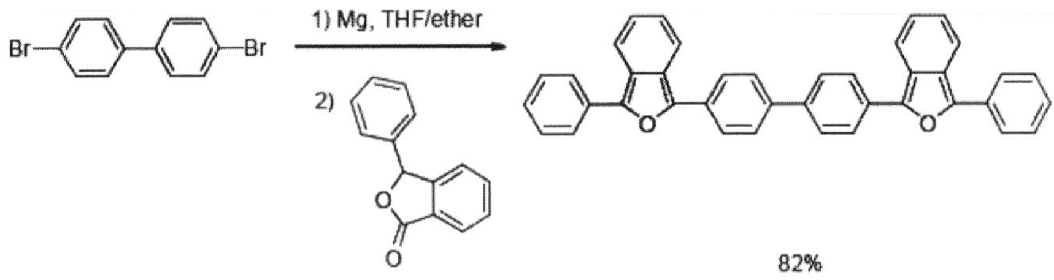

Figure 12. Synthetic route to 4.

SOLUTION PHOTOPHYSICS OF THE WEAKLY COUPLED DIMERS 2 AND 3

We start this section by commenting on the behavior of 2 and 3 as polycrystalline solids. Since we noted above that irradiation of polycrystalline 1 produces triplets fairly efficiently, almost certainly by SF it is not surprising that solid 2 and 3 do so as well. From the point of view of application in a solar cell, they however offer no obvious advantage, and are harder to make. The interesting question in their case is whether their dimeric nature permits efficient SF within a single molecule, and this section therefore focuses on their photophysics in solution.

The issues to be addressed are (i) is singlet or triplet excitation in these dimers localized in one of the constituent chromophores, or delocalized as it undoubtedly is in a crystal, (ii) is the coupling between the two chromophores contained within 2 and 3 strong enough for SF to compete with fluorescence and other possibly present channels that depopulate the S_1 state, (iii) is the coupling of the chromophores weak enough not to perturb the state energies excessively, or does it shift the excitation energy of the S_1 state below twice that of T_1, making the formation of a pair of triplets from the excited singlet endothermic, hence SF too slow.

We have examined the spectral properties of 2 and 3 similarly as those of 1. Figure 13 compares the absorption and fluorescence spectra of 14 and Figure 14 compares their triplet–triplet absorption spectra obtained by anthracene sensitization in DMSO. The spectra of 2 and 3 are only about 20 nm red shifted relative to 1, similarly as in crystalline 1, leaving no doubt that the coupling between the two chromophores in the dimeric molecules is only weak, as anticipated. This is reasonable, considering that there is only hyperconjugative interaction between the two halves of the dimer in 2, and that there is very limited conjugation between them when the ortho methyl groups force them to be nearly perpendicular to each other in 3.

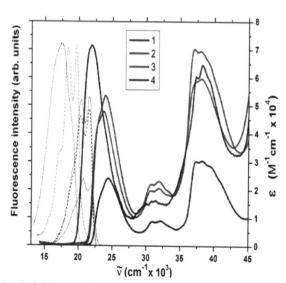

Figure 13. Absorption (solid line) and fluorescence (brokenline) spectra of the ground state of 1, 2, 3, and 4 in acetonitrile. The fluorescence spectrum of 4 in toluene is also shown.

The spectra are compatible with the localization of excitation on one of the chromophores, presumably rapidly jumping from one to the other. The TD-DFT calculations of optimized S_1 and T_1 geometries (Table 1) support this interpretation.

The photophysical properties of 2 and 3 are collected in Tables 5 and 6. In cyclohexane solution, there is no indication that the dimers 2 and 3 undergo SF as isolated species. Their photophysical properties are nearly identical with those of 1 in the same solvent. The quantum yield of fluorescence is still very close to unity and its lifetime is 56 ns as before. No triplet–triplet absorption is detectable after direct irradiation of the solution without a sensitizer, nor is any other transient absorption other than that due to the $S_1 \rightarrow S_n$ process. This absorption decays with the same lifetime as fluorescence, and the decay follows a single exponential.

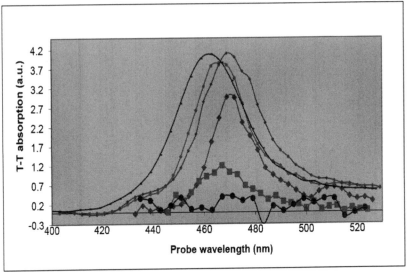

Figure 14. Triplet–triplet absorption of 1, 2, 3 sensitized with anthracene (small dots) and obtained upon direct excitation (large dots, intensity multiplied by 10) in DMSO.

Table 5. Photophysical characteristics of 2 and 3a.

Solvent	$\tilde{V}_{abs}$ $(cm^{-1} \times 10^3)$	$\tilde{V}_g$ $(cm^{-1} \times 10^3)$	$\varnothing_F$	$\varnothing_T$	V_{T-T} $(cm^{-1} \times 10^3)$
2 CH	23.77 ± 0.05	21.92 ± 0.05	0.97 ± 0.03	$<0.01 \pm 0.005$	21.75 ± 0.10
2 AN	23.72	21.60	0.19	0.012	21.55
2 DMSO	23.42	21.60	0.13	0.026	21.30
2 DMF	23.57	21.61	0.11	0.023	21.45
3 CH	23.57	21.78	0.92	<0.01	21.60
3 AN	23.61	21.55	0.32	0.025	21.40
3 DMSO	23.24	21.47	0.17	0.057	21.25
3 DMF	23.40	21.41	0.28	0.049	21.30

^{a} For symbol definition see Table 3. $\varnothing_T$ is the triplet quantum yield. CH: cyclohexane, AN: acetonitrile, DMSO: dimethylsulfoxide, DMF: dimethylformamide.

Table 6. Multiexponential fluorescence decay of 2 and 3 in solution[a].

Solvent	A_1	τ_1(ns)	A_2	τ_2(ns)	A_3	τ_3(ns)
2 AN	0.41	0.222 ± 0.005	0.58	1.93 ± 0.12	0.01	5.0 ± 2.6
2 DMF	0.40	0.231 ± 0.004	0.60	2.13 ± 0.10	--	--
2 DMSO	0.23	0.215 ± 0.005	0.75	0.81 ± 0.04	0.02	3.5 ± 0.6
2 THF	--	--	--	--	1.0	4.53 ± 0.02
2 CH	--	--	--	--	1.0	4.59 ± 0.02
3 AN	0.23	0.150 ± 0.003	0.75	1.72 ± 0.03	0.02	4.4 ± 1.4
3 DMF	0.30	0.128 ± 0.003	0.70	1.69 ± **0.02**	--	--
3 DMSO	0.34	0.134 ± 0.004	0.62	2.51 ± 0.06	0.04	6.8 ± 1.0
3 THF	--	--	--	--	1.0	3.75 ± 0.02
3 CH	--	--	--	--	1.0	3.93 ± 0.02

[a] A_i are the amplitudes associated with the lifetimes τ_i shown. CH: cyclohexane, THF: tetrahydrofuran, AN: acetonitrile, DMSO: dimethylsulfoxide, DMF: dimethylformamide.

The situation changes when the solvent is highly polar. In DMF or DMSO, and to a lesser degree in acetonitrile, the fluorescence quantum yield is considerably reduced (Table 5) and fluorescence lifetime is shorter and multiexponential (Table 6). Triplet–triplet absorption after direct excitation without a sensitizer becomes quite pronounced and the quantum yield of T_1 is 0.010.06. These measurements were repeated at various temperatures and the highest observed quantum yield of T_1 was about 0.09. However, time-resolved absorption measurements show that the faster decay of the S_1 species is not due to its conversion into T_1. Instead, in a few hundred ps this decay produces a new species S* with a broad absorption peak from 500 to 600 nm and another broad peak near 670 nm. Our search for a fluorescence from this species has been fruitless. The species S* decays to form the triplet on a time scale of a few ns (Figure 15). The lifetime for the decay of S_1 is equal to the rise time of S* and the decay time of S* is equal to the rise time of T_1. Global analysis of transient absorption in the 450–700 nm region at times up to 8 ns yielded the absorption spectra of S_1, S*, and T_1, evidence for stimulated emission from S1, and a set of rate constants for the transformation of S_1 to S* and back and for the transformation of S* to T_1 in 2 and 3 (Figure 16 and Table 7). The inefficiency of T_1 formation is attributed to the conversion of S* and probably also T* to S_0. The overall reaction scheme is shown in Figure 17 (for simplicity we do not show explicitly the direct path from T* to S_0).

The structural assignment of S*, which is only formed in polar solvents, to a dipolar species in which a radical cation of 1 is covalently attached to the radical cation of 1 is secured by comparison of its absorption spectrum with those of the radical cation and radical anion of 1 (Figure 8). A good precedent for this type of intramolecular charge transfer in a polar solution is provided by 9,9'-bianthryl [12]. The existence of T*, the triplet state of the dipolar species S*, is postulated to account for the temperature dependence of the triplet yield. Its absorption spectrum is expected to be essentially identical with that of S* and it is believed to be only slightly more stable than S*, because the two unpaired electrons are well separated in space.

At room temperature the vertically excited state S_1 is rapidly converted by electron transfer into an equilibrium mixture with the dipolar species S*, which can undergo back electron transfer to yield S_1. From temperature dependence of the equilibrium constant we find that S* is more stable than S_1 by 1.8 kcal/mol. At low temperatures, the electron transfer in S1 to yield S* becomes rate limiting and has an activation energy of 3 kcal/mol. At the same time, the quantum yield of T_1 formation increases, and this is attributed to a shift of the equilibrium between S* and T* in favor of the lower energy species T*. The available data are not good enough to permit a quantitative evaluation of their energy difference.

Upon the disappearance of one molecule of S*, 1.2 ± 0.4 molecules of T_1 are formed. Within the error margin, this can be attributed fully to ISC, and SF need not be invoked, although we cannot exclude that it makes a minor contribution to the formation of T_1.

The nature of the inter-chromophore coupling in 2 and 3 is distinctly different than the stacking π-face–π-face interaction in crystalline 1, but the effects on absorption spectra are similar, a red shift of about 20 nm. This suggests that the contrast between the efficient SF in polycrystalline 1 and non-existent SF in the isolated dimers 2 and 3 is not due to a vastly different strength of the inter-chromophore coupling. Instead, we believe that it is related to the delocalized nature of the S_1 excitation in the crystal and localized nature of S_1 excitation in the dimers, along the lines discussed above in connection with Figure 1. It is of interest to direct future efforts in the direction of covalent structures in which the excitation would also be delocalized, perhaps stacked dimers or oligomers and polymers instead of dimers.

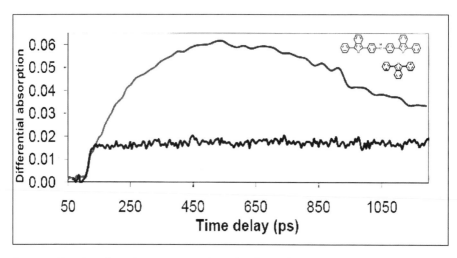

Figure 15. Transient absorption at 575 nm of 1 and 2 obtained upon direct excitation at 400 nm in DMSO.

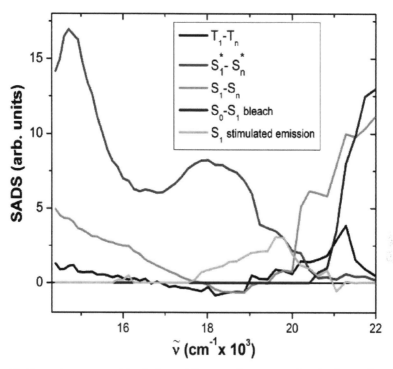

Figure 16. Absorption spectra of the S_1, S^*, and T_1 states of 2, transient bleach of S_0, and stimulated emission from S_1 obtained from global analysis.

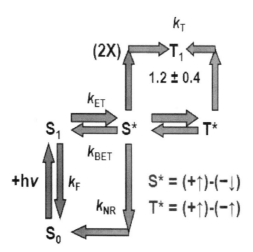

Figure 17. Reaction Scheme for the Photophysics of 2 and 3 in DMSO.

Table 7. Rate constants[a] for the photophysics of 2 and 3 in DMSO solution.

RT rate constants (1/ns)	k_F	k_{ET}	k_{BET}	k_{NR}	k_T
2 DMSO	0.23	1.8	2.8	~1.0	~0.1
3 DMSO	0.25	3.8	4.9	~0.8	~0.09

[a] See Figure 17 for the meaning of symbols.

SOLUTION PHOTOPHYSICS OF THE STRONGLY COUPLED DIMER 4

It is clear in Figure 13 that the first transition in 4 is not just doubled in intensity relative to that of 1 and otherwise nearly unchanged, like those of 2 and 3. Instead, it is significantly red shifted and intensified. This is true for the solid spectrum as well and is not surprising, given the quite strong direct conjugation when the two halves of the molecule are nearly coplanar. This structure is best viewed as a single chromophore conjugated throughout, even though it still has the capability to accommodate two triplet excitations after the two halves are uncoupled by twisting around the central bond. Our investigations of this dimer are just now in course as a part of the follow-up contract, but some photophysical results can already be stated here (Table 8).

In the solid state, and also in solution, T_1 formation is easily detectable, but only at higher excitation energies. The triplet formation action spectrum does not follow the absorption spectrum, as it did in solid 13. Instead, in solution and in the solid, it shows a clear threshold at about twice the triplet energy of 1. Clearly, in 4, the coupling of the two halves is too strong for the condition $E(S_1) \geq 2\ E(T_1)$ to be satisfied. Even above the threshold, the triplet formation efficiency is low, only 3%, undoubtedly because SF has to compete with fast vibrational deactivation and/or internal conversion. It appears that in 4 the coupling is strong enough to undergo SF in solution and in neat crystalline material, but that it also is strong enough to lower the singlet excitation energy below twice the triplet excitation energy.

Table 8. Photophysical characteristics of 4 in solution[a].

Solvent	$\tilde{V}_{abs}$ (10³×cm⁻¹)	$\tilde{V}_F$ (10³×cm⁻¹)	θ_F	θ_T
4 CH	22.17 +/-0.05	19.64	0.25	0.021 +/-0.005
4 TOL	21.79	19.43	0.42	0.031
4 AN	22.08	17.48	0.35	--
4 DMSO	21.50	17.55	0.59	0.013
4 DMF	21.55	17.73	0.41	0.009

[a] For symbol definition see Tables 3 and 5. CH: cyclohexane, TOL: toluene; AN: acetonitrile, DMSO: dimethylsulfoxide, DMF: dimethylformamide.

CONCLUSION

We have considered the potential benefits offered by using SF sensitizers in photovoltaic cells and identified two key issues involved in the search for such sensitizers. One of these is thermodynamic and deals with the arrangement of the electronic levels of the chromophore, and the other is kinetic and deals with the strength of coupling between the two or more chromophores that need to be present in a SF sensitizer. Considerable theoretical and experimental progress in both has been described, and it has been pointed out that the two criteria impose contradictory demands.

A set of guidelines for the search for an optimal SF chromophore has been elaborated, and has led to the first observation of what we believe to be SF in a rationally designed polycrystalline neat solid sensitizer, 1, and its covalent dimers 24. The triplet yield has not yet been measured accurately, but in 1 itself it appears to be at least 10%. The behavior of the covalent dimers 24 in solution is different. In nonpolar solvents, the weakly coupled dimers 2 and 3 form no triplets, and the coupling between the constituents is clearly too weak. The thermodynamic criterion is met, but the kinetic one is not. In strongly polar environments, 2 and 3 form triplets in yields that range up to ~9%, but this occurs primarily or exclusively by ISC in a dipolar intermediate. Although this is interesting photophysics, it does not lie on the path to more efficient SF. In dimer 4, the coupling is strong enough for SF to take place with a triplet yield of ~3%, and the kinetic criterion is satisfied. However, the coupling is so strong that the thermodynamic criterion is no longer satisfied and the SF takes place only upon excitation above a threshold that lies considerably above the 00 transition into the lowest excited singlet state. Competition with vibrational deactivation and/or internal conversion is then inevitable and is responsible for the low triplet yield.

The paths to further progress are clear: (i) work with nanocrystals of the monomer, (ii) find a mode of covalent coupling that causes the initial singlet excitation to be delocalized (this may involve stacking interactions or perhaps going to oligomers and polymers), (iii) identify a different chromophore in which the excitation energy of the lowest triplet T_1 is significantly smaller than half the excitation energy of the lowest excited singlet S_1, such that even after the required strong coupling in a dimer the S_1 excitation energy remains larger than twice the T_1 excitation energy. These routes are being pursued in a current follow-up project.

KEYWORDS

- **Covalent oligomers**
- **Electron energy loss**
- **Nanocrystals**
- **SF chromophores**
- **Singlet fission**

Chapter 8

Detection and Selective Destruction of Breast Cancer Cells

Yan Xiao, Xiugong Gao, Oleh Taratula, Stephen Treado, Aaron Urbas, R. David Holbrook, Richard E. Cavicchi, C. Thomas Avedisian, Somenath Mitra, Ronak Savla, Paul D. Wagner, Sudhir Srivastava, and Huixin He

INTRODUCTION

Nanocarrier-based antibody targeting is a promising modality in therapeutic and diagnostic oncology. Single-walled carbon nanotubes (SWNTs) exhibit two unique optical properties that can be exploited for these applications, strong Raman signal for cancer cell detection and near-infrared (NIR) absorbance for selective photothermal ablation of tumors. In the present study, we constructed a HER2 IgY-SWNT complex and demonstrated its dual functionality for both detection and selective destruction of cancer cells in an *in vitro* model consisting of HER2-expressing SK-BR-3 cells, and HER2-negative MCF-7 cells.

The complex was constructed by covalently conjugating carboxylated SWNTs with anti-HER2 chicken IgY antibody, which is more specific and sensitive than mammalian IgGs. Raman signals were recorded on Raman spectrometers with a laser excitation at 785 nm. The NIR irradiation was performed using a diode laser system, and cells with or without nanotube treatment were irradiated by 808 nm laser at 5 W/cm^2 for 2 min. Cell viability was examined by the calcein AM/ethidium homodimer-1 (EthD-1) staining.

Using a Raman optical microscope, we found the Raman signal collected at single-cell level from the complex-treated SK-BR-3 cells was significantly greater than that from various control cells. The NIR irradiation selectively destroyed the complex-targeted breast cancer cells without harming receptor-free cells. The cell death was effectuated without the need of internalization of SWNTs by the cancer cells, a finding that has not been reported previously.

We have demonstrated that the HER2 IgY-SWNT complex specifically targeted HER2-expressing SK-BR-3 cells but not receptor-negative MCF-7 cells. The complex can be potentially used for both detection and selective photothermal ablation of receptor-positive breast cancer cells without the need of internalization by the cells. Thus, the unique intrinsic properties of SWNTs combined with high specificity and sensitivity of IgY antibodies can lead to new strategies for cancer detection and therapy.

Although significant progress has been made in both the understanding and treatment of cancer during the last 30 years, it remains the second leading cause of death in the US. Non-invasive detection of cancer in its early stages is of great interest since early cancer diagnosis, in combination with precise cancer therapies, could significantly increase the survival rate of patients. Nanomedicine, an emerging research area that integrates nanomaterials and biomedicine, has the potential to provide novel diagnostic tools for detection of primary cancers at their earliest stages, and to provide improved therapeutic protocols. Research in nanomedicine will also lead to the understanding of the intricate interplay of nanomaterials with components of biological systems.

Attaching antibodies or other targeting agents (such as receptor ligands) to the surface of nanocarriers to achieve specific targeting of cancerous cells is a promising modality for therapeutic and diagnostic oncology [1]. Improved therapeutic efficacy of targeted nanocarriers has been established in multiple animal models of cancer, and currently more than 120 clinical trials are underway with various antibody-containing nanocarrier formulations [2]. The most commonly explored nanocarriers include polymer conjugates, polymeric nanoparticles, lipid-based carriers such as liposomes and micelles, and dendrimers [1]. Recent developments in nanotechnology have engendered a range of novel inorganic nanomaterials, such as metal nanoshells [3] and carbon nanotubes [4], offering unique opto-electronic properties compared with conventional organic nanocarriers [3, 4].

The SWNT is a novel nanomaterial that exhibits unique structural, mechanical, electrical, and optical properties that are promising for various biological and biomedical applications, such as biosensors [5], novel biomaterials [6], and drug delivery transporters [7-11]. Water-solubilized SWNTs have been shown to transverse the cell membrane via endocytosis to shuttle various cargoes into cells, including proteins [12], nucleic acid such as plasmid DNA [13, 14] and short interfering RNA [15], without causing cytotoxicity. Two unique intrinsic properties of SWNTs can be exploited to facilitate cancer detection and therapy. The SWNTs have very strong resonant Raman scattering [16] that can be harnessed for cancer cell detection [17-19]. The SWNTs absorb NIR light in the 700–1100 nm spectral window to which biological systems are transparent; continuous NIR irradiation of SWNTs attached to cancer cells produces excessive heat in the local environment that can be utilized to achieve selective destruction of these cells without harming normal cells [7, 20-22].

To achieve specific targeting of tumor cells for photothermal ablation, SWNTs have been either conjugated to folate to target folate receptors in folate positive cancer cells [7, 22] or attached noncovalently (through adsorption) [20] or indirectly via streptavidin-biotin interaction [21] to antibodies targeting specific receptors on cancer cells. Direct covalent attachment of antibodies to SWNTs for specific tumor targeting has also been reported [23], however, using such antibody-SWNT conjugates for specific photothermal ablation of cancer cells with NIR light has not been reported.

All of the antibodies in clinical use today for cancer cell targeting are mammalian IgG monoclonal antibodies [24]. Recently, there has been renewed interest in using avian IgY antibodies as IgG substitutes in immunoassays and clinical applications [25]. The IgYs, distinct from IgGs in molecular structure and biochemical features, have many attractive biochemical, immunological and production advantages over

IgGs and are suitable for further development [25]. We have recently demonstrated the advantages of using anti-HER2 IgY antibody in detecting breast cancer cells [26]. The IgY antibodies provide specific and more sensitive detection of breast cancer cells compared with commercial IgG or IgM antibodies. Coupled with quantum dots, anti-HER2 IgY antibodies have the potential to give quantitative biomarker measurements [26].

In an effort to improve breast cancer detection and therapy, we have developed a novel method which combines the advantages of anti-HER2 IgY antibody with the unique properties of SWNTs. We constructed a HER2 IgY-SWNT complex by directly functionalizing SWNTs with the anti-HER2 IgY antibody through covalent bonding, explored the Raman and NIR optical properties of the complex, and tested its feasibility for detection and selective destruction of cancer cells.

MATERIALS AND METHODS

Preparation of the HER2 IgY-SWNT Complex

Purified HiPco SWNTs were purchased from Carbon Nanotechnologies (Houston, TX) and solubilized by carboxylation using a microwave-assisted functionalization method described previously [27]. In a typical reaction, ~1 mg of as-received carbon nanotubes were added into 2 ml of a 1:1 mixture of 70% nitric acid and 97% sulfuric acid aqueous solutions in a plastic beaker. The mixture was then subjected to microwave radiation for 2 min. Afterwards, the mixture was diluted with deionized water and centrifuged at 2,000 g for 15 min to remove insoluble materials. The supernatant was filtered through a Microcon YM-50 centrifugal filter unit (Millipore, Billerica, MA) and rinsed thoroughly with 100 mm MES buffer in order to adjust pH to 4.5. For covalent attachment of HER2 IgY antibody onto SWNTs, 2.0 mg N-(3-Dimethylaminopropyl)-N′-ethylcarbodiimide hydrochloride (EDC), 88.3 mg N-Hydroxysuccinimide (NHS) and 100 μL MES buffer solution (100 mm, pH 4.5) were added to the microwave-functionalized SWNT solution and incubated for 60 min at room temperature. The mixture was then centrifuged in Microcon YM-50 centrifugal filter unit and rinsed with a 100 mm MES buffer solution (pH 6.3) to remove excess EDC, NHS, and the byproduct urea. The purified, activated carbon nanotubes on the filter were re-dispersed into a 100 mm MES buffer solution (pH 6.3). Thereafter, 60 μl (1.0 mg/ml) chicken anti-HER2 IgY antibody, prepared as described previously [26], was added into the above solution and reacted for 2 hr Finally, the solution was centrifuged at 25,000 g for 20 min to remove the unreacted materials. The collected precipitate was resuspended in PBS buffer (100 mm, pH 7.4) and used for further studies. The concentration of antibody conjugated to SWNTs was determined using BCA protein assay (Pierce, Rockford, IL) following the manufacturer's instructions. The SWNT concentration in the solution was estimated from the absorbance spectrum at 808 nm acquired with a Cary-500 UV-visible-NIR spectrophotometer (Varian, Palo Alto, CA) in double-beam mode.

Cell Culture and Treatment

Breast carcinoma cell lines SK-BR-3 and MCF-7 were obtained from ATCC (Manassas, VA) and cultured under conditions as recommended by the supplier. Cells were grown for 24 hr to reach ~30–40% confluence, then treated with the HER2 IgY-SWNT complex or SWNT or antibody alone at the final nanotube concentration of 4 mg/l for 24 hr under the same culture condition. The cell culture was washed three times with fresh medium before NIR irradiation to remove unbound nanotubes, antibodies or the antibody nanotube complex.

Atomic Force Microscopy (AFM)

The SWNTs before and after conjugation with anti-HER2 IgY antibody were imaged with a tapping mode Nanoscope IIIa atomic force microscope (Veeco, Chadds Ford, PA). In order to image the SWNTs, 5 μl of the prepared solutions were deposited on freshly cleaved mica. After a 3–5 min incubation, the mica surface was rinsed with three drops of deionized water four times and dried under a flow of nitrogen. During imaging, a 125 μm long rectangular silicon cantilever/tip assembly was used with a spring constant of 40 N/m, resonance frequency of 315–352 kHz and a tip radius of 5–10 nm. The images were generated by the change in amplitude of the free oscillation of the cantilever as it interacts with the sample.

Dispersive Raman Spectrometric Analysis

Raman spectra for the HER2 IgY-SWNT complex solution were collected on a S1000 Raman spectrometer from Renishaw (Hoffman Estates, IL) coupled to a DM LM microscope from Leica (Bannockburn, IL) using a 50× objective. The source was an Ar-ion pumped tunable Ti:sapphire laser from Coherent (Santa Clara, CA) operating at 785 nm. Laser power was 7 mW measured at the sample.

Senterra dispersive Raman spectrometer from Bruker Optics (Billerica, MA) was used to collect Raman spectra from cell cultures. The system consisted of laser excitation at 785 nm focused on the samples through an optical system, producing spectra of Raman shifts, which were evaluated to identify and determine the presence and location of the nanomaterial. Measurements were made using a 10× objective lens, with laser powers ranging from 1 to 25 mW, and exposure times of 10–60 sec. Various settings were tried in order to determine the optimum signal-to-noise ratio and to avoid damaging the samples from over-heating by the laser.

NIR Irradiation and Temperature Measurement

A Spectra-Physics diode laser from Newport (Irvine, CA) at a wavelength of 808 nm mounted on a heat sink was employed to heat the samples. The heat sink was controlled by a driver and a temperature controller. Two thermocouples made from 80 μm diameter type K wire from Omega (Stamford, CT) were positioned outside of the beam path to record the temperature response. A program written in LabVIEW (National Instruments, Austin, TX) was used to control the timing and power of the laser and to record the temperature of the two thermocouple junctions. The laser exposure was conducted at a current of 3.0 A in the laser diode which produced an output of 5 W/cm^2, and the duration was set to 120 sec.

Samples were contained in cells with a 1 cm × 1 cm glass bottom. A removable Teflon cap was fitted over the cells. A 4.5 mm hole was drilled through the cap for the laser to pass through (the beam diameter was 4.4 mm). Two smaller holes were drilled through the cap for the thermocouple wires to pass through. One of the thermocouples was extended below the cap into the fluid and positioned just at the edge of the illuminated area. The second was positioned approximately 1 mm outside the edge of the beam path. The beam passed through the sample's glass bottom and entered a 3 cm diameter beam dump located just below the glass bottom. The beam dump serves to prevent radiation passing through the sample from making its way back to the sample. A diagram for the laser exposure arrangement is shown in Figure 1.

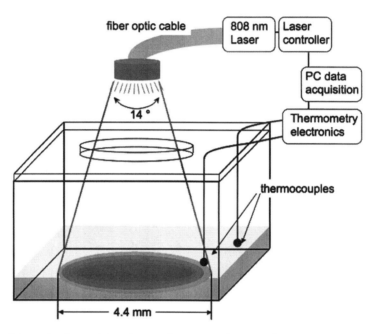

Figure 1. Schematic of experimental setup for NIR irradiation and temperature measurement. Dimensions not to scale.

Cell Viability Assay

Ten min after heating with NIR irradiation, cell viability was examined by the calcein AM/ethidium homodimer-1 (EthD-1) staining [28]. The LIVE/DEAD Viability/Cytotoxicity Kit from Molecular Probes (Eugene, OR) was used and protocols provided by the manufacturer were adopted. Cells showing green fluorescence were considered alive; while dead cells showed red fluorescence. Results were expressed as percentage of live cells relative to the number of cells on a control slide that did not go through treatment or NIR irradiation.

SWNTs Localization Study by Immunohistochemistry (IHC)

Cells were grown on tissue culture chamber slides (Nunc, Rochester, NY) at a density of 30,000 cells/cm^2 and then treated with the HER2 IgY-SWNT complex at the final nanotube concentration of 4 mg/l for 24 hr. Cell monolayers were subsequently fixed in 10% neutral-buffered zinc formalin (Fisher, Pittsburgh, PA), and were pre-blocked with 5% (w/v) nonfat dry milk in TBST (50 mm Tris-HCl, 150 m, NaCl, 150 mm Tween 20), 20°C, for 20 min. For detection, slides were robotically prepared (reaction with secondary antibody and fluorescent detection reagents) with a Benchmark XT workstation (Ventana, Tucson, AZ) [29]. Anti-IgY biotinylated antibody (GenWay, San Diego, CA) was used as the secondary antibody and was detected by fluorescence microscopy with streptavidin-Qdot655 (Invitrogen, Carlsbad, CA). Imaging systems for analysis of fluorescence signals from quantum dots and integration of the signal with an imaging system were described elsewhere [30, 31].

Confocal laser scanning microscopy images were obtained on a TCS SP5/DM6000 from Leica using an HCX Pl Apo oil immersion 63× coverslip corrected objective. A 405 nm Diode laser was used as the excitation source while the emission bands were set to 440–480 nm (DAPI, channel 1), 640–660 nm (QDs, channel 2), and diffraction (cells, channel 3). Zoom functions between 1× and 6× were used as needed.

Data Analysis

All experiments were repeated at least three times with at least three replicates each time. For comparative studies, one-way ANOVA tests (with Bonferroni post test if $p < 0.05$) were used for statistical analysis. Differences were considered statistically significant if a p value of < 0.05 was achieved.

RESULTS AND DISSCUSSION

Preparation and Characterization of the HER2 IgY-SWNT Complex

The HER2 IgY-SWNT complex was prepared by first carboxylating HiPco SWNTs using a microwave-assisted functionalization method published previously [27]; the carboxylated SWNTs were then activated by EDC and NHS and reacted with HER2 IgY antibody to form the covalent complex, through amidation between the carboxyl groups on the SWNTs with primary amines on amino acid residues such as lysine and arginine on the antibody (Figure 2A) [32]. Free unconjugated antibodies were removed through ultracentrifugation. The SWNTs used consisted of short, straight fragments (with average diameter and length being 1.17 ± 0.28 nm and 88.00 ± 43.68 nm, respectively) and exist as individual tubes and small bundles rather than large aggregates as evidenced by AFM image (Figure 2B and C). After antibody attachment, the diameter of the nanotubes increased to 4.02 ± 0.82 nm (Figure 2D and E). Based on the concentrations of the carbon nanotubes and the IgY antibodies used, it was estimated that on average about 10 IgY antibody molecules were attached to each nanotube. The nanotube complex solutions were highly stable in PBS buffer, without forming aggregates for several months when kept at 4°C.

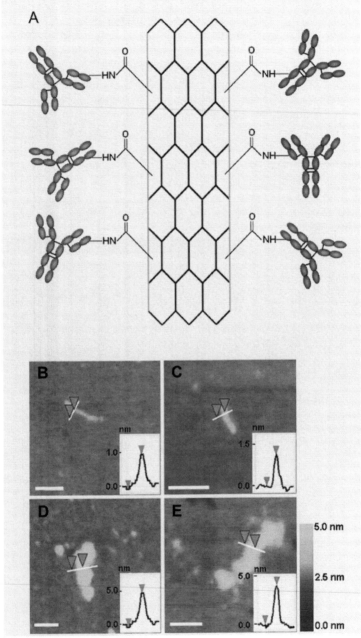

Figure 2. The HER2 IgY-SWNT complex. (A) Schematic representation of SWNTs covalently functionalized with anti-HER2 IgY antibody. (B-E) atomic force microscopy (AFM) images of carboxylated SWNTs prior to conjugation (B and C) and after conjugation (D and E) to anti-HER2 IgY antibodies. Insets shows AFM cross-section analysis indicating the changes in height of SWNTs prior to and after conjugation with anti-HER2 IgY antibodies. The height differences on the surface are indicated by the shades shown on the right. Scale bars represent 60 nm.

The optical properties of the freshly prepared HER2 IgY-SWNT complex were tested. The Raman spectra (Figure 3A) of the complex showed a number of well characterized resonances such as the radial breathing mode (RBM) region between 100 and 300 cm^{-1} and the tangential (G-band) peak at 1,590 cm^{-1}. A narrow G- feature was also visible in the G-band region, confirming the presence of semiconducting SWNTs in the sample. The spectra also contained the disorder-induced D band around 1300 cm^{-1}. The UV-visible-NIR spectra (Figure 3B) indicated that the HER2 IgY-SWNT complex has fairly strong absorbance in the NIR region (700–1,100 nm spectral window), even though the interband absorption peaks, originating from electronic transitions between the first and second van Hove singularities of the nanotubes [33, 34] were smeared out during the microwave dispersing and IgY functionalization process. Thus, SWNTs covalently functionalized with antibody retained a significant portion of their optic properties that are potentially useful for biomedical applications.

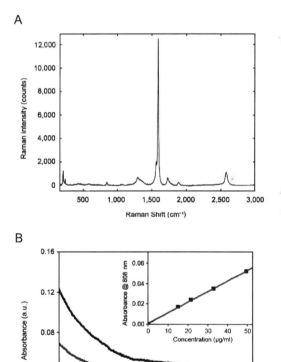

Figure 3. Optical properties of the HER2 IgY-SWNT complex. (A) Raman spectra. (B) The UV-visible-NIR spectra at different nanotube concentrations (from top to bottom: 49.70, 33.02, 21.61, and 15.11 µg/ml). Inset shows the linear relationship of the absorption at 808 nm versus concentration (optical path = 0.3 cm).

Raman Spectrometric Detection of Cancer Cells Using the HER2 IgY-SWNT Complex

We first explored the feasibility of harnessing the characteristic ~1,590 cm^{-1} Raman band for *in vitro* specific detection of cancer cells. Breast carcinoma SK-BR-3 cells, which have high HER2 expression [26], were treated with the HER2 IgY-SWNT complex for 24 hr Raman spectroscopy collected at single-cell level from randomly selected cells showed the characteristic G band at ~1590 cm^{-1} (Figure 4). The Raman signal from the complex-treated breast cancer cells resulted from the specific binding of the IgY antibody moiety of the complex to the HER2 receptor on the cancer cells, as the same cells treated with SWNTs alone did not exhibit Raman scattering. In addition, MCF-7, which are negative for HER2 expression [26], did not exhibit Raman signals when treated with the HER2 IgY-SWNT complex. Thus the characteristic Raman band at ~1,590 cm^{-1} from the HER2 IgY-SWNT complex differentiated HER2-expressing SK-BR-3 cells from the receptor-negative MCF-7 cells.

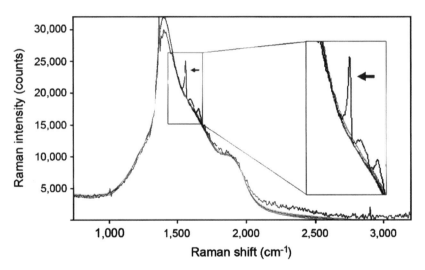

Figure 4. Raman spectra of breast cancer cells treated by the HER2 IgY-SWNT complex. A 785 nm laser diode was used for excitation at 1–25 mW through a 10x objective lens on randomly selected cells. Raman spectra of a representative cell from each sample are shown. Blue line, SK-BR-3 cell treated with the HER2 IgY-SWNT complex; red line, SK-BR-3 cell treated with SWNT alone; magenta line, untreated SK-BR-3 cell; green line, MCF-7 cell treated with the HER2 IgY-SWNT complex. The Raman signal indicated by the arrow is the characteristic G band at ~1590 cm^{-1}. Inset shows higher resolution spectrum in the area around 1590 cm^{-1}.

NIR Irradiation-induced Heating of the HER2 IgY-SWNT Complex Suspension

To demonstrate the heating effect of the HER2 IgY-SWNT complex upon NIR irradiation, we carried out a control experiment in which an aqueous solution of the HER2 IgY-SWNT complex in PBS at a concentration of 4.0 mg/l was irradiated for 2 min using a laser diode with a wavelength of 808 nm at 5.0 W/cm^2 (Figure 5). The temperature rose rapidly after a short lag of a few seconds then increased constantly with time. The maximum temperature increase was ~14°C. On the other hand, PBS solution without SWNTs showed very little temperature rise (<1°C), indicating the solution is transparent to the 808 nm NIR light.

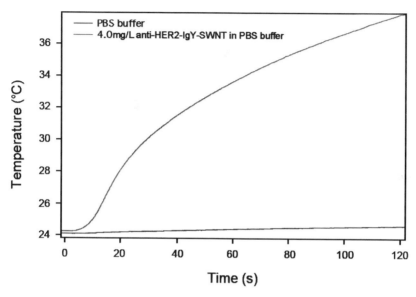

Figure 5. Temperature measurement during NIR irradiation. The green line records the temperature evolution of the HER2 IgY-SWNT complex solution at a concentration of 4 mg/l during continuous irradiation by a 808-nm laser at 5.0 W/cm² for 2 min. The blue line records that for the phosphate-buffered saline (PBS) solution.

Selective Photothermal Ablation of Cancer Cells Using the HER2 IgY-SWNT Complex

Next, we explored the feasibility of using the HER2 IgY-SWNT complex for *in vitro* selective destruction of breast carcinoma SK-BR-3 cells (Figure 6). We conducted the NIR irradiation with a 808 nm laser at 5 W/cm² for 2 min. SK-BR-3 cells treated with the HER2 IgY-SWNT complex showed extensive cell death after heating with NIR irradiation (Figures 6D and G); in stark contrast, negligible cell death was observed with SK-BR-3 cells treated with SWNTs alone (Figure 6B and G) or untreated (Figure 6A and G), and in MCF-7 cells treated with the HER2 IgY-SWNT complex (Figure 6F). These results clearly demonstrated the high transparency of biosystems to NIR light in the vicinity of 808 nm, and at the same time indicated that the specific binding of the IgY antibody moiety of the complex with HER2 receptors on the SK-BR-3 cells is essential for the selective thermal ablation of tumor cells. On the other hand, the SWNT moiety is equally indispensable for the hyperthermia effect, as cell death observed in SK-BR-3 cells treated with the IgY antibody alone (5.9%; Figure 6C and G), although statistically significant ($p = 0.040$), was to a much less extent than in cells treated with the complex (97.7%, $p = 3.38 \times 10^{-7}$; Figure 6D and G).

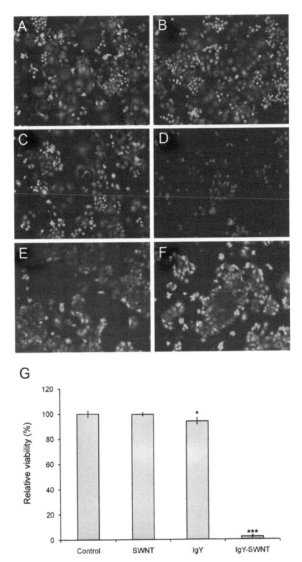

Figure 6. Cell viability after treatment with the HER2 IgY-SWNT complex followed by NIR irradiation. Cell viability was examined by calcein AM/EthD-1 fluorescence staining and representative images from each sample are shown. Cells with green fluorescence were considered alive, whereas those with red fluorescence were dead. (A) Untreated SK-BR-3 cells. (B) The SK-BR-3 cells treated with SWNT alone. (C) The SK-BR-3 cells treated with anti-HER2 IgY antibody alone. (D) The SK-BR-3 cells treated with the HER2 IgY-SWNT complex. (E) Untreated MCF-7 cells. (F) The MCF-7 cells treated with the HER2 IgY-SWNT complex. All treatments were for 24 hr, followed by NIR irradiation with a 808 nm laser at 5 W/cm² for 2 min. Magnification for all the images (A-F) was 10x. (G) Bar graph showing the percentage of live cells in each sample of SK-BR-3 cells following NIR irradiation. Cells were counted under microscope for three randomly selected view fields, and the total number was used for the calculation. The experiment was repeated for three times (n = 3). Error bars represent standard deviation. Cells that did not go through treatment and irradiation were used as controls (852 ± 20, 100%). * p < 0.05 and *** p < 0.001 versus control.

Localization of the HER2 IgY-SWNT Complex on the Cell Membrane

To localize the HER2 IgY-SWNT complexes in the cancer cells, we first performed an immunohistochemical experiment using quantum dots as detection agent. As shown in Figure 7A, most of the HER2 IgY-SWNT complexes were localized on the membrane of the SK-BR-3 cells forming a shell-like shape, with little detected inside the cells. Fluorescence signal computed from 10 randomly selected cells shows that the intensity ratio of fluorescence on the cell surface to that inside the cell is (563 ± 35): 1. No fluorescence signal was detected in receptor-free MCF-7 cells (Figure 7B), suggesting that binding of the complex onto SK-BR-3 cells resulted from the anti-HER2 activity of its antibody moiety.

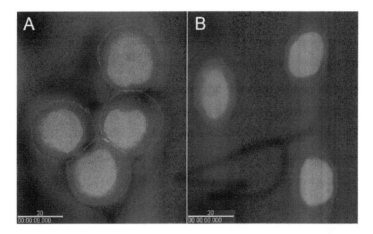

Figure 7. Localization of the HER2 IgY-SWNT complex on the cell membrane. (A) The SK-BR-3 cells. (B) The MCF-7 cells. The HER2 IgY antibody on the complex was probed by biotinylated anti-IgY antibody and detected by fluorescence microscopy with streptavidin-Qdot655 fluorophores. Bars represent 20 μm.

To confirm the above result, we performed additional experiments using confocal microscopy for imaging. The high-resolution images shown in Figure 8 clearly demonstrated that the HER2 IgY-SWNT complexes were localized on the membrane of the SK-BR-3 cells and were not internalized by the cancer cells.

The first problem to tackle for biomedical applications of SWNTs is to solubilize and disperse carbon nanotubes in aqueous solutions and functionalize them with biomolecules such as proteins/antibodies, nucleic acids, and carbohydrates. Past studies using SWNT-antibody conjugates for specific photothermal ablation of cancer cells attached antibodies to SWNTs either noncovalently through adsorption [20] or indirectly via streptavidin-biotin interaction [21]. Direct adsorption of antibodies to SWNTs is simple to execute but the weak interaction between the antibody and the nanotubes raises the possibility of loss of the targeting function of the antibodies. Indirect conjugation via streptavidin-biotin interaction involves an additional step of preparing the antibody-biotin complex. A method for direct covalent attachment of antibodies to

SWNTs for specific tumor targeting has been reported [23] that involves four reactive steps. Here, we used a simpler method for direct covalent conjugation of antibody to SWNTs. The HiPco SWNTs were first dispersed in water through microwave-assisted carboxylation, activated by EDC and NHS, and reacted with HER2 IgY antibody to form the covalent complex. Microwave-assisted functionalization has several advantages over conventional chemical techniques, such as rapidness and environmental friendliness [27]. However, the functionalization process causes some changes in the optoelectronic properties of the SWNTs, such as increase in the disorder mode (D-band) at ~1300 cm^{-1} (Figure 3A) and loss of interband transitions between van Hove singularities in the absorption spectrum (Figure 3B). Similar changes have been reported previously for covalently functionalized carbon nanotubes [27, 35, 36]. Nevertheless, the resultant IgY-SWNT complexes retain a significant portion of the optic properties of SWNTs, as evidenced by the fairly strong Raman and NIR absorbance.

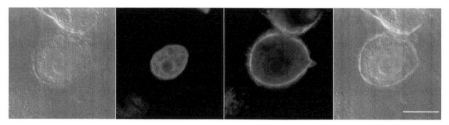

Figure 8. Confocal microscopic visualization confirming the localization of the HER2 IgY-SWNT complex on the cell membrane of SK-BR-3 cells. From left to right, the panels are the brightfield image, DAPI channel image showing the nucleus, quantum dot (QD) channel image showing fluorescence from the HER2 IgY-SWNT complex, and an overlay of the three. The white bar represents 10 μm.

The characteristic G band at ~1590 cm^{-1} was detected in HER2-expressing SK-BR-3 cells treated with the IgY-SWNT complexes (Figure 4) but not in the similarly treated receptor-negative MCF-7 cells, indicating the ability of Raman spectroscopy to specifically detect cancer cells *in vitro*. As a nondestructive optical spectroscopic technique that does not require extrinsic contrast-enhancing agents, the use of Raman spectroscopy has seen a remarkable increase during the last decade in its application to the field of medicine [37]. In particular, Raman spectroscopy has shown great promise as a new tool for detection of malignant and premalignant tissues and as a real-time guidance tool during oncosurgical procedures [38]. However, most of these studies are based on spectral differences between normal and neoplastic tissues that result from compositional changes in the affected tissues, and thus, in most cases, the detection is not highly specific and only possible at later stages of tumor progression. In the current study, characteristic Raman signals (at ~1590 cm^{-1}) are collected at the single-cell level from cancer cells targeted by the IgY-SWNT complexes, thus opening the possibility of using Raman spectroscopy for targeted molecular detection of tumors at the incipient stage. An added advantage of Raman spectroscopy lies in its potential for *in vivo* applications for which limited penetration depth is a fundamental barrier. Until recently, Raman spectroscopy has been generally restricted to probing surface or near-surface areas of biological tissues with penetration depth of only several hundred microns into tissue. This limitation mainly stems from the diffuse scattering nature of tissue which leads to random propagation of photons within its matrix and prevents the formation of sharp images required to discriminate signals emerging from deeper areas. Several methods have been developed recently for the retrieval of Raman signals from deep areas thus enhancing tissue penetration of Raman spectroscopy. These deep Raman techniques discriminate between Raman signals emerging from different depths within the sample using temporal or spatial gating [39]. For instance, combining spatially offset Raman spectroscopy (SORS) with three-dimensional tomographic imaging, it was possible to image a canine hind limb section of a thickness of up to 45 mm using transmission Raman spectroscopy [40, 41]. Therefore, combined with advances in Raman spectroscopic technologies for deep tissue imaging [39], SWNTs functionalized with antibody specific for tumor cell receptors may be exploited for *in vivo* specific detection of cancer cells at early stages.

The present study demonstrates very high specificity of the HER2 IgY-SWNT complexes for HER2-expressing cancer cells, indicating the potential usefulness of the IgY antibody for selective targeting of cancer cells. The IgY antibodies offer many advantages over their mammalian IgG counterparts in terms of both production and biochemical and immunological properties. The IgY antibodies can be isolated in large quantities from egg yolk using simple separation methods; the non-invasive production method also brings the great benefit concerning the welfare of the immunized animals [42]. The IgY antibodies can also be used to avoid interference in immunological assays caused by the human complement system, rheumatoid factors, human anti-mouse IgG antibodies (HAMA), or human and bacterial Fc-receptors [43]. Similarly, for clinical use as antibody-based therapeutics, they neither activate mammalian complement nor interact with mammalian Fc receptors that could mediate inflammatory responses [44]. Despite these advantages, the application of IgY antibodies in research and medicine has been very limited [45]. Oral administration of IgY antibodies have shown great promise as immunotherapy for the prevention and treatment of enteric, respiratory, and dental infections in humans and animals [44-47]. As eggs are normal dietary components, there is practically no risk of toxic side effects of oral administration of IgY antibodies [44, 46]. However, the phylogenetic distance between birds and mammals implies potential concerns over the immunogenicity of IgY antibodies in human. So far, there has been no report on intravenous administration of IgY antibodies in human and the associated immune responses. Nevertheless, concerns over IgY immunogenicity in human should be completely cleared out before any clinical application of IgY should be attempted. The results presented here and in a previous study [26] may bring more attention to this class of antibodies and promote studies on the immunogenicity of IgY preparations in human.

Temperature measurement of the IgY-SWNT complex solution at the nanotube concentration of 4 mg/l showed an increase of ~14°C in the bulk solution, indicating the temperature rise of the surrounding environment would not cause harm to normal cells that do not bind to the SWNT-containing complex in the short time period (2 min). On the other hand, the same result also hinted that the thermal destructive effect to cancer cells must be microscopic rather than macroscopic. We hypothesize that temperature rise in the nanoscale vicinity of individual nanotubes can be dramatic. The sharp local temperature increase may cause damage to subcellular structures such as cell membranes ultimately leading to cell death. The ability to directly measure temperature of an individual nanoparticle will help to validate the hypothesis, and such an endeavor is currently underway [48].

The method described here for selective cancer cell destruction differs from the previously published ones [7, 20-22] in that our method does not require internalization of SWNTs into tumor cells. The HER2 is a transmembrane glycoprotein with the receptor motif extended outside the cell membrane [49]. The reason for the lack of internalization of the SWNT complex by the cancer cells after binding to the cell surface receptor is not known; however, it is likely due to the surface chemistry of the SWNTs used here [27, 32]. It has been reported the surface chemistry has a profound impact on the cellular uptake of nanoparticles such as quantum dots [50]. Although the exact mechanism may differ for various nanoparticles, the surface dependent cellular uptake may be a common phenomenon for all nanoparticles [51]. It is very important to note that the functionalization method used in the current study is different from those published previously where internalization of SWNTs after binding to the cell surface receptors have been reported [7, 20, 22]. In the study by Chakravarty et al. [21], cellular localization of SWNTs after incubation with cancer cells was not reported.

The method described here for selective photothermal ablation of cancer cells without the need of internalization by the cells has the advantage of being more easily extended to other types of cancer cells over agents that need internalization, as cellular internalization is not always achievable with all cancer types. Many cancer cells overexpress specific tumor markers (receptors) on their surface for which IgY antibody with high specificity and sensitivity can be developed. Thus, the IgY-SWNT complex, as exemplified in this study by the anti-HER2 IgY antibody, has the potential to become a novel, generic modality for detection and therapy of various cancer types. Our next step is to evaluate the pharmacokinetics, biodistribution, cytotoxicity, and activity of such IgY-SWNT complexes *in vivo* using animal models.

CONCLUSION

Our current work exploited two unique optical properties of SWNTs-very strong Raman signals and very strong NIR absorbance. We constructed a HER2 IgY-SWNT complex by covalently functionalizing SWNTs with anti-HER2 IgY antibody to impart to SWNTs the high specificity and sensitivity of the IgY antibody. The resultant complex was successfully used *in vitro* for both detection and selective destruction of HER2-expressing breast cancer cells. Raman signal from cancer cells was detected at the single-cell level. A uniqueness of this dual-function agent is that it does not require internalization by the cancer cells in order to achieve the selective photothermal ablation, thus offering the advantage of being more easily extended to other types of cancer cells. However, further research is needed before these findings can be translated into clinical trials.

KEYWORDS

- **Atomic Force Microscopy**
- **Immunohistochemistry**
- **Nanomedicine**
- **Near-infrared**
- **Single-walled carbon nanotubes**

AUTHORS' CONTRIBUTIONS

Yan Xiao conceived of the study, designed, and carried out most of the experimental work, coordinated the project, analyzed the data, and drafted the manuscript. Xiugong Gao participated in the design of the study, in the IgY antibody design and production, performed data analysis, and drafted the manuscript. Oleh Taratula carried out HER2 IgY-SWNT complex preparation and characterization, and analyzed the data. Stephen Treado and Aaron Urbas participated in the Raman spectrometry studies. R. David Holbrook participated in confocal imaging studies. Richard E. Cavicchi and C. Thomas Avedisian participated in NIR irradiation and temperature measurement studies. Somenath Mitra prepared the SWNT samples. Ronak Savla participated in HER2 IgY-SWNT complex preparation. Paul D. Wagner and Sudhir Srivastava participated in the design of the study and critically revised the manuscript. Huixin He participated in the design of the study, supervised the HER2 IgY-SWNT complex preparation and characterization, and helped to draft the manuscript. All authors read and approved the final manuscript.

ACKNOWLEDGMENTS

We thank Joseph Hodges and Kristen Steffens of NIST for their help with the laser radiation apparatus. This work was supported by NCI-Early Detection Research Network joint NIST-Biochemical Science Division Interagency Agreement #Y1 CN5001 and in part by National Science Foundation (NSF) Grant #CHE-0750201. The contribution of CTA was supported partly by the New York State Office of Science, Technology, and Academic Research. Certain commercial equipment or materials are identified in this chapter in order to specify adequately the experimental procedures. Such identification does not imply recommendation or endorsement by the National Institute of Standards and Technology, nor does it imply that the materials or equipment identified are necessarily the best available for the purpose.

COMPETING INTERESTS

The authors declare that they have no competing interests.

Chapter 9

Electrochemical Screen-printed (Bio)Sensors

Elena Jubete, Oscar A. Loaiza, Estibalitz Ochoteco, Jose A. Pomposo, Hans Grande, and Javier Rodríguez

INTRODUCTION

Screen-printing technology is a low-cost process, widely used in electronics production, especially in the fabrication of disposable electrodes for (bio)sensor applications. The pastes used for deposition of the successive layers are based on a polymeric binder with metallic dispersions or graphite, and can also contain functional materials such as cofactors, stabilizers, and mediators. More recently metal nanoparticles (NPs), nanowires, and carbon nanotubes (CNTs) have also been included either in these pastes or as a later stage on the working electrode. This review will summarize the use of nanomaterials to improve the electrochemical sensing capability of screen-printed sensors. It will cover mainly disposable sensors and biosensors for biomedical interest and toxicity monitoring, compiling recent examples where several types of metallic and carbon-based nanostructures are responsible for enhancing the performance of these devices.

INTRODUCING SCREEN PRINTING TECHNOLOGY AND THE APPLICATIONS OF NANOMATERIALS IN SCREEN PRINTED ELECTRODES (SPEs)

The SPEs is a low-cost thick film process that has been widely used in artistic applications and more recently in the production of electronic circuits and sensors. In the 80s the process was adapted to the production of amperometric biosensors [1-4], making their commercialization much easier. This was due to the multiple advantages that the technology offers including reduced expense, flexibility, process automation, reproducibility, and wide selection of materials. A huge number of successful electrochemical devices have been built using this technique [5].

The process of screen printing is rapid and simple; it consists of squeezing an ink or paste through a patterned screen onto a substrate held on the reverse of the screen. Successive layers can be deposited by this procedure and repeat patterns can be designed onto the same screen to enhance production speed. The substrate needs to be an inert material, most commonly PVC [6], polycarbonate [7], polyester [8], or ceramic [9], although nitrocellulose [10], and glass fiber [11] are also employed. Each layer is deposited through the corresponding mask providing a specific pattern. These masks are prepared by a photolithographic technique with photosensitive gels and nylon, polyester or, stainless steel meshes.

There are mainly two types of pastes that can be used in SPE production: conductive or dielectric inks. The conductive inks give rise to the formation of conductive tracks on the electrodes. They are based on an organic binder where gold, silver, platinum or, graphite are dispersed at high loads as conducting fillers. Recently water-based inks have also been employed [12-15]. Functional materials can also be part of the formulation such as cofactors, stabilizers and mediators, and more recently nanosized metals, and CNTs. Dielectric inks are often based on polymers or ceramics and form the encapsulating layer of the sensor, delimiting the working area, and electric contacts.

Regarding biosensors production via screen printing, the biological component (e.g., enzyme, antibody, nucleic acid) can be added in different ways, giving rise to different preparation alternatives; the deposition by hand or the electrochemical entrapment after a multilayer deposition process is the most used alternatives. Another possibility is to introduce the biological material in a printing paste, printing it as a last layer on the electrode surface or as a one step deposition layer forming a biocomposite with the rest of the ingredients [16]. However, the incorporation of the biomolecules in screen printing pastes depends on their nature and is not always possible due to the paste drying conditions that can often bring the denaturalization of the biocomponent. Some of the most commonly used configurations for screen-printing enzymatic biosensors were reviewed by Albareda-Sirvent et al. [17]. More recently, arrays have been developed with multiple working electrodes on the same printed strip, for simultaneous electrochemical detection of different analytes including phenol/pesticides [18-20] or several target sequences of hybridization in genosensors, with up to eight working electrodes on the same strip [21].

The design of new nanoscale materials has found a wide range of applications in the field of sensors, revolutionizing this field. They have also started to find their place in screen printed devices, bringing uncountable benefits. Among them, CNTs, nanowires, and metallic NPs have lately become favorite tools in the sensor area, since they can promote the electron transfer reactions of many molecules, lower the working potential of the sensor, increase the reaction rate, improve the sensibility, or in case of biosensors, contribute to a longer stability of the biocomponent. In genosensors and aptasensors the metallic NPs can serve also as electroactive labels for electrochemical stripping techniques (e.g., stripping voltammetry, stripping potentiometric detection). Nanowires have also received considerable attention in nanoscale electronics and sensing devices [22-26], due to their high aspect ratios, capability of multi-segmented synthesis, and surface modification compatibility. Recent research supports that nanowires can be applied in biofuelcells [27], adaptive sensors [28], and enzymes-based electrochemical sensors [29, 30]. They are also interesting tools for magnetic control of electrochemical reactivity or to adapt on demand (bio)electrocatalytic transformations as it was shown for ethanol/methanol [29] and glucose detection [27, 31].

The incorporation of nanomaterials in SPEs can be done following different alternative strategies. In most of the cases the addition of NPs or nanotubes into screen printed inks, although possible [32], is not an easy task. In spite of not suffering from temperature stability problems in the same extent as enzymes or nucleic acids, they are insoluble in many solvents that constitute the matrix of screen printing pastes. For these reasons, other methods have been developed that are applied after the printing. These postprinting modifications include drop casting of the working electrode with nanotubes dispersed in DMF/water [33], Nafion [34], polyethylenimine [35], or DMSO [36], or electrodeposition of metallic particles [37]. More examples will be discussed in the following sections.

There are many good reviews concerning the preparation and application of nanomaterials in electrochemical sensors [38-50]. However, there are scarce references on them to screen printed devices as electrodic material to support these nanomaterials. On the other hand, there are articles reviewing advances in SPE sensors [16, 51-57], but their content on applications of nanomaterials into these types of sensors is limited. In any case, there is no review, to the best of the authors' knowledge, dedicated exclusively to the application of nanotechnology to SPS. The following sections will be addressed to meet this need, covering examples of practical applications of screen printed sensors in the clinical and environmental field, explaining their basic principles and recent improvements with the use of nanotechnology.

SPE IN CLINICAL DIAGNOSIS

Glucose

Due to the prevalence of diabetes in the developed nations, 85% of the current market of biosensors is aimed to glucose monitoring, resulting in more than $5 billion expense [58]. Disposable screen-printed biosensors are widely employed to address this need of frequent glucose monitoring in diabetics, and they are also used in food industry for quality control. They show a superior performance compared to reflectance devices since they give a rapid and accurate answer using disposable strips with no risk of instrument contamination. This shift toward electrochemical sensing has already been accounted for companies like Roche Diagnostics, Lifescan, Abott, and Bayer giving rise to more than 40 blood glucose meters on the market.

The majority of these devices are based on screen-printed carbon electrodes modified with the enzyme glucose oxidase (GOX), which oxidizes glucose to gluconic acid. In these systems, the presence of a mediator is needed to achieve direct electron exchange between the electrode and the redox centers of GOX, since these centers are situated in the interior of an insulating glucoprotein shell which prevents the direct process [59]. There are two major types of mediators: hydrogen peroxide oxidation mediating reagents and enzymatic glucose oxidation mediating reagents. The first type is employed in sensors where the oxygen participates actively in the oxidation of glucose catalyzed by GOX forming gluconic acid and H_2O_2. Reduction or oxidation of H_2O_2 occurs at high potential in nonmediated electrodes. Therefore, mediators are employed in this type of glucose biosensors to lower such potentials. This is the case of Prusian blue (PB) [60-64]). Instead of these mediators, metalized carbon can also be used in the working electrode as the dispersed metal particles have shown favorable catalytic activity to oxidation and reduction of H_2O_2. The second type of mediators, the artificial mediating reagents, offers the advantage of not requiring oxygen in the system, which is the limiting reagent in the first type of systems and thereby lowering their sensitivities.

The use of nanotechnology has served to improve both types of systems. As mentioned previously, dispersed metal particles can be used to diminish the oxidation potential of H_2O_2 in glucose sensing without the need of mediators. If these particles are in the "nano" range, cannot only the potential be decreased but also the sensitivity enhanced. Shen et al. [32] reached this effect very recently by adding iridium NPs on a screen-printable homemade carbon ink based on hydroxyethyl cellulose, polyethylenimine, and a commercial carbon material. They first made a study of feasibility applying this ink on the working electrode of a 3-electrode configuration. Over the printed working electrode the enzyme (GOX) was covalently attached via glutaraldehyde. They passed subsequently to successful mini disposable electrodes with a 3-electrode configuration, with a working diameter of 1 mm. These minielectrodes responded linearly to glucose between 0 and 15 mm and needed as little as 2 µl Sample volume.

Zuo et al. [65] used a silver NPs-doped silica sol-gel and polyvinyl alcohol hybrid film on a PB-modified screen-printed electrode to immobilize GOX. Although they did not avoid the use of the mediator for H_2O_2 detection (PB), they doubled the sensitivity of the sensor comparing with the biosensor without NPs. The immobilized GOX remained with a 91% activity for 30 days in buffer.

The synthesis, characterization, and immobilization of PB NPs of 5 nm diameter have been reported [66] as mediators in indium tin oxide (ITO) electrodes for the amperometric detection of H_2O_2. A similar strategy could be applied to SPE for the H_2O_2-based glucose detection.

Some other examples of nanomaterials for glucose sensing are applied to biosensors that do not require the formation of H_2O_2. Guan et al. [67] dispersed multi-walled carbon nanotubes (MWCNTs) within mineral oil following the procedure described by Rubianes and Rivas [68]. They mixed the MWCNT dispersion with an enzymatic solution of glucose GOX in citrate buffer and potassium ferrocianide and deposited a drop of the mix over the working electrode of screen-printed electrodes. After the drop dried they tested these electrodes (several types varying the mixing time for enzyme/MWCNT) and compared them with the corresponding type without MWCNT. It was shown that the best response toward glucose was obtained in the systems where the MWCNTs were present and had been mixed with the enzymatic solution during 30 min before deposition. Longer mixing time would lead to axial electron transfer. A wider linear response range and higher sensitivity was reached when the MWCNTs were present.

Lu and Chen [69] drop-coated also the working electrode of their screen-printed strips with a solution containing magnetite NPs (Fe_3O_4) with ferricyanide. In this case, the enzyme (GOX) was added in a later stage, after the NP-containing drop had dried. Sensitivity of 1.74 µA mm-1 was achieved. Rossi et al. [70] prepared and functionalized Fe_3O_4 NPs with amino groups to link them covalently with GOX. The GOX-coated magnetite maintained the enzymatic activity for up to 3 months. Although they could have applied this modified enzyme in screen-printed sensors, they opted for quantifying the oxygen consumption from the transformation of glucose to gluconic acid catalyzed by this enzyme by measuring the increase of the steady state fluorescence intensity of Ru(phen)3. This consumption of oxygen increased with the glucose concentration so glucose was able to be monitored without any surface of sensing material but directly in the solution. In spite of being an elegant approach, a digital fluorescence imaging system was required, which is a more sophisticated piece of equipment than a small portable potentiostat.

Gao et al. [71, 72] built and patented a nanocomposite membrane to be screen printed into a carbon strip using an aqueous slurry ink of a diffusional polymeric mediator (polyvinyl ferrocene coacrylamide) on a PVPAC binder and alumina NPs. The nanoparticulate membrane served not only as biosensing media but also for analyte regulating functions.

Wang et al. [27] reported the possibility of modulating the electrochemical reactivity toward glucose and methanol of a screen-printed working electrode using nickel nanowires. A screen-printed carbon strip served as the working electrode and was limited by a glass cylinder to form an electrochemical cell. An Ag/AgCl and platinum wire were used as reference and counter electrodes, respectively. The nanowires were previously grown by electrodeposition into nanopores of alumina membranes, then removed from the templates, washed and stored in a KOH solution until their use. After being magnetically separated from the storing solution, they were dispersed in a NaOH solution used as electrolyte and placed in the homemade electrochemical cell. The modulation of the magnetic field was performed by placing a small magnet under the working electrode surface. The magnetic properties of nickel and its catalytical action toward aliphatic alcohols and carbohydrates entitled this action, and an enhancement of the electrochemical signal was observed when the nickel nanowires were vertically oriented with a magnetic field. Similarly, when the nanowires were magnetically orientated in a horizontal position the enhancement was produced to a lesser extent. Moreover, such modulated redox transformation was observed multiple times, upon repetitive changes of surface orientation.

Cholesterol

The alarming increase of clinical disorders such as hypertension, heart related illnesses, cerebral thrombosis, arteriosclerosis, and coronary artery disease. Due to abnormal levels of cholesterol in blood have stimulated the development of biosensors with the purpose of quantifying the levels of this compound. Besides, the quality control and nutritional labeling of foods in the food stuff industry is another application for the measurement of cholesterol [73, 74].

The biosensing element most commonly used in cholesterol biosensors is cholesterol oxidase (ChOx), which can be immobilized in the working electrode of screen-printed sensor, catalyzing the conversion of cholesterol in presence of oxygen and water into four cholestene-3 one and hydrogen peroxide. As in the case of glucose biosensing, amperometric measurements of hydrogen peroxide are often monitored, and here the high potential also causes interferential problems (ascorbic acid, uric acid, and other easily oxidizable species); so mediators are also required. Some commonly used mediators in cholesterol sensing are cobalt phtalocianine [75], ferrocene derivatives [76], and phenothiazine derivatives [77]. Another possibility is also to add a mediator for cathodic determination of H_2O_2 such as PB [78], titanium dioxide [79], or metal hexacyanoferrate [80]. As in glucose sensing, there is also a possibility to avoid the route of H_2O_2 production; in this case peroxidase (POD) can be combined with ChOx with potassium ferrocyanide. The drawback of this last route can be that the air oxidation of ferrocyanide is taking place as a competitive reaction of the enzymatic oxidation since it can affect the system.

Several matrices have been employed to construct cholesterol sensors including glassy carbon or gold electrodes, graphite-Teflon, tungsten wire, ITO-coated glass, and porous silicon. Advances implemented on these electrodes for cholesterol detection have been recently reviewed for Arya et al. [81], including the application of nanomaterials. However, there are only a few reports that focus on the development of disposable cholesterol biosensor. For example, in Arya's review only three out of 100 cited references were based on screen-printed electrodes, two of them containing NPs. This is an example of how the application of nanomaterials on screen-printed cholesterol sensors is a field that is yet starting.

A collaboration between Italian and Russian scientists has led to applications of AuNPs [82] and very recently MWCNT [83] on screen-printed rhodium graphite electrodes for cholesterol detection. In both cases they did not use ChOx but opted for another enzyme, cytochrome P450scc, for its specific catalysis of cholesterol side chain. An interesting review on applications of this enzymatic family in biosensors can be found [84].

The same scientists had previously built a cholesterol sensor with this enzyme in SPE but without NPs, by immobilization of cytochrome P450scc biomolecule with glutaldehide or agarose hydrogels over the rhodium-graphite working area [85]. This sensor needed the use of a mediator for electronic transfer: riboflavin. By drop coating the working area with AuMPs suspended in chloroform, they converted the electrode into a nonmediated system, since the roughness of the surface was enough to penetrate the protein matrix, reaching a sensitivity for cholesterol of 0.13 µA µm^{-1}. The addition of MWCNT to the electrode prior to the deposition of the electron transfer, P450scc enzyme, once the enzyme had been immobilized, would increase sensitivity more than 17 times with respect to bare electrodes, or 2.4 times with respect to the electrode containing gold nanoparticles (AuNPs). This catalytic effect is shown in Figure 1. Although the sensitivity to cholesterol was of the same order as using the AuNPs, the linearity in the response improved significantly in the range of 10–80 µm.

Li et al. [86] also proved the electron transfer improvement with MWCNT in an electrode containing ferrocyanide, POD, ChOx, and cholesterol estearase. They performed clinical trials in blood of 31 patients with the biosensor showing fairly good correlation between this method and the results obtained by a clinical blood analyzer.

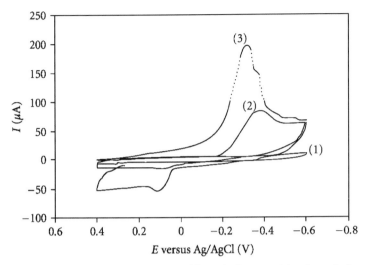

Figure 1. Cyclic voltammograms of screen-printed bare rhodium-graphite electrode for cholesterol detection (1), on electrodes modified with Au nanoparticles and P450scc (2), or with multi-walled carbon nanotubes and P450scc (3). Experiments were performed under aerobic conditions, 100 mm phosphate buffer, 50 mm KCl, pH 7.4, and the scan rate was 50 mV s^{-1}. With permission from [83].

Hybridization Sensors

The detection of specific sequences of DNA is a booming field due to its applications for diagnosis of pathogenic and genetic diseases, forensic analysis, drug screening, and environmental testing. Different strategies can be used for the detection of DNA in sensors, among them the most useful tools are the intrinsic electroactivity of nucleic acids [87], the use of DNA duplex intercalators [88], the labeling with enzymes [89], or the addition of electroactive markers [90].

Metallic NPs have emerged as appealing electroactive markers in electrochemical sensors, especially in stripping voltammetry. This technique is cheap, simple and fast in comparison with optical methods in which commercial DNA chips are based. Another advantage of the use of NPs in DNA hybridization sensors is their multiplexing capability, being able to recognize different molecules in the same sample due to the distinct voltammetric waves produced by different electrochemical tracers [91]. Additionally, their life cycle is much longer than other markers, making their use even more attractive.

Although most of the work of application of metallic NPs for DNA recognition events is performed in other types of electrodes (e.g., gold disks [92], glassy carbon electrodes [93], graphite-epoxy composite electrodes [94], pencil graphite electrode [95]) some examples have emerged regarding the use of SPE in combination with elements from nanotechnology and will be studied in the following paragraphs.

Wang et al. [96] developed a hybridization assay employing a combination of electrodes: a probe-modified gold surface and an SPE. The method was based on the electrostatic collection of silver cations along the DNA duplex, the reductive formation of silver nanoclusters along the DNA backbone, the dissolution of the silver aggregate with a nitric acid solution and the stripping voltammetry detection of the dissolved silver with the SPE. A scheme of the working protocol is shown in Figure 2.

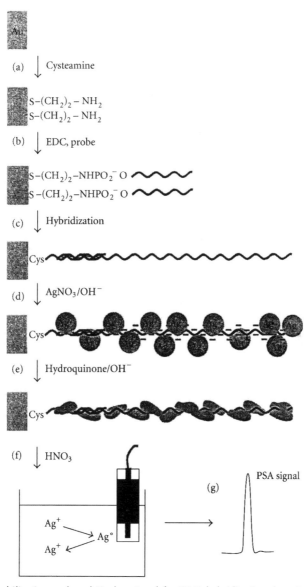

Figure 2. Immobilization and analytical protocol for DNA hybridization detection. (a) Formation of self-assembled cysteamine monolayer, (b) immobilization of ssDNA probe, (c) hybridization of complementary target, (d) "loading" of the silver ion to DNA, (e) hydroquinone-catalyzed reduction of silver ions to form silver aggregates on the DNA backbone, (f) dissolution of the silver aggregates in nitric acid (50%) and transfer to the detection cell, (g) stripping potentiometric detection with SPE. From [96] with the permission of Elsevier.

The DNA segments related with the BRCA breast-cancer gene were detected to concentrations as low as 200 ng/ml with an original strategy developed also by Wang et al. [97]. It was based on the use of magnetic particles as tools to perform DNA hybridization. The assay involved the hybridization of a target oligonucleotide to probe-coated magnetic beads, followed by binding of the streptavidin-coated AuNPs to the capture target and catalytic silver precipitation on the gold-particle tags. The DNA-linked particle assembly was then magnetically collected onto a screen printing electrode surface with a permanent magnet positioned bellow. This way a direct contact between the silver tag with the surface was managed and the solid-state electrochemical transduction was enabled. This silver aggregate did not form in the presence of only noncomplimentary DNA. The described method did not require acidic solution or metal deposition of the silver, so the time needed for the assay was reduced. This technique of combined magnetism and metal detection is now used frequently for DNA hybridization detection not only for screen-printed sensors but also with other electrodic supports.

Suprun et al. [98] recently published the design of an SPE with AuNPs included in its surface as an electrochemical sensing platform of interactions from the protein thrombine and the thrombin binding aptamer (APT). The nanostructrured DNA aptasensor had the APT immobilized to the AuNP by avidin-biotin linkages. Detection of a binding between APT and thrombin was performed by introducing the aliquots of the targets and binding-buffer onto the electrodes. The difference between cathodic peak areas in the system SPE/AuNP/APT/ thrombin and in the SPE/AuNP/APT/ buffer was measured in a stripping voltammetry with E_{ox} = +1.2 V, and a calibration curve was built for different thrombin concentrations. The thrombin detection limit was 10^{-9} M.

The same detection limit (10^{-9} M) was also found for a thrombin biosensor designed by Kerman and Tamiya [99]. They developed an APT-based sandwich assay where the primary APT was immobilized on the surface of the SPE and the secondary APT on the AuNP. The electrochemical reduction current response of the AuNPs was monitored to quantify detection of thrombin.

The AuNPs and stripping voltammetry were employed by Authier et al. [95] for the quantitative detection of amplified human cytomegalovirus (HCMV) DNA. In this case it was only the oligonucleotide probe which was marked with AuNPs. The detection was permitted after the release of the gold metal atoms anchored on the hybrids by oxidative metal dissolution, given rise to a response with anodic stripping voltammetry at a sandwich type screen printed microband electrode. With this technique it was possible to detect 5 pM-amplified HCMV DNA fragment.

Drugs Determination

New applications for SPE are emerging for determination of drugs in the pharmaceutical and biomedical fields. Recently Shih et al. [100] determined codeine, an effective analgesic and antitussive agent in pharmaceutical preparations. They developed a nontronite clay-modified screen-printed carbon electrode that detected codeine in urine by square wave stripping voltammetry. The codeine quantification was achieved by measuring the oxidation peak after background subtraction in voltammograms run between 0.6 and 1.3 V at a square wave frequency of 15 Hz and amplitude of 45 mV. Under these conditions they found linearity for codeine detection in the range of 2.5–45 μm.

Burgoa Calvo et al. [101] developed a silver NP-modified carbon SPE to detect lamotrigine (LTG), a new generation antiepileptic drug for treatment of patients with refractory partial seizures or without secondary generalization. They determined LTG by differential pulse adsorptive stripping voltammetry with a detection limit of 3.7 × 10⁻⁷ M. The SPE used was modified with silver NPs that had been electrodeposited from $AgNO_3$ in an acidic Britton–Robison solution by accumulating potential during a time under stirring. They studied different parameters of silver deposition to optimize the intensity of the reduction peak at −1.06 V needed for the LTG quantification. Good agreement was found between the level stated by the manufacturer of commercial capsules and the one measured by the biosensor.

Enzymatic amperometric sensors with AuNPs have just been developed [102] for the determination of Phenobarbital, a first generation of anticonvulsant drug widely used to treat epilepsy. Different electrode preparation methods were evaluated to immobilize covalently the enzyme, cytochrome P450 2B4. The best results were obtained in gold SPE modified with electrodeposited AuNPs and with the cytochrome attached covalently by Mercapto Propionic Acid/N-hidroxysuccinimide with N-(3-dimethylamoinopropyl)-N′-ethylcarbodiimide hydrochloride, or in carbon SPE functionalized with diazonium salt. The former covalent attachment in gold SPE without NPs did not give any response to Phenobarbital. The same research group detected another antiepileptic drug, leveticeratum, by carbon SPE [103] but in this case without modification by NPs, using POD immobilization by pyrrole electropolimerization.

Martinez et al. [104] designed an MWCNT-modified SPE for Methimazole (MT) determination in pharmaceutical formulations. The MT is used as a drug to manage hyperthyroidism associated with Grave's disease, but it has side effects as possible decrease of white blood cells in the blood. The designed sensor consisted of a rotating disk together with an MWCNT-modified graphite SPE (the working electrode was drop casted with a dispersion of the MWCNT in a mixture of methanol, water and Nafion). The rotating disk contained tyrosinase immobilized in its surface, which catalyzed the oxidation of catechol (C) to o-benzoquinone (BQ). The back electrochemical reduction of BQ was detected on MWCNT-modified graphite SPE at −150 mV versus Ag/AgCl/NaCl 3 M. Thus, when MT was added to the solution, this thiol-containing compound participate in Michael type addition reactions with BQ to form the corresponding thioquinone derivatives, decreasing the reduction current obtained proportionally to the increase of its concentration. This method made possible the determination of MT for concentrations from 0.074 to 63.5 μm with a reproducibility of 3.5%.

Ethanol Quantification

Quantification of ethanol is useful not only in clinical diagnostic analyses but also in fermentation and distillation processes. The amperometric biosensing response of ethanol can be based in two approaches: using alcohol oxidase (AOX) or alcohol dehydrogenase (ADH) as catalytic enzyme [52]. In the first case AOX is employed to catalyze the formation of aldehydes and H_2O_2 by oxidizing low molecular alcohols with O_2. The electrochemical response would be given by the mediated oxidation of H_2O_2, as in other biosensoric detections (glucose, cholesterol). This is the case of the screen-printed sensor designed to determine ethanol in beer, built by Boujtita et al. [105], using cobalt phthalocyanine as mediator.

The second approach uses the action of the enzyme ADH catalyzing the oxidation of ethanol or other primary alcohols (excepting methanol) following Scheme 1. This mechanism was used by Liao et al. [107] to built an ethanol biosensor with ferricyanide-magnetite NPs as mediator. They used the two-step immobilization method that Lu and Chen [69] had previously employed for glucose sensing. The method involved drop coating the carbon working electrode with a mix of Fe_3O_4 and ferrricyanide, drying this layer at high temperature, and posterior addition of the enzyme in buffer (in this case ADH from baker's yeast (YADH) and NAD^+). The NAD^+-YADH/Ferri-Fe_3O_4 based biosensor worked at 200 mV and showed excellent sensitivity for ethanol in buffer: 0.61 μA mm^{-1}.

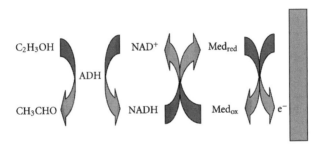

Scheme 1. Detection of ethanol at the biosensor surface catalyzed by alcohol dehydrogenase (ADH).

The use of nickel nanowires with SPE for ethanol/glucose detection was previously revised. Additionally, multi-segmented nickel-gold-nickel nanowires were recently employed in the detection of ethanol [29]. The principle of use was similar to the procedure reported for glucose/ethanol [27], but in this case the electrodic support to control magnetically the orientation of the nanowires was not a SPE but a glassy carbon disk.

SPE FOR DETECTION OF ENVIRONMENTAL POLLUTANTS

Hybridization Sensors

The use of organophosphorus and carbamate pesticides in agriculture has risen exponentially in the last decade causing public concern regarding the environment and food safety. For this reason, many examples of SPE have been proposed for detection of pesticides to substitute other techniques such as HPLC that require trained personnel and can be time consuming and tedious. Most of the screen-printed biosensors are enzymatic systems based on cholinesterase (ChE) inhibition, alone or in combination with choline oxidase (CHO), or noninhibition systems based on organophosporus hydrolase (OPH). However the applications of the later are quite limited since OPH is not a commercial enzyme. Few examples have also been reported using tyrosinase as biocomponent.

Andreescu and Marty [108] compiled in a good review the advances in ChE biosensors, describing immobilization procedures, different designs, and configurations including SPE and practical applications. However, there is only one reference related to the use of nanomaterials on pesticide detection with SPE. The appearance of more work in this area will entitle us to complete that work.

Lin et al. [109] modified in 2003 the classical system of amperometric bienzymatic biosensor for pesticide detection that was used from the late 1980s (see Scheme 2) applying MWCNTs covalently linked to the enzymes. The MWCNT were dispersed in DMF and dried over the carbon working area of the screenprinted electrode; they created carboxylic groups in their surface. Both enzymes were then immobilized by forming amide linkages with the MWCNT using 1-ethyl-3-(3-dimethylaminopropyl) carbodiimide (EDC) as coupling agent. Performing amperometric detections at 500 mV parathion was detected in buffer, without the use of mediators obtaining a linear calibration curve from 50 to 200 μm and a low detection limit of 0.05 μm.

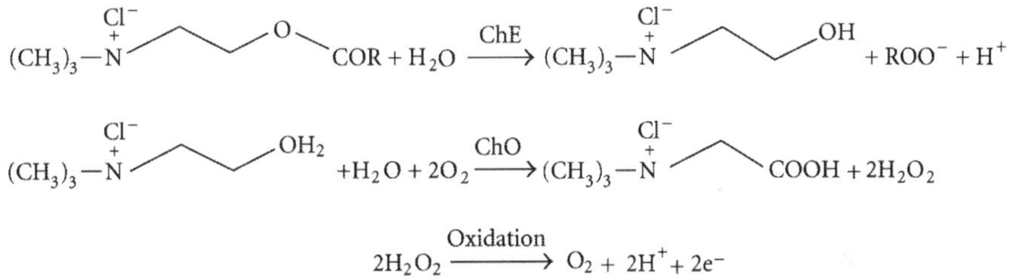

Scheme 2. First generation bi-enzymatic ChE/ChO amperometric biosensor [108]. With permission from Elsevier.

Cai and Du [110] reported very recently the detection of Carbaryl using an MW-CNT-based composite in screen-printed carbon electrodes. The nanotube containing composite was used to modify the other classic approach of pesticide detection: the monoenzymatic approach using ChE to catalyze the hydrolysis of thiocholine (see Scheme 3). Since thiocholine is already electroactive, its oxidation is the reaction to be studied, without the need of a second enzyme. When pesticides are present they inhibit the catalysis of thiocholine formation and a decrease in the signal is observed. Classically, this approach is used with mediators in the system (CoPh, PB) and the amperometric study of thiocholine oxidation can be performed at a low oxidation potential such as 100 mV. However, Cai and Du, avoided the use of mediators with the use of the MWCN-cross-linked cellulose acetate composite, with the ChE covalently bounded to it. This was possible due to the catalytical activity of the nanotubes toward the oxidation of the enzymatically produced thiocholine. The percentage of inhibition of the thiocholine oxidation signal for different concentrations of Carbaryl was obtained by quantifying the peak current at the 535 mV oxidation peak that appeared in cyclic voltammetry before and after inhibition. The detection limit of this sensor for Carbaryl was 0.004 µg/ml concentration (equivalent to a 10% decrease in signal).

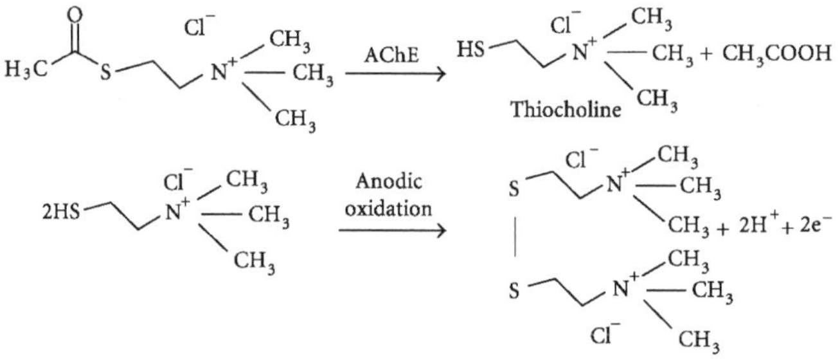

Scheme 3. Approach used in monoenzymatic electrodes. In presence of pesticides the catalytic formation of thiocholine is partially inhibited and less electrochemical response is obtained for the oxidation of thiocholine.

A CNT-modified screen-printed sensor combined with a flow-injection system has been built very recently [106] for the assessment of salivary ChE enzyme activity as an exposure biomarker of pesticides. The modification of the carbon screen-printed electrode was performed by drop casting of an aqueous dispersion of MWCNT, leaving it to dry naturally over the working electrode. A diagram of the flow injection system with the incorporated SPE is shown in Figure 3. A quick and noninvasive approach was reached to determine pesticide exposure by measuring the activity of ChE in rat saliva via the electrochemical monitoring of oxidation of thiocholine production (see Figure 3). In this case the enzyme is not immobilized in the SPE but present in the saliva.

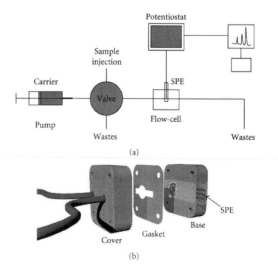

Figure 3. (a) Schematic diagram of the flow-injection sensing system. (b) The 3D image of the electrochemical flow-cell containing the modified SPE. With permission from [106].

Metals

Although essential metals play an integral role in the life processes of living organisms being catalysts in biochemical reactions or essential nutrients, some other metals have no biological role (such as silver, aluminium, cadmium, gold, lead, and mercury). High concentrations of most metals, regardless of being essential or nonessential are toxic for living cells [111]. Thus, there is a growing demand for rapid, inexpensive, and reliable sensors for measurement of metals not only in the environment but also in biomedical and industrial samples. In this sense, SPE can be of great use for metal detection since it has been proved that they give comparable results to those obtained by more expensive, laboratory-based techniques [112].

In the past 5 years, research has been developed on the use of the enzyme urease as biorecognition element in screen-printed biosensors for the detection of metal ions [113-115]. However, no records have been found for urease inhibition-based SPE containing metallic- or carbon-based nanomaterials.

On the other hand, there are many other examples of screen-printed sensors without including any enzyme or biorecognition element (i.e., sensors, not biosensors) for metallic ion detection (see Figure 4) and some of them have seen improvements by the incorporation of nanomaterials. This section will focus on them. In most of the cases they are based on voltammetric stripping analysis, a technique that traditionally was undertaken by Hg electrodes and that is one of the most sensitive alternatives for metal ion determination.

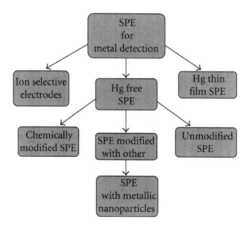

Figure 4. Different approaches for detection of metals with screen-printed electrodes.

Domínguez-Renedo et al. [37] have recently published the detection of chromium (VI), a strong carcinogenic and toxic species, by means of an SPE modified by electrochemical deposition of metallic NPs. Both gold and silver NP-modified electrodes were tested toward Cr (VI) by differential pulse voltammetry. For the electrode preparation, the silver deposition was similar to the procedure described by this group for

LTG detection [101]. For the electrodes that were modified with AuNPs, a solution of 0.5 M H_2SO_4 containing 0.1 mm of AuClO4 was used for the electrochemical deposition of gold at the graphite working electrode, at a potential of 0.18 V during 200 sec. While the best results in terms of sensibility (4×10^{-7} M) and reproducibility (%RSD = 3.2) were obtained by the gold nanomodified sensors, the silver-modified ones offered no interference in presence of any tested metallic ion (gold-modified sensors showed interference with Cu(II) in concentrations higher than 10^{-5} M).

The same preparation techniques for silver and AuNPs in SPE were employed a year earlier to detect Sb (III) [116, 117]. In that case, the detection was performed using anodic stripping voltammetry. A sensitivity of 6.8×10^{-10} M was reached in the case of silver NPs, while 9.4×10^{-10} M was measured in the case of AuNPs. Common interferants in anodic stripping voltammetry (such as bismuth) did not affect the electrochemical response of these sensors.

Nanogold-modified SPEs were also employed to detect As(III) [118]. In this case the working electrode was previously treated with Triton X-100 solution. The AuNPs were obtained dissolving poly (L-lactide) in THF containing $HAuCl_4$ with posterior addition of $NaBH_4$. The poly-L-lactide-established NPs were then drop coated on the pretreated working electrode. Differential pulse anodic stripping voltammetry was also used here obtaining a linear calibration curve up to 4 ppb of As(III) with a detection limit (S/N = 3) of 0.09 ppb. Using the same types of electrodes, an indirect method to detect traces of hydrogen sulphide was developed [119] by measuring the inhibited oxidation current of As(III). The detection limit for hydrogen sulphide was 0.04 μm.

Other Pollutants

Nitrites can contaminate water, foodstuff, and environmental matrices by conversion into carcinogenic nitrosamines. Its quantitative determination is therefore of increasing interest. There is a large number of electrochemical sensors developed with this purpose [120-127], some of them including nanomaterials [122-130] but only one example [131] has been encountered using SPE and nanotechnology. The later work consisted of the immobilization of hemoglobin (Hb) into SPE containing colloidal AuNPs incorporated into carbon ink. An unmediated sensor was thus created with sensitivity for nitrites of 0.1 μm. The colloidal AuNPs decreased the background current, improved the conductivity, amplified the electrochemical signal, helped to retain the bioactivity and accelerated the electron transfer rate.

Hydrazine and its derivatives are potential reducing agents of environmental and toxicological significance. The application of SPE for the detection of this compound was performed already in 1995 for Wang and Pamidi [132]. They printed only the working electrode as a strip containing cobalt phtalocianine or modified with mixed valent ruthenium and detected hydrazine spiked samples with concentrations of 10^{-5} M by amperometric and voltammetric measurements. No biomolecule was required. More recent work [133] includes the application of copper-palladium alloy NP plated screen-printed electrodes in a flow injection analysis system, reaching a linear detection range of 2–100 μm and a detection limit of 270 nm. The SPE, prepared under successive electrochemical deposition of Cu and Pd, showed and enhanced hydrazine electrocatalytic response at low detection potentials in neutral media.

CONCLUSIONS AND PERSPECTIVES

This review has summarized recent applications of nanotechnology in thick film electrochemical sensors. Apart from a revision of the state of the art for this printing technique, the article has gathered different types of screen printing sensing devices by application topic. In every topic a brief introduction has been given to explain the mechanism of detection, followed by examples of nanotechnology applications, emphasizing on the preparation of the (bio)sensor and its response. The reported examples have shown situations where metallic NPs and CNTs have been valuable for screen-printed sensors in different ways.

1. To substitute the use of other mediators, as it was demonstrated with MWCNT or iridium NPs for H_2O_2 detection, substituting PB in glucose sensors, or riboflavin in cholesterol sensors. The same was demonstrated with MWCNT in pesticide detection.

2. As labels or electroactive markers for stripping voltammetry. Metallic NPs were used to improve detection in SPE genosensors, to test drugs, or to detect metallic pollutants in SPE sensors.

3. To retain the bioactivity of enzymes once they have been immobilized in the sensor, as it was the case of AuNPs with hemoglobin in nitrite sensors.

4. To eliminate interferences, as the Au and AgNPs in Sb sensors, where the interference of bismuth is not observed.

5. To improve the sensitivity of the devices. The increase in the surface area of the working electrode generally leads to more sensitive responses.

As summary, the coupling of electrochemical screen-printed sensors with nanoscale materials is being used as a tool for improved sensitivity, longer stability of the bioelement in biosensors and new detection possibilities. The application of nanomaterials in this type of electrochemical devices is still in progress and literature continues to grow and broaden. Although the use of nanotechnology in screen-printed sensors may not be the only key needed to solve some problems faced occasionally by these sensors (e.g., selectivity in some applications), it is certainly a step forward, and the implemented improvements are multiplying the future possibilities of these disposable sensor devices.

KEYWORDS

- **Cholesterol oxidase**
- **Glucose oxidase**
- **Metallic nanoparticle**
- **Multi-walled carbon nanotubes**
- **Screen printed electrodes**

Chapter 10

Bulk Nanocrystalline Metal Creation

David D. Gill, Pin Yang, Aaron C. Hall, Tracy J. Vogler,
Timothy J. Roemer, D. Anthony Fredenburg, and Christopher J. Saldana

INTRODUCTION

Nanocrystalline (nc) materials are super-materials with grain size under 100nm. This small grain size causes the materials to exhibit very high hardness and strength which could be advantageous for many applications. Additionally, the fine grain structure might also create more homogenous material properties for small features on meso-scale parts. These features are often small enough that they consist of only a few grains across their width causing the material to behave differently than would be predicted by analyses using bulk material properties.

There are several size scales that all exhibit different amounts of strengthening based on grain size. Figure 1-1 shows the terminology that corresponds to different ranges of grain size for the microstructure of fine grained materials. This project has utilized powders and materials with microstructure ranging from ultra-fine grained (UFG) down to nanocrystalline. Because of the methods used to make the materials, many of the grain structures used during this project actually fell into different categories depending on the orientation of the sample. The project included methods that created lamellar structures that were tens of nanometers in two dimensions with the third dimension being 100s of nanometers long. So, the range of grain size in each general definition is important, as is a description of the morphology of the particle and the morphology of the grains as well.

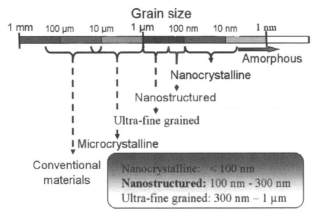

Figure 1-1. The terminology and grain size ranges used to describe the microstructure of fine grained materials.

The purpose of the study was to evaluate several methods of creating nanocrystalline metal, investigate means of consolidating the powdered nanocrystalline metals into bulk material, and then to perform testing and analyses on these materials to understand the microstructure and the mechanical properties.

Creating Nanostructured Material

There are numerous means of creating nanocrystalline metal and each has some advantages and some disadvantages. Figure 1-2 shows a number of the most researched means of creating nanocrystalline materials. This project used cryogenic ball milling because of its ability to create high density nanocrystalline material from high strength materials of interest to weapons designers. These materials include Inconel which was studied in the third year of the project.

The film methods were not used due to the lack of an ability to create bulk structures. Similarly, equal channel angular extrusion and high pressure torsion were not used due to their inability to create nanocrystalline structures in high strength materials like Inconel.

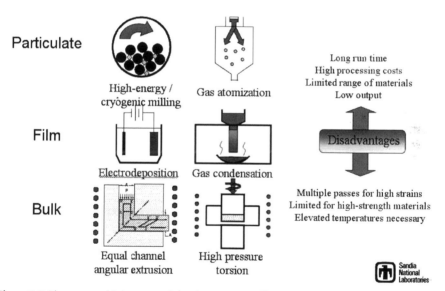

Figure 1-2. There are multiple means of creating nanocrystalline materials depending on the material of interest, the size of material required, and the desired material properties.

In addition to the methods shown in the figure, two new methods of creating nanocrystalline materials were also researched and evaluated during the course of this project. These methods were modulation assisted machining (MAM) and large strain extrusion machining (LSEM). The MAM and liquid nitrogen ball milling, processes that create powdered metal that must later be consolidated by some means, were studied for their ability to create fully nanocrystalline material, their cost, and their ability to create powders from materials of interest to the weapons community.

Modulation Assisted Machining

The MAM is a process in which a single point tool is oscillated at a high rate while cutting a workpiece. This oscillation produces a higher strain on the machined chip than is achieved in normal machining. Also, in a turning operation, the rate of oscillation can be adjusted in conjunction with the rate of rotation of the workpiece to tune for different strain and to tune for different particle (chip) morphology. The team investigated three morphologies including platelet, needle, and equiaxed. It was hoped that this method would be good for quickly creating nanocrystalline materials at a cost much lower than ball milling. However, research into cutting methods showed very slow chip production with hours of work producing just a few grams of material. Material was purchased from M4 Science of West Lafayette, IN. This material was prohibitively expensive for the average application. The project team investigated the material produced by this method and the results of this analysis are given.

Liquid Nitrogen Ball Milling

Liquid nitrogen or cryogenic ball milling is a method used to induce significant cold work into metal powder. In the process, the powder and some type of milling balls are put into a cylinder and rotated around a horizontal axis. The repeated impact of the balls on the material induces cold work and the introduction of liquid nitrogen both makes the material more brittle and also prevents heating from the deformation process. It is important to prevent heating so that the grains in the material will not grow. The project utilized cryomilled powder which was purchased from Novemac, LLC of Dixon, CA. This powder was used in both the cold spray and the shock consolidation processes and showed great promise for both applications.

CREATING BULK NANOSTRUCTURED MATERIAL

The nanocrystalline metal powders created by the methods detailed in the previous sections are often not practically usable due to the small particle size. Common methods of compacting or consolidating particles rely on high pressure or high temperature applied over a long time period which makes them unusable for the consolidation of nanocrystalline material. Extended time under high temperature or high pressure causes grain growth in the material and the unique nanocrystalline properties are lost. To address these significant challenges, the project team conducted research and experimentation on three potential methods of creating bulk material. The first method, LSEM creates material directly from large machining chips. This method is only usable for meso-scale devices because of the difficulty in achieving machining chips large enough for macroscale parts. The next two methods rely on short duration events at high pressure. Cold spray uses the high speed impact of particles on a surface to achieve a coating. Shock consolidation uses a very short time frame impact created by a gas gun. Each of these methods was evaluated for its ability to create or maintain nanocrystallinity in the material, its ability to create fully dense bulk material, and the microstructural and mechanical properties of the material created.

Large Strain Extrusion Machining

The LSEM is a process that uses a machine tool such as a lathe to create a machining chip or foil. The lathe tool has a special gate arrangement at the back of the tool such that the foil coming off the rotating workpiece is fed through the gate. The gate helps to straighten the foil and to smooth one side somewhat. For the process, the workpiece is brought to speed on a powerful lathe and the tool is fed into the part at a high rate. A continuous foil comes out quickly through the gate at the back of the tool and feeds out rapidly until the lathe either loses power or the tool is backed away from the workpiece. The long foil created by this process has nanocrystalline properties due to the high strain event that it experienced. This foil can be lapped or machined to be flat and then meso-scale parts can be cut from the foil. The strain in the foil is controlled by the machining parameters, and the strain predicts the grain size in the final foil material.

The project team evaluated the LSEM process for making foils from which meso-scale parts could be machined. As would be expected, there is significant stress in the material and this can lead to problems in utilizing the foil for precise parts with meso-scale features. The activities completed for creating and understanding this method of creating nanocrystalline material are covered below.

Cold Spray Consolidation

Cold spray, a new member of the thermal spray process family, can be used to prepare dense, thick metal coatings. It has tremendous potential as a spray forming process. In the cold spray process, finely divided metal particles are accelerated in an inert gas jet to velocities in excess of 500 m/s. When accelerated to velocities above a material-dependent critical velocity, V_c, metal particles will bond to the substrate and form a dense, well-adhered deposit, this is the foundation of the cold spray process. The cold spray process is capable of preparing deposits of low-oxide content in air at near-room-temperature conditions. This makes cold spray very attractive for many applications. Some of those applications, such as spray forming, require the cold-sprayed material to exhibit a modest amount of ductility. In their as-sprayed state, cold-sprayed metals tend to be brittle because the particles are heavily cold worked during the deposition process and interparticle bonding tends to be incomplete.

For this project, the stock metal powder used in the cold spray process was nano-structured, nanocrystalline, or UFG, depending on the material. The team tested powders created by liquid nitrogen ball milling and by MAM. The tests showed the ability of cold spray to consolidate materials into a bulk material sample with relatively large dimensions. The team evaluated the use of liquid nitrogen cooling of the cold spray substrates to remove any built-up heat from the process. Figure 1-3 shows a schematic of the cold spray process as well as a sample of nanocrystalline material cold sprayed on the end of an aluminum substrate.

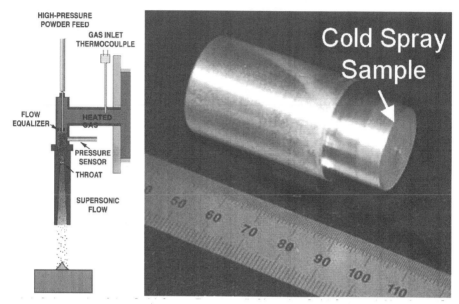

Figure 1-3. Schematic of the cold spray process (left) and a cold sprayed aluminum sample deposited on the right-end of the bar (right).

Shock Consolidation

The shock consolidation process was used due to the high rate of impact which prevents excessive heat build-up. The shock testing was performed at Sandia National Laboratories and utilized the two gas guns shown in Figure 1-4 as well as an 80 mm gas gun at Georgia Tech University. The remainder of this section of text has been adapted from the Ph.D. proposal of team member D. Anthony Fredenburg.

Figure 1-4. These gas guns were used in the shock consolidation of nanocrystalline metals.

Shock Compression of Solid and Porous Media

When a stationary material is impacted under intense dynamic loading conditions, large compressive stress waves are induced in both the impacting material and the body initially at rest. The stress waves produced can take several forms [1-1], and are largely dependent upon the impact velocity and mechanical properties of the material through which the wave is travelling. A strong shock is characterized by a stress pulse with a constant speed and near instantaneous rise to peak pressure; the front is relatively thin and discontinuities exist across the front. Relationships have been developed to correlate the state of material ahead of and behind the discontinuous front through conservation of mass, momentum, and energy. They are termed the Rankine-Hugoniot relationships and serve to connect the particle velocity U_p, shock velocity U_s, specific volume $V = 1/\rho = 1/$density, pressure P, and specific internal energy E [1-2]:

$$U_p - U_0 = U_s(1\text{-}V/V_0) \text{ (Mass Conservation)} \tag{1}$$
$$P - P_0 = (U_s/V_0)(UP\text{-}U_0) \text{ (Momentum Conservation)} \tag{2}$$
$$E - E_0 = {}^{3/2}(P+P_0)(V_0\text{-}V) \text{ (Energy Conservation)} \tag{3}$$

where the subscript 0 indicates initial values for the uncompressed state. The assumptions inherent in the above relationships require that the pressure, volume, and energy states measured are in thermodynamic equilibrium, and that the amount of compression for a given applied pressure is the same as would be produced by a hydrostatic pressure of equal magnitude [1-2]. The first assumption requires a very narrow shock front, and the second is valid for applied pressures that far exceed the yield strengths of the solids; effectively restricting the application of the above relationships to strong shocks. In addition to the above equations, a fourth relationship, commonly referred to as the equation of state (EOS) is required. For inert solids a linear relationship between U_s and U_p has been found to be valid for many materials at pressures below ~100 GPa. The EOS for a non-reacting solid material in response to a strong shock takes the form [1-2]:

$$U_s = C_0 + S_1 U_p \tag{4}$$

where C_0 is the sound speed of the material and S_1 is a material constant. Once any two of the unknown variables are experimentally determined, it is possible to derive all other unknowns through these four equations.

Using these relationships one can construct a compressibility curve for the material, termed the "Hugoniot," which shows the response of a material in pressure-volume (P-V) space. The curve, illustrated in Figure 1-5, is not a continuous P-V path, but rather the location of attainable shock states. The line joining the initial and final states of the shocked material represents the actual path of the material under shock loading, referred to as the Rayleigh line, and is proportional to shock speed.

The energy imparted during impact is the area under the Rayleigh line, and during subsequent release to ambient pressure the energy absolved is the area under the release isentrope, which is often approximated as the Hugoniot curve. Thus for a solid system, the total energy retained (dashed area in Figure 1-5) is the area between the Hugoniot and the Rayleigh line. With the passage of a strong shock, stress values in a porous media can reach values near or above the "crush strength" (point C in Figure 1-5), an applied stress that will compress a material to full density. As such, upon release to ambient pressures the energy absolved is often approximated by the area under the corresponding solid Hugoniot. The amount of energy retained by a porous material is much greater than that of a solid for similar pressures, as shown in Figure 1-5, where the release isentrope for the porous sample is separated from the solid Hugoniot and follows the path A' → C → O.

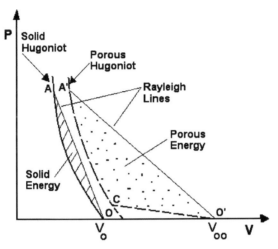

Figure 1-5. Stress-volume Hogoniot for non-reactive solid and porous materials. The V_{oo} and V_o indicate the initial specific volume for porous and solid material, respectively.

In order to develop the Hugoniot of porous materials, various EOS have been proposed to relate material properties in the porous and condensed states. One such EOS is the Mie-Grüneisen EOS, relating the pressure and internal energy difference between the porous and solid states. The Mie-Grüneisen EOS is [1-2]:

$$P - P_K = (\gamma/V)(E - E_K) \qquad (5)$$

where γ is the Grüneisen parameter and P_K and E_K represent the pressure and specific internal energy at zero Kelvin, respectively, such that γ can be assumed as only a function of volume. Though useful in establishing compressibility curves for heterogeneous mixtures, application of the Mie-Grüneisen EOS has shown that substantial deviations occur between predicted and experimental results for porous media due to the constant volume assumption and large dependency on obtaining an accurate γ function.

Shock Compaction of Powders

Transforming a powder from its initial loosely associated state to final solid density state is a complicated process, and becomes even more involved when dealing with heterogeneous powder mixtures. Several models have been developed that describe the compaction process for homogenous powders and will be subsequently discussed; however, a complete model describing the compaction of heterogeneous powders (mixtures) is still lacking.

One such model developed for ductile distended or porous materials is the so called "P-α" model proposed by Herrmann [1-3] to provide a description of the compaction behavior at low stresses and the correct thermodynamic behavior at higher stresses. The complete EOS relates the pressure P, specific volume V, and specific internal energy E through [1-3]:

$$P = f(V/\alpha, E) \tag{6}$$

$$\alpha = g(P, E) \tag{7}$$

where α is the porosity defined as the ratio of the specific volume of the porous material to the specific volume of the analogous solid density material ($\alpha = V_{oo}/V_o$). Defining α as such allows the change in volume due to bulk compression and pore collapse to be separated. This equation is based on the assumption that the relation between pressure, specific volume, and specific internal energy for the matrix material is the same for the porous and nonporous states, and that the surface energy of the particles is negligible. Herrmann suggests that the initial compression of a highly porous material is elastic and is a result of the buckling of the particle walls. Once the elastic limit of the particles is reached, plastic deformation begins to occur as the particle walls permanently deform and the pores collapse. As porosity decreases toward unity and approaches that of a solid material, the bulk of compression will be due to volume compression of the material, with little change in porosity. This concept is illustrated schematically in Figure 1-6 for multiple loading and unloading schemes.

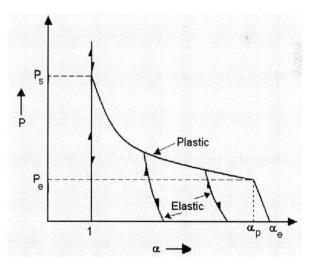

Figure 1-6. Illustration of elastic and plastic region of compression in P-α model for loading and unloading.

Modifying Herrmann's P-α model, Carroll and Holt [1-4] included an additional term supposing the volume average of the stress in the matrix is a pressure P_m, and not P, such that:

$$P_m = \alpha P \text{ and } P = \alpha^{-1}f(V/\alpha, E). \qquad (8)$$

Building upon the above relation and extending the work of Mackenzie [1-5] to the dynamic regime, Carroll and Holt [1-6] developed a model for the compaction of a porous material assuming pore geometry is only a function of initial porosity and pore diameter. This pore collapse relation is expressed:

$$\alpha = 1 + (\alpha_p - 1)\{(P_S - P)/(P_S - P_E)\}^2, \qquad (9)$$

where α_p is the distention achieved during elastic compression, P_E is the elastic compression pressure, and P_S is the crush strength. This spherical shell model assumes that pores in a material collapse in the same manner as would a hollow sphere of the matrix material, and that during pore collapse there is incompressibility of the matrix. Using a finite difference computer code they have shown that compression of aluminum hollow spheres occurs in three distinct phases, a purely elastic phase, an elastic-plastic phase, and a purely plastic phase where change in porosity occurs almost exclusively in the plastic phase. These distinct phases are observed during both static and dynamic compaction.

Similar observations are made for extremely distended ceramic powders under static compression; suggesting three defining stages in the compaction process [1-7]. At low initial pressures the first stage is characterized by particle rearrangement with little to no fracture occurring. During this stage intimate contact is established with neighboring particles such that voids or pores are formed between the particles. The second stage, characteristic of an increased applied load, will exhibit either deformation or fracture of the particles, and allows for additional rearrangement. At higher loads still, the particles or granule fragments undergo further rearrangement. It is supposed that all three stages occur simultaneously with one being the dominant process for a specific pressure range [1-7].

It has been proposed by Raybould [1-8, 1-9] and supported by others [1-10–1-12] that, if the applied pressure is high and the width of the shock front short, during dynamic compaction particles are thrown against one another so violently that oxide layers can be broken. Heat is also generated at interparticle contacts as a result of friction and deformation of moving particles, and temperatures can reach that of the melt. For melting at the particle surfaces to occur, Raybould [1-13] has introduced a criteria for shock rise-time which requires the rate at which the energy of the shock is deposited must be greater than the rate of heat conduction to the particle interior, and has shown this criteria can be reached for lead, aluminum, and steel powders at impact velocities of 500, 1,000, and 1,000 m/s, respectively [1-8]. In an attempt to quantify the process of energy deposition and resultant microstructural changes during dynamic compaction a model has been proposed by Gourdin [1-14–1-17]. This model, shown in Figure 1-7, relates the energy flux at the particle surface F_o to the specific energy behind the shock E, the rise time τ, and the specific area of the powder A, and provides a quantitative link between shock conditions, powder characteristics, and the final microstructure of the compact. Good agreement between model and experiment has been found for 4330V steel, aluminum-6% silicon, and copper powders. Separate investigations by Roman, Nesterenko, and Pikus [1-18] and Gourdin [1-19] on how particle size effects melting have been conducted and show that for a homogenous monodispersed powder the amount of dynamic densification and subsequent melting is increased as the particle diameter is increased.

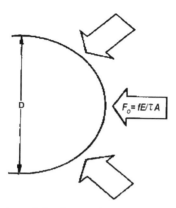

Figure 1-7. Schematic illustration of the flux model developed by Gourdin, where the flux is represented as F_o.

Particle Packing

The dynamic compression of heterogeneous powder particles is a complicated process and is far from well understood. However, with a solid theoretical understanding of the packing of powder particles and the use of computer simulations to observe the meso-mechanics of deformation and flow during compaction, one can begin to estimate, and even predict, the behavior of heterogeneous powders under dynamic compression. It is intuitive to begin with the ideal scenario for packing of powder particles, then consider some of the physical interactions that occur between particles, followed by an examination of the actual process of compaction.

It is well established [1-19] that a narrow bimodal distribution of powder particle size will lead to a higher density than that of a unimodal distribution. According to Cumberland and Crawford [1-20] a dense mixture of heterogeneous spheres can be formed in two ways. First, one can start with a large sphere system in dense random packing and insert small spheres into the existing voids without disturbing the arrangement of the large spheres. If random dense packing is assumed for both the large sphere matrix and the small sphere interstitials, then the fractional solid content (FSC) for both the large spheres, v1, and the small spheres, v_2, is $v_1 = v_2 = 0.625$. The minimum fraction by weight of the coarse particles, Cmin, such that density is maximized is then determined through:

$$C_{min} = v_1/(v_1 + v_2 - v_1 v_2) \tag{10}$$

and is equal to 0.73. The optimal FSC, V'_{opt}, for the mixture is accordingly:

$$V'_{opt} = v_1/C_{min} = v_1 + v_2 - v_1 v_2. \tag{11}$$

For a random dense packing of large spheres with a similar packing of small spheres in the intersticies, the optimal FSC for the mixture is 0.86. The other option for obtaining a dense mixture is to start with a matrix of small spheres and subsequently add large spheres until the number of large spheres increases to the point where there is a matrix of large spheres in dense random packing and small spheres in the intersticies. Similar to the previous scenario, the weight fraction of large spheres to small spheres that maximizes density is $C_{min} = 0.73$ and the optimal FSC for the mixture is 0.86. This approach assumes that both large and small spheres have the same density. However, if the densities of the large and small spheres are not the same, then the weight fraction of large spheres that maximizes density is:

$$C_{min} = v_1\rho_1/(v_1\rho_1 + (1-v_1)v_2\rho_2) \qquad (12)$$

where ρ_1 and ρ_2 are the densities of the large and small particles, respectively. In practice however, densities achieved by mixing of heterogeneous powders are almost always lower than theoretical prediction.

When real powder particles are mixed, additional considerations must be taken into account when attempting to achieve dense packing, especially as particle size is decreased. It has been found that as particle size decreases and the surface area to volume ratio of the particles increases, friction, adhesion, and other surface forces begin to play a more important role in the arrangement of particles [1-21]. For particle sizes less than 100 μm, gravity is no longer the dominant force and body forces such as van der Waals and electrostatic forces begin to dominate and become more influential as particle size decreases, with electrostatic forces only becoming significant for particle sizes less than 1 μm [1-22]. These attractive forces can lead to lower coordination and an increase in porosity as particle size decreases, as well as the formation of agglomerates. Surface effects also become more dominant as particle size decreases. As the surface becomes increasingly more rough, the sliding and rolling friction coefficients increase which inhibit the motion of particles, leading to more porous structures [1-23]. Irregularity of particle shape, though decreasing the van der Walls attractive force, can increase the porosity as an irregular shaped particle can bridge the gap between two adjacent particles, leading to the formation of voids beneath the particle [1-24].

Die filling can also have a substantial effect on the initial density of the particles [1-21]. It has been observed that as the intensity of particle deposition increases, that is the number of spheres falling per unit time per unit area, in the range between 100 and 1400 particles/in²-s the packing efficiency improves. Conversely, increasing the height at which the particles are dropped tends to decrease the packing efficiency; where there is an optimal height of drop for a given intensity of deposition that will produce the closest packing. Cumberland and Crawford [1-21] also reported that a large container to particle size ratio is necessary to ensure the most close-packed particle arrangement, on the order of ~50:1. High frequency, low amplitude vibration during packing has also been found to increase the packing efficiency, especially for irregular shaped particles because they can fit into the indentations and irregularities of neighboring particles. Location and arrangement of powder particles during mixing or filling is important, especially in heterogeneous powder mixtures, because significant segregation can occur which can lead to density gradients and non-uniformity in compacts.

Consolidation of Particles

The mechanics of compaction for the quasi-static and dynamic regime vary greatly and produce compacts with significantly different characteristics. However, a general relationship has been developed that can be used to predict the pressures necessary to produce a certain density compact in both the quasi-static and dynamic regimes. This relationship, known as the Fishmeister-Artz relationship [1-25], is given by:

$$P_y = 2.97\rho^2(\rho-\rho_0)/(1-\rho_0)\sigma_0 \tag{13}$$

where P_y is the pressure needed to achieve a density ρ, starting from an initial density of ρ_0, where the powder mixture has an effective yield strength of σ_0. This model, originally developed for the compaction of homogeneous mono-sized spheres, can also be used as a rough approximation for predicting the pressures necessary to compact heterogeneous powder mixtures. In an effort to better understand the different phenomena occurring during quasi-static and dynamic compaction, simulations on homogeneous mono-sized spheres have been carried out at different impact velocities [1-26]. The authors have shown that during quasi-static compaction, plastic strain is generally initiated and spread throughout the contact regions between particles, whereas during dynamic compression plastic strain initiates near the impact surface and travels in the direction of stress propagation. The development of plastic strain from an array of mono-sized spheres modeled in two dimensions is shown in Figure 1-8. Note that at the low impact velocity of 300 m/s the strain develops at the contact points between particles and grows outwardly, showing quasi-static compression characteristics and leads to a hexagonally shaped particle. For the higher impact velocity of 1000 m/s the plastic strain travels in the direction of wave propagation, leading to very low plastic strains in the particles ahead of the stress front and high plastic strains behind the front. This causes the lower particles to behave as rigid bodies, such that the upper particles will deform on to the top of the lower particles [1-26], as shown in Figure 1-9. Other important parameters investigated [1-26] are interparticle friction, strain hardening, and particle size. It has been found that interparticle friction effects increase under dynamic compaction because of the increased particle shearing that occurs along the particle surfaces. An increased amount of strain hardening in a particle will not change the overall shape of the particle, but rather increases the temperatures generated during dynamic compaction. Decreasing particle size causes an increase in temperature when both small and large particles are subjected to the same particle velocity because there is an increase in strain rate for the smaller particle [1-26].

The preceding analysis deals with an idealized system of homogeneous mono-sized spheres, and it is expected that for heterogeneous systems composed of ductile and brittle components with varying particle sizes and volume fractions, behavior during compaction will be quite different. For simulated mixtures of ductile and brittle particles [1-27] preferential deformation of the ductile particles occurs during compaction. If the volumetric percentage of the ductile component is high enough, the brittle particles can become surrounded by a ductile matrix without undergoing much, if any, deformation or fracture [1-27]. However, when the amount of ductile component is limited, the strength of the compact is dominated by the brittle component and compaction occurs by fracture and rearrangement of the brittle particles. In a compaction study on Cu-W, a system composed of ductile and brittle components, optimal mechanical properties were observed for mixtures containing 20–50% of the brittle tungsten particles [1-28]. The effect of particle size, shape, strength, density, and so forth, is expected to be very influential on dynamic compaction characteristics.

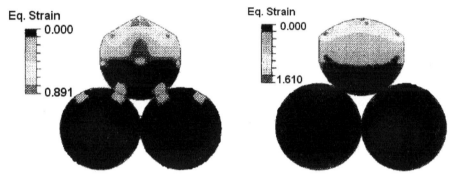

Figure 1-8. Equivalent plastic strain for two-dimensional cylinders compacted at 300 m/s (left) and 1000 m/s (right).

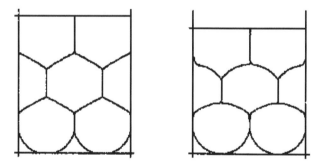

Figure 1-9. Resultant shape from two-dimensional compaction of cylinders at 300 m/s (left) and 1000 m/s (right).

FABRICATION OF NANOSTRUCTURED INCONEL 718 BY LARGE STRAIN EXTRUSION MACHINING

In an effort to create low cost nanocrystalline material, the team investigated the use of material from the LSEM process. This work, done in conjunction with team members from Purdue University, characterized LSEM material, created parts from LSEM material, and evaluated its performance. The text in this section is from a paper by Pin Yang that will be submitted at a later time. Co-authors on the paper include David Gill and Michael Saavedra of Sandia National Laboratories and Christopher Saldana,

James B. Mann, Willie Moscoso, Srinivasan Chandrasekar, and Kevin P. Trumble of Purdue University.

Abstract for Large Strain Extrusion Machining

This study investigated the feasibility of fabricating nanostructured engineering alloys by a low-cost, severe plastic deformation (SPD) process. Special focus was placed on LSEM of solution treated and precipitation hardened Inconel 718 alloy as a precursor material for manufacturing of meso-scale parts. This approach and its generalization are shown to be well suited for production of small components and structures of high-aspect ratio from nanostructured alloys. Correlations between the amount of strain induced by the extrusion machining and microstructure evolution, as well as the mechanical performance were established. In addition, the hardness evolution and thermal stability of a solution treated, strain refined nanostructured Inconel were investigated. Prototyping of meso-scale parts with Nanostructured Inconel super alloy was demonstrated.

Introduction for Large Strain Extrusion Machining

Meso-scale components play an important role in system miniaturization. The size reduction from a mechanical clock to a wrist watch is a practical example of the past. Emergent needs ranging from stronglink components, satellite flexures, microelectro-mechanical systems (MEMS), to compact biomedical devices are just a few examples looming. These meso-scale components vary from a few millimeters (10^{-3} m) in size to the finest feature of few microns (10^{-5} m), and normally carry higher mechanical stresses, experience severe sliding contact, and sustain greater torque [2-1]. Therefore, high strength-to-weight ratios are critical for the miniaturization of structural components. Sometimes, thermal stability is also required during high temperature brazing or diffusion bonding to assure hermeticity of the micro-systems and to protect these components from hazardous environments. In addition, feature sizes in micro-scale components are often on the same order as the grain size when conventional microcrystalline alloys are used as precursor materials, a potentially undesirable mechanical design condition [2-2]. For these reasons, UFG and nanostructured alloys with enhanced strength and hardness are attractive candidates for micro-system applications.

In the 1990s, LIGA (German acronym for Lithographie, Galvanoformung, Abformung) emerged as an important process for making small-scale components out of electro-formable metals such as nickel. The electrochemically-deposited metal is typically nanocrystalline, with high strength and hardness. Perhaps, a more versatile manufacturing approach for making 3-D micro-scale components is by direct fabrication from bulk nanostructured materials using conventional micro-machining processes such as electro-discharge machining (EDM), electro-chemical machining (ECM), micro-milling, and femto-second laser beam processing [2-1]. These processes have the advantage of being applicable over a broad parameter space both in terms of materials and component geometries [2-1, 2-3], and are capable of overcoming the low machining efficiency imposed by these high strength materials through traditional machining process. For example, micro-EDM is used to make 3-D structures and features down to a few micrometers in size and of high-aspect ratios for fuel system components, jet-engine turbine blades, and micro-power systems [2-1]. However, central to this approach is the availability of high-strength nanostructured materials in bulk forms.

The challenge of creating bulk nanostructured materials from high-strength alloys can now be at least partially addressed with SPD techniques [2-4, 2-5]. Conventional SPD techniques like equal channel angular pressing (ECAP) are well developed to the stage where bulk nanostructured materials can be routinely produced from alloys of initial low or moderate strength. A few high-strength, nanostructured alloys can also be made by these techniques by heating the precursor bulk material to elevated temperatures during the SPD, with some attendant trade-off in the level of microstructure refinement [2-6]. Recently, a new class of SPD processes based on large strain machining has emerged; a constrained variant, LSEM is used in the present work. These machining processes are well suited for SPD of difficult-to-deform alloys (e.g., Inconel, Ti alloys, stainless steels, Ta) by chip formation, without pre-heating of the bulk microcrystalline material [2-7–2-10]. The LSEM, in particular, is capable of not only affecting microstructure refinement by imposition of large plastic strains, but is also amenable for producing bulk chips with controlled geometry (e.g., foil, sheet, wire, bar) and UFG and/or nanocrystalline microstructures.

In this study, a low-cost production of nanostructured Inconel 718 in bulk foil/ sheet forms using LSEM is first demonstrated. This approach and its generalization are shown to be well suited for production of small components and structures of high aspect ratio from nanostructured alloys. In the next few sections, a brief background on the LSEM process is reviewed and various microstructures developed under different processing conditions (shear strain induced by LSEM process) and their mechanical performances are reported. Fabrication of micro-scale components from high-strength Inconel foils were then demonstrated, using a wire micro-EDM process. In addition, the effect of thermal aging (or precipitation hardening) on the microhardness of nanostructured Inconel is compared.

Background for Large Strain Extrusion Machining

Imposition of large plastic strains is the most effective method to produce bulk metals with UFG and nanocrystalline microstructure [2-9, 2-11 - 2-13]. Materials with the refined microstructure generally possess a significantly improved strength and hardness as compared with their microcrystalline counterparts [2-8, 2-13]. Large plastic strains can be introduced by SPD methods such as wire drawing [2-11], ECAP [2-12, 2-13], high-pressure torsion [2-4, 2-14], and rolling [2-15]. Most of these processes involve multiple passes of deformation to achieve a desired microstructure and enhanced mechanical properties. However, this practice becomes progressively more difficult as material is cumulatively strengthened by each deformation cycle. This is particularly true for high-strength materials where tooling difficulties have usually precluded using these methods without some preheating above ambient levels. These constraints make the aforementioned methods cumbersome, time-consuming and expensive to scale up for large volume production.

Simplicity of chip formation in plane-strain machining enables grain refinement by SPD at near-ambient temperatures in the full range of commercial alloys, including high-strength alloys which are difficult to process by conventional SPD [2-7–2-10]. The LSEM is the experimental configuration used to create bulk foil samples under controlled conditions of strain and strain rate [2-9]. Figure 2-1 illustrates an LSEM process wherein a bulk foil/sheet sample (chip) of controlled thickness is produced by machining and extrusion imposed in a single step using a specially designed tool. The tool consists of two components—a bottom section with a sharp cutting edge inclined at a rake angle (α) and a wedge-shaped top section that acts as a constraining edge. The tool moves radially into a disk-shaped workpiece rotating at a constant peripheral speed, V. The chip is simultaneously forced through an "extrusion" die formed by the bottom rake face and the top constraining edge, thereby effecting dimension and shape control. The velocity of the material at the exit of the tool is $V_c = V_t/t_c$, where t and t_c are the undeformed and deformed thickness of the chip, respectively. The LSEM, unlike conventional machining, involves chip formation by constrained deformation where the deformed chip thickness is controlled a priori. Thus, microstructure refinement by large strain machining is effected in combination with shape and dimensional control.

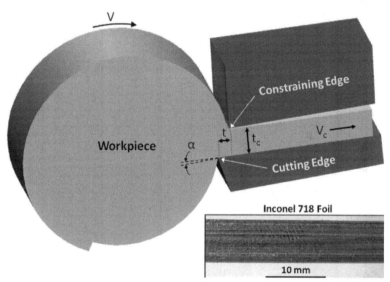

Figure 2-1. Schematic of large strain extrusion machining (LSEM). Insert at the bottom right shows a typical Inconel 718 foil strip created using this process.

The LSEM process produces bulk UFG and nanostructured materials at small deformation rates that suppress in situ heating and microstructure coarsening [2-10]. The LSEM can be considered a generalization of ECAP with the exception that the thickness of the material at the tool exit can be different than that at the inlet. This ensures imposition of strain levels that are sufficiently high to induce formation of an UFG microstructure in a single-step deformation process. The shear strain (γ) imposed in the foil/sheet sample is determined by the extrusion (chip thickness) ratio ($\lambda = t_c/t$) and the rake angle, α, as [2-10, 2-16]:

$$y = \frac{\lambda}{\cos\alpha} + \frac{1}{\lambda\cos\alpha} - 2\tan\alpha \cdot \qquad (1)$$

A wide range of strain [1-10] can be imposed in the foil/sheet samples by varying α and λ in a single pass of the cutting tool, while strain rate can be varied by varying the velocity of LSEM; to first order, the strain rate varies linearly with velocity. Most importantly, bulk foil/sheet samples with UFG and nano-scale microstructures can be created even from high-strength alloys at near-ambient temperatures [2-6–2-8]. The process produces good surface quality, uniform properties over several meters of ex-truded chip [2-10], and is versatile for fabrication of bulk materials in the form of foils, bars and wires [2-11]. Therefore, LSEM is an excellent candidate to make foil sheets that can serve as precursor materials for the meso-scale components.

Experimental Procedure for Large Strain Extrusion Machining
The LSEM was used to produce bulk Inconel foils having UFG microstructure for manufacturing of small-scale components and mini-tensile specimens for mechanical evaluation. For this purpose, the as-received, precipitation hardened bulk Inconel 718 alloy was solution treated at 1035°C for 1 hr followed by a rapid air quench, in order to ensure that any precipitates that were initially present went into solution. The initial grain size in the solution-treated Inconel 718 was ~50 µm, as measured by the line in-tercept method, and the Vickers hardness after the solution treatment was ~2.16 GPa. Additional hardness and elastic modulus were determined by a MTS nanoindenter XP system, based on the force-displacement curve. To account for the influence of elastic unloading on the projected area of the indentation impression, the Oliver-Pharr method [2-17] was used. The elastic stiffness (S) was obtained from the slope of the unloading curve. Elastic modulus can then be determined by the contact area (Ac), Ge-ometry ratio for Berkovich indentation tip ($\beta = 1.024$), and Poissons ratio (ν) through the following expression,

$$E = \frac{\sqrt{\pi}}{2\beta(1-v^2)} \frac{S}{\sqrt{A}} \cdot \qquad (2)$$

Chips in the form of bulk foils were created from the solution-treated material by LSEM at a speed of 40 m/min with tungsten carbide tools having rake angles of 0° and +15°. The undeformed chip thickness was ~130 µm and the width of cut was 6 mm. Using Equation 1, the strain in the foils was estimated as 2.1 ($t_c = 260$ µm) and 3.5 ($t_c = 415$ µm) for the +15° and 0° rake angle tools, respectively. Foils of different shear strains were made by a CNC Fryer lathe. Hardness and microstructure of the foils were characterized by Vickers indentation and transmission electron microscopy (TEM). For the hardness measurements, 20 indents, each at least 50 µm in size, were made on three samples for each condition. The microstructure of the foils was examined using a JEOL 2000FX TEM operating at 200 KeV. The TEM samples were prepared by mechanical thinning of the chips to ~150 µm, followed by electrolytic thinning in a Struers Tenupol-5 using a solution of 25% HNO_3 and 75% CH_3OH.

Tensile testing was performed by dog-bone shape mini-tensile specimens fabricated from LSEM foils by micro-EDM process on a Sandia developed test system (see Figure 2-2). Prior to the micro-EDM process, the LSEM foils were first flatten and then about 10 μm surface layers (the secondary deformation arising from tool-chip friction) were removed by lapping. Sample was loaded and aligned in a custom built fixture before testing. The system was controlled by the Teststar programmable controller (MTS Inc, Mineapolis, MN) under a strain rate of 25.4 μm (or 0.001″)/second. Strain was measured with an optical extensometer (Keyence LS7000 LED shadow edge detection).

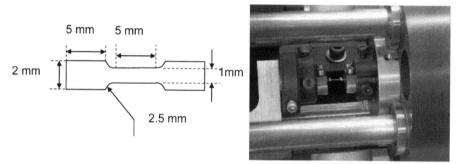

Figure 2-2. The schematic of physical dimensions of a mini-tensile test specimen and tensile test system with an optic extensometer.

Prototype micro-scale components—spur gears with ~2 mm diameter—were produced from the foils by micro-EDM. This component geometry was selected because it incorporates sufficiently complex features of high-aspect ratio typical of 3-D micro-components [2-1, 2-18]. A wire electrical discharge machine (Agie AC Vertex 2F) was used to machine the gears from the foils. Machining was carried out using low-energy discharges with a high pulse rate, conditions that are ideal for machining of micro-scale features. Feature sizes as small as 40 μm were achieved as determined by OGP Avant video measurement system. De-ionized water was used as the dielectric medium. The EDM process created about 1 μm thick recast layer with rough appearance on the surface. The gears were subsequently electro-polished using a mixture of sulfuric acid (73 Vol. %), glycerol (20 Vol. %), and water (7 Vol. %) to obtain a smooth surface finish [2-19]. The electro-polishing was performed with 220 mA current for 60 sec with a custom designed tungsten spring as electrical contact. The gears were characterized using a scanning electron microscope (SEM, JEOL JSM-6490LA).

Results and Discussion for Large Strain Extrusion Machining
Shear Strain and Hardness
Figure 2-1 (inset) shows a typical Inconel foil produced by LSEM of the solution treated material. The foils were approximately 600 mm long and 6 mm wide. The straightness of the foils facilitated subsequent processing. The hardness values of the foils created with shear strain (γ) of 2.1 and 3.5 were 4.84 + 0.22 GPa and 5.51 + 0.11 GPa,

respectively. These values, determined by the Vicker hardness, are much greater than the hardness (2.17 + 0.07 GPa) of the solution treated bulk microcrystalline Inconel, suggesting a significant level of microstructure refinement during LSEM. Indeed, the foil hardness values are nearly that of aged (or precipitation hardened) microcrystalline Inconel 718 (4.12 ± 0.05 GPa). In comparison to a LSEM foil created from a bulk precipitation hardened Inconel 718 (shear strain unknown), the hardness value of this material reached 5.46 ± 0.11 GPa which is 32.6% increase for the bulk microcrystalline material. Uniaxial tensile test results from the foils showed increases in strength that are directly proportional to the increases in hardness [2-20].

These experiments were repeated with four different strain values (up to 4.1) to establish correlations between processing conditions with the hardness and microstructure by nano-indentation and TEM. Indentation impressions and the force-displacement curve for a chip deformed to strain of 1.5 is given in Figure 2-3, with indentation depth controlled at ~1 μm and indents ~20 μm apart. The average hardness and modulus were obtained from 15 indents for each condition. Results of the measured hardness and modulus data are given in Table 2-1. Results indicate that the hardness increases while the modulus decreases with the increase of shear strain. The increase in hardness can be attributed to microstructural refinement induced by SPD where dislocations are hindered by the grain boundary and plastic deformation becomes more difficult, as shown by a classic Hall-Patch behavior. Most modulus values determined from the nanoindentation technique are lower than the bulk modulus of microcrystalline Inconel 718 (~215 GPa). Many researchers have attribute to the decrease in modulus to the increase of grain boundary areas in these hard UFG materials.

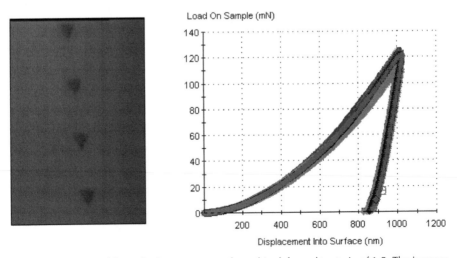

Figure 2-3. Indents and force-displacement curve for a chip deformed to strain of 1.5. The increase and decrease in loading with respect to the displacement correspond to the loading and the unloading conditions respectively.

Table 2-1. Hardness and modulus of solution treated extrusion machined inconel 718 as measured by the nano-indentation technique.

Hardness (GPa)				
Rake angle, °	−20	−5	+5	+15	+45
Shear Strain†	4.1	3.9	2.7	2.3	1.5
Average	6.7	6.1	6.3	5.9	5.5
Std Dev	0.35	0.31	0.21	0.28	0.25
Elastic Modulus (GPa)				
Rake angle, °	−20	−5	+5	+15	+45
Shear Strain†	4.1	3.9	2.7	2.3	1.5
Average	163	173	159	181	192
Std Dev	6.5	7.0	3.9	5.5	8.7

† The amount of shear strain was controlled by the raking angle in a LSEM process.

Microstructure of Large Strain Extrusion Machining

Direct measurements of strain and strain rate in chip formation using Particle Image Velocimetry (PIV)[9] have shown the deformation to be uniformly distributed across the chip thickness. Thin surface layers (~10 μm thick, see Figure 2-4 (a) on the right hand size of the cross section, separated by a dotted line) on both faces of the chip are, however, affected by secondary deformation arising from tool-chip friction. The uniform distribution of strain is consistent with observed uniformity of the hardness through the chip volume. Additional support for homogeneity of the deformation comes from electron backscattered diffraction (EBSD) investigations that have revealed a uniform microstructure through the foil thickness [2-21].

In this investigation, we used an electron beam channeling contrast imaging technique to examine the plastic deformation induced by the LSEM process from precipitation harden Inconel 718 alloy at the microscopic level. Figure 2-4a shows the electron beam channeling image of a cross section of finely polished LSEM foils and Figure 2-4b shows a TEM micrograph from the secondary deformation region (immediately adjacent to the cutting tool) sliced by focused ion beam (FIB) technique. The arrow direction shows the rolling direction during the machining. TEM results indicate that the secondary deformation area is truly nanocrystalline that is, average grain size less than 100 nm with fine carbides (majority are the niobium rich carbides as determined by EDS technique) uniformly dispersed within the microstructure. Electron beam channeling reveals an elongated microstructure in Figure 2-4a which is consistent with TEM observation in the bulk LSEM foil. The aligned carbides and the elongated microstructure in the electron beam channeling image reveal the curved plastic deformation flow lines (or flow plane for simple shear) in front of the secondary deformation area at the cutting tool edges (dotted lines for guiding the eyes only). Microstructural observations indicate that during the SPD process, titanium carbides, mostly in a cubic shape, remain coherent to the Inconel matrix (as shown by the dark particles in Figure 2-5a), while the niobium rich carbides (white particles in the image in Figure 2-5b) tend to crack and sometimes turn into smaller particles that elongate along the plastic deformation direction, leaving microscopic defects in the materials. These defects play an important role on final failure during tensile test.

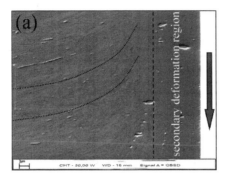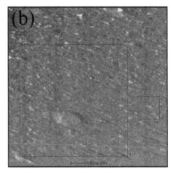

Figure 2-4. Microstructure of a severely plastic deformed precipitation hardened inconel 718 after LSEM. The micrograph (a) of a cross section of deformed foils along the rolling direction (shown by arrow), and (b) TEM image of the secondary deformation area.

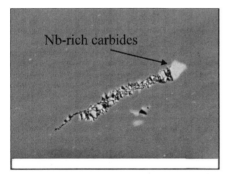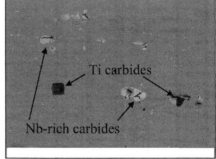

Figure 2-5. Plastic deformation around precipitated carbides. Black titanium carbides retain their integrity while the large Nb rich carbides tend to crack or deform along the plastic deformation direction.

Shear Strain and Microstructure Refinement

Direct evidence for ultrafine microstructures of the foils is seen in the TEM images and selected-area diffraction (SAD) patterns shown in Figure 2-6 - 2-9, with shear strains from 1.5 to 3.9. The micrographs illustrate increased microstructure refinement and formation of nano-scale and UFG structures with increasing strain, which is consistent with the well-known dependence of microstructure on strain. At lower shear strains (1.5 and 2.3), the appearance of elongated microstructural features is evident. Diffraction patterns that indicate the presence of operating reflections in the crystal lattice are observed. The elongated nature of these structures, as well as the details in the diffraction patterns, suggest that these features are deformation twins in the Inconel. This will be verified later with the indexing of the diffraction patterns of these structures.

Figures 2-6 and 2-7 show the bright and dark field TEM images of LSEM foil with shear strain of ~1.5 and ~2.3, respectively. Many regions show elongated features resembling twinned structures. Diffraction patterns of these areas indicate operating reflections in the lattice, which could be indicative of twinning. Other regions show dense accumulation of dislocations. Deformation twinning is the more dominant mode of deformation in certain material systems at higher strain rates and lower strains and has been observed to occur in Inconel 718 in the literature. Deformation twins are lower energy grain boundaries and may play a role in enhanced ductility and thermal stability.

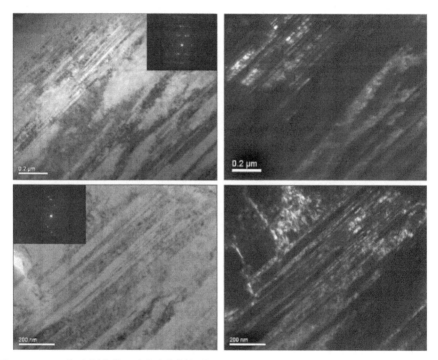

Figure 2-6. Bright-field (left) and dark-field (right) TEM micrographs of inconel 718 foils created with shear strain of ~1.5, showing the nano-size microstructure.

As the shear strain increases, these elongated structures are no longer evident as twinning is presumed to no longer be the prevalent mode of deformation. Figures 2-8 and 2-9 depict the microstructure changes in the LSEM Inconel 718 as the shear strain increases up to ~2.7 and ~3.9. Both figures show the absence of the elongated structures. The microstructure evolves from one characterized by dense areas of dislocations to one with sub-500 nm grain sizes. With increasing strain, the diffraction patterns also transform from a smeared single crystal pattern to one that is more ring-like and indicative of high angle misorientation between grains. This has also been confirmed by sample-tilting experiments in the electron microscope.

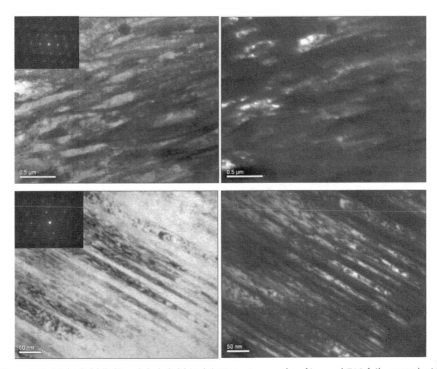

Figure 2-7. Bright-field (left) and dark-field (right) TEM micrographs of inconel 718 foils created with shear strain of ~2.3 showing the nano-size microstructure.

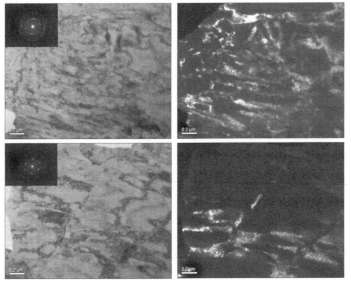

Figure 2-8. Bright-field (left) and dark-field (right) TEM micrographs of inconel 718 foils created with shear strain of ~2.7 showing the nano-size microstructure.

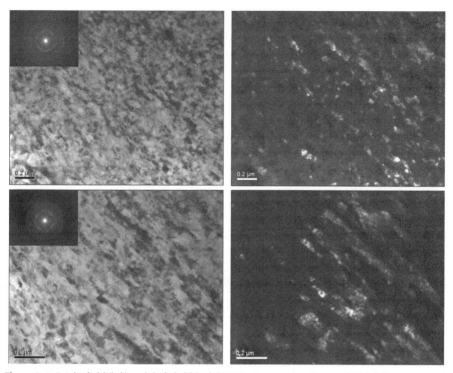

Figure 2-9. Bright-field (left) and dark-field (right) TEM micrographs of inconel 718 foils created with shear strain of ~3.9 showing the nano-size microstructure.

Shear Strain Versus Mechanical Properties

The stress-strain responses for LSEM Inconel 718 samples are given in Figure 2-10. Mini-tensile specimens fabricated from a bulk precipitation hardened (ppt) Inconel 718 alloy were also tested (green curve) for comparison purposes. Yield stress is determined at 0.2% engineering strain on the stress-strain curve. Results show that there is a significant increase in yield strength for the LSEM specimens. A 50% increase in the yield strength for a precipitation hardened LSEM sample (strain ~ 3.1) was observed in comparison to the microcrystalline bulk material; however, the maximum strain to failure (ε_f) was also significantly reduced (from 13% to below 2%). For solution treated (ST) samples, both yield strength (σ_{ys}) and maximum strength (σ_{max}) increase with the amount of strain induced during the extrusion machining process, while the average strain to fail (based on five test specimens) decreases from 3.5 to 2.5% and eventually to 1.4% as the strain values in LSEM process increased from 1.5, 2.7–3.9. When the maximum strain to failure exceeds 2%, necking at gauge region is readily observed, suggesting ductile deformation is dominant before failure. A 2% of strain before failure might not be suitable for large structure applications; however, it could be adequate to provide a sufficient reliability for meso-scale components. Data indicate that strengthening is more effective in the precipitation hardened samples than these solution treated specimens; however, the amount of ductility will be forfeited.

Therefore, there is a moderate trade-off between strength and ductility, either by machining conditions or initial microstructure.

Elastic moduli estimated from the slope of the stress-strain curve (dσ/dε) show that these LSEM samples have a lower modulus in comparison to the bulk materials. The observation is consistent with nano identation measurements. However, the estimated modulus increases with the shear strain induced by the LSEM process, which contradict the indentation results. It is plausible that errors in estimating the slope of the stress-strain curve were introduced since these samples are all slightly warped after extrusion machining. In parallel, we suggest to continue future investigation on this subject to obtain a better understanding of this issue.

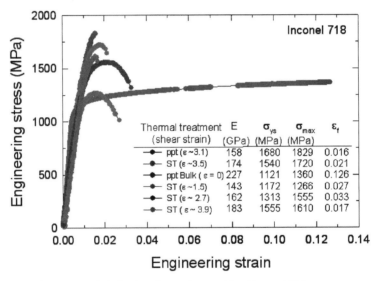

Figure 2-10. The stress strain behavior of extrusion machined inconel 718.

Microstructure of the fracture surface for precipitation hardened, extrusion machined specimens was examined because of the dramatic increase in its yield strength and the significant loss in its ductility. Fracture surface were examined by SEM. A typical fracture surface obtained by SEM for these specimens is given in Figure 2-11. The fracture surface exhibits regions with ductile and brittle failure morphologies. The dimpled rupture on the top is characteristic of a ductile failure; while the bottom region with many brittle cracks associated with niobium-rich carbides (determined by electron microprobe analysis) shows evidence of a brittle fracture. Observation suggests that fracture might be initiated from these large cracked niobium rich carbides (see Figure 2-5) during the extrusion machining process. Upon pulling under a tensile stress, cracks can quickly propagate through the sample as the material reaches its yield point, therefore contributing to the shorter elongations observed in these specimens

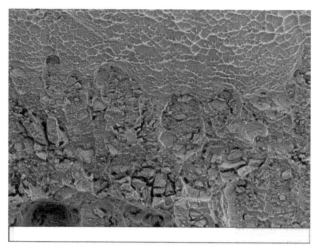

Figure 2-11. Microstructure of the facture surface for a precipitation hardened, extrusion machined inconel 718 alloy. The top portion exhibits a dimpled rupture and the bottom shows brittle fracture from cracked carbides ($\gamma = 3.1$).

Thermal Stability of Extrusion Machined Inconel 718 Alloy

The thermal stability of precipitation-hardenable alloys such as Inconel 718 is largely determined by the stability of the precipitate phase [2-22]. This is even more important for metals with a nano-scale matrix, as the precipitates serve to pin grain boundary motion, hindering coarsening and softening of the fine-grained material. An aging study was conducted to investigate the precipitation kinetics (precipitate formation and ripening) in the microstructure (or grain size) refined Inconel 718 by LSEM process. The thermal aging study will help to determine the feasibility of the use of nanostructured Inconel 718 in micro-component applications at elevated temperature (i.e. $> 650°C$).

The original plan was to study the kinetics of the precipitation process of extrusion machined Inconel at different strain conditions [2-23] by SEM, TEM, x-ray diffraction [2-24, 2-25] (phase volume fraction, precipitate phases (γ', γ," and δ) [2-26], and their lattice parameters), and differential scanning calorimetry (DSC) [2-27]. Results will be compared to the tensile testing data and microhardness measurements. Follow-up will also include a precipitation strengthening study to investigate the effect of over-ageing (OA) and peak-ageing (PA) on the strength, ductility and thermal stability of Inconel 718. The correlation of hardness data with the data from DSC and XRD testing will provide a good foundation for understanding precipitation kinetics and thermal stability, while the tensile data will give insight into ductility of nanostructured Inconel 718. However, due to the time constraint, only the microhardness experiment was completed and will be reported here.

The hardness evolution at 680°C for bulk and extrusion machined Inconel 718 is given in Figure 2-12. Results show that the time required for peak aging of the nano-structured Inconel is much quicker than in the bulk material. The bulk material peak hardens to a value of about 4.73 GPa after aging at 200 hr. A peak in hardness (6.24 GPa) is seen after a treatment of 5 hr for a shear strain of 3.9. With longer exposure up to 300 hr at 680°C, the strength of the highly nanostructured Inconel is seen to be stable at ~5.85 GPa. As the shear strain during the extrusion machining is reduced, the peak hardness falls and the time to peak hardness increases, as illustrated by sample with $\gamma = 1.5$.

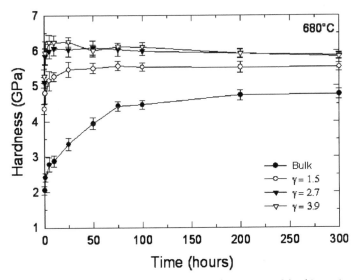

Figure 2-12. Hardness evolution for solution treated inconel 718 at 690°C (melting point = 1300°C).

The level of strengthening seen in the bulk material and nanostructured materials is consistent with the formation of γ' and coherent γ'' precipitates in the matrix [2-26]. The primary mode of strengthening in Inconel 718 is the γ'' precipitate. Softening of Inconel 718 has been shown to correspond to the transformation of the second phase from γ'' to the δ phase, this does not appear to be the case for the range of conditions tested here due to the lower temperature.

Of particular interest is the stability of the nanostructured Inconel 718 with extended time at 680°C, which is greater than half the melting point of alloy (Tm ~ 1300°C). Materials with grain sizes in the nanometer regime typically suffer from low thermal stability, with re-crystallization and grain coarsening occurring at temperatures as low as 0.3–0.4 Tm. These preliminary results are promising in this regard, suggesting extrusion machined Inconel alloys might have a great potential for high temperature applications.

The aging curves for 750°C are markedly different than those at 680°C (see Figure 2-13). The time to peak hardness for the bulk material is much quicker in this case (in comparison to 680°C) and the peak hardness of the high strain samples is quickly lost with time at this temperature. The bulk material also begins to lose strength with treatment beyond 100 hr, exhibiting a similar softening trend to that of the nanostructured material. One observation to note is that the extrusion machined materials are able to retain about 10% higher strength than the bulk up to 300 hr.

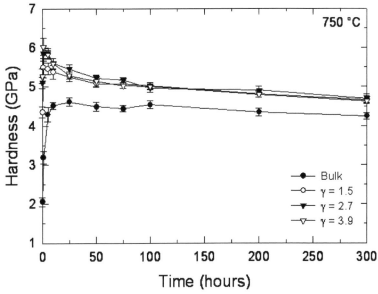

Figure 2-13. Hardness evolution for solution treated inconel 718 at 750°C (melting point = 1300°C).

The level of strengthening seen in the bulk material is once again consistent with the formation of γ′ and coherent γ″ precipitates in the matrix [2-26]. The softening of the undeformed bulk and the extrusion machined materials suggests the loss of thermal stability of the γ″ precipitate phase in both the microcrystalline bulk and the nanostructured Inconel in this time/temperature regime. The decreasing hardness for the nanostructured Inconel 718 samples with time, therefore, could be attributed to the ripening of the γ″ phase or the fast transformation of the γ″ to the δ phase at this temperature. Further characterization of the precipitate phases by high-resolution TEM in the future will shed light on these ideas. Nevertheless, these results indicate that the nanostructured Inconel is able to maintain strength improvement over the bulk material and exhibit a higher hardness value.

Prototype Meso-Scale Component Fabrication

Production of small components from the nanostructured Inconel foils was carried out using micro-EDM. These foils were extrusion machined from a precipitation hardened Inconel 718. Prior to the micro-machining, about 10–20 μm of material was taken off from each of the two extended foil surfaces by mechanical polishing to remove layers deformed by the tool faces. Micro-EDM produced a rough surface finish with ~1 μm thick recast layer (see Figure 2-14). The recast layer has a coarse grain structure with residual tungsten (from the wire used in the EDM machine) and complex oxides on the surface (Nb-Cr-Ti-Al-O, determined by EDX), presumably due to oxidation in the electro-discharge process. Figure 2-15 shows an SEM picture of micro-scale gears created by micro-EDM and electro-polishing. Measurements showed the form accuracy and tolerances to be excellent and the surface finish (Ra) value on the gear teeth to be ~0.1 μm. The high-aspect ratio features associated with the gear teeth are sharply demarcated in the picture, demonstrating the capability of micro-EDM to create a complex micro-scale part from pre-cursor nanostructured materials. Electron microscopy of the gear surfaces showed an almost complete absence of recast layers (see insert in Figure 2-15). However, white niobium-rich carbides can still be observed on the surface. The absence of the recast layers is undoubtedly a consequence of the low discharge energies used to effect material removal. Thus, the nano-scale microstructure of the Inconel foil, with its enhanced mechanical properties, is retained in the gear.

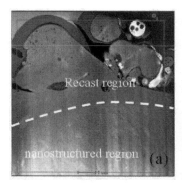 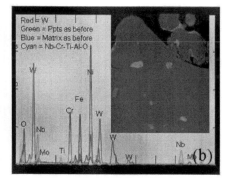

Figure 2-14. TEM micrgraph of the micro-EDM surface of an inconel component (a), and the EDX analysis of the recast region (γ=3.1).

Figure 2-15. Meso-scale gears produced by micro-EDM and electro-polishing of precipitation hardened inconel foil (γ=3.1).

The combined LSEM/micro-machining approach offers exciting opportunities for creation of a new-class of small-scale components with enhanced structural performance. This manufacturing approach is flexible and applicable to a variety of high-strength nanostructured material systems such as titanium alloys and stainless steels. The nanostructured alloys created by LSEM can serve as pre-cursor materials for conventional micro-machining or micro-forming processes. This offers, potentially, a new paradigm for micro-systems manufacturing that incorporates nano-scale microstructure control. Ongoing work is aimed at developing this manufacturing framework for a broader class of nanostructured alloys and micro-scale components.

Conclusions for Large Strain Extrusion Machining

The LSEM has been used to create foils of Inconel 718 with nano-scale microstructure. The nanostructuring is a consequence of the large plastic strains imposed in the Inconel 718 by SPD. Although the LSEM is a relatively low cost manufacturing process to fabricate nanostructured precursor materials for meso-scale components, small warp, and residual stress could be an issue in fabrication of high precision components. The evolution of the microstructure with increasing shear strain is evident from TEM study. At low shear strain, the appearance of elongated microstructural features, as well as the details in the diffraction patterns, suggest that these features are deformation twins in the Inconel, presumably due to lower energy grain boundaries which may play a role in enhanced ductility. As the shear strain increases, these elongated structures disappear and are replaced by dense areas of dislocation to one with sub-500 nm grain sizes and high angle misorientation between grains becomes evident. These microstructural changes are reflected by the increased hardness and yield strength, as well as by reduced elongation during tensile testing. Fracture surface exhibits regions with both ductile and brittle failure morphologies. Microhardness evolution with thermal aging at 680°C shows that the time required for peak aging of the nanostructured Inconel is much quicker than in the bulk material. With longer exposure up to 300 hr at 680°C, the strength of the highly nanostructured Inconel is seen to be stable at~ 5.85 GPa, which is 23.6% higher than the bulk microcrystalline material. The peak hardness slightly decreases for bulk and highly strained Inconel after thermal aging at 750°C for 100 hr. However, the extrusion machined materials are able to retain about 10% higher strength than the bulk up to the 300 hr. Prototype 3-D micro-scale components from the nanostructured Inconel foils have been demonstrated using micro-EDM and electropolishing. This approach is applicable to a range of alloys and micro-machining processes.

Acknowledgments for Large Strain Extrusion Machining

This work was supported in part by an NSF Graduate Fellowship (to Christopher Saldana at Purdue University.). The majority of the funding and work are support by Sandia's LDRD program. We are grateful to Professor S. Chandrasekar and C. Saldana (both from Purdue University) for their work on extrusion cutting, TEM, and thermal aging characterizations, Michael Saavedra for assistance with micro-EDM, Gilbert Benavides for encouraging development of this micro-scale manufacturing approach, David Schmale and Thomas Buchheit for mini-tensile test evaluations, Michael Rye and Joseph Michael for electron beam channeling contrast imaging, Richard P. Grant for SEM study, and Paul G. Kotual for the TEM investigation.

FABRICATION OF NANOSTRUCTURED ALUMINUM 6061-T6 PARTICULATES BY MODULATION ASSISTED MACHINING

A major focus of the project was to utilize particles created by MAM as a potentially inexpensive source of nanocrystalline materials. This work was done with project partners from Purdue University. Below is a paper yet to be submitted which details the project efforts toward the creation, analysis, and utilization of MAM particles. The chapter is authored by Pin Yang, David Gill, Luke Brewer, Bonnie McKenzie, and Joseph Michael of Sandia National Laboratories, with co-authors Christopher Saldana and Srinivasan Chandrasekar of Purdue University's Center for Materials Processing and Tribology.

Abstract for Modulation Assisted Machining

This study investigated the feasibility of fabricating nanostructured particulates of aluminum 6061-T6 alloy by a low-cost, MAM method. These particulates were used for low temperature consolidation processes, including cold spray and shockwave consolidation, to provide high-strength nanostructured materials in bulk forms for meso-scale component fabrication. Small particulates, less than 100 μm with controlled morphology, were produced by MAM during turning. Particle size, morphology and distribution were studied by optical and scanning electron microscopy (SEM), as well as particle size analyzer. Microstructure of these particulates was analyzed by SEM, electron channeling contrast imaging, electron backscatter diffraction, and TEM.

Introduction for Modulation Assisted Machining

Despite all the extraordinary properties [3-1–3-3] of nanocrystalline metals reported in the literature for more than 20 years, the transition of these materials from laboratory curiosity to viable engineering materials has not yet been realized. To deliver the promised properties of nanostructured metals and unleash their full potential for military and commercial applications, two major obstacles including the development of an effective consolidation process and the cost [3-4] and availability of nanostructured materials have to be overcome. Recent advances show that chips generated from machining operations are nanocrystalline, and cost 1/100th of current production methods. In addition, a variety of nanocrystalline metals and alloys can be produced by this technique for specific engineering applications. We used cold spray and shockwave compaction as low temperature forming processes to consolidate these particulates and form bulk materials that can be machined into practical meso-scale components (e.g., gears and cams) without compromising their unique microstructure.

It has been known for some time now that the application of a superimposed, low frequency modulation to conventional machining processes can result in controlled "chip breakage." This observation has been exploited to develop a MAM method for producing micron-sized particulate, that is chip particles, of varied morphology and controlled size distribution. Furthermore, since machining imposes large plastic strains in chips it results in significant refinement of the microstructure [3-5, 3-6].

These particulates are expected to be nanostructured. This prediction has been directly supported by the high hardness and strength observed in chips created by large strain machining [3-7–3-10]. This strengthening mechanism is best illustrated in the pioneering work of Hall [3-11] and Petch [3-12] who showed the yield strength of metal varies as the inverse square root of the grain size. The present study seeks to exploit these observations by incorporating two aspects of MAM to create nanostructured particulates: (1) the ability to control the geometry and size of the particle and (2) the ability to impose severe plastic strains during machining that affect microstructure refinement. It is anticipated that microstructure, morphology, chip size, and distribution of these MAM produced particulates can significantly affect packing efficiency and final density during low temperature consolidation processes, and will be the focus of this investigation. Special emphasis is placed on the microstructure of these machined chips as it has never been reported in the literature due to the complexity of sample preparation for TEM study.

Background for Modulation Assisted Machining

Aluminum 6161-T6, a precipitation hardenable AL-Mg-Si alloy, was used for this study. This material has been strengthened by a heat treatment at 175°C for 8 hr (temper process, T6) after solution treatment and water quench. During the tempering process, precipitates such as β'' (needles), β' (rods or spheres), β (Mg_2Si), and FE-Mn-Cr-Si precipitates, homogeneously form from an oversaturated solid solution and strengthen the aluminum alloy. The strength of this alloy can be further increased by strain hardening. Imposing severe plastic strains during MAM would, therefore, refine the microstructure and strengthen the particulates, as evidenced by a 39.2% increase in the hardness in comparison to the stock material [3-13].

The MAM in turning is used for production of micro-sized particulates using a conventional lathe machining configuration as shown in Figure 3-1. A sinusoidal modulation, $\Delta z(t)$, is superimposed in the feed direction (i.e., direction of undeformed chip thickness), which is perpendicular to the cutting velocity (v_c). Particulate production is possible via interrupted cutting of the workpiece at the fine-scale due to sinusoidal motion of the tool. The formation of this particulate via a direct cutting process capitalizes on the SPD that causes microstructural refinement. This is advantageous due to the aforementioned enhancements in material characteristics seen in metals that have undergone SPD. Thus, the capabilities of this technology include (1) the ability to directly control the size and shape of this particulate, (2) the ability to generate particulates in a wide variety of metals and alloys of varying mechanical properties, and (3) the ability to generate particulate with a fine grain size in the submicron regime, a consequence of SPD imposed during the material removal process. These features enable alternative ways to make nanostructured bulk materials by low temperature consolidation processes. Unlike other SPD routes, such as equal channel angular extrusion [3-14] and high pressure torsion [3-15], this approach can produced desirable microstructure without multiple passes, expensive tooling and inability to process materials of high initial strength without the use of elevated temperatures that may lead to coarsening of the microstructure.

In a turning MAM process, the tool is fed linearly into the rotating workpiece with the modulation superimposed. Particulates are created by the intersection of the successive machining passes on the workpiece. An example of this is depicted in Figure 3-1 as the intersection of passes n and n-1. Tool motion, which defines the morphology of the particulate created from the cutting process, is a function of modulation and machining parameters, which are summarized in Table 3-1.

Each parameter influences the final geometry of the particulate. Modulation frequency (f_m) and workpiece rotational frequency (f_w) determine the contact time of the tool with the workpiece and affect the resultant particulate length. This is evident, as increased workpiece rotation frequencies required increased modulation frequencies to create particles of equivalent length. Modulation amplitude (A) and feed rate (s_o) describe the axial position of the tool at any point of time, defining the thickness and cross-sectional shape of the resultant undeformed particulate. Similarly, increased feed rates require increased modulation amplitudes to ensure interrupted cutting and deformation of particulate. Any determination of particulate morphology must account for each of these variables. A two dimensional model of the process depicted in Figure 3-1 was developed [3-13] to quantitatively determine the particle's undeformed dimensions as a function of machining and modulation parameters. The geometric characteristics of the undeformed particulate (length, width, and thickness) are determined from this model. Details of the mathematical modeling and the validation of the shape and morphology with respect to machining parameters were documented in Saldana's thesis [3-13]. The present work utilizes his code to fabricate aluminum 6061-T6 particulates with equiaxed, needle, and platelet shapes for low temperature consolidation processes.

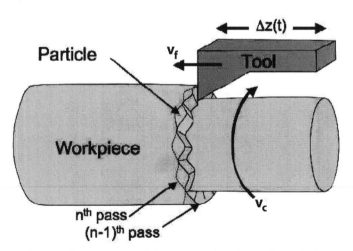

Figure 3-1. A Schematic illustration of modulation assisted machining for particulate production.

Table 3-1. Modulation assisted machining (MAM) parameters.

Type	Parameter	Symbol	Units
Machining	Feed Rate	S_0	mm/rev
	Particle Width	W_0	mm
	Workpiece Rotational Frequency	f_w	RPM
	Workpiece Diameter	d	mm
Modulation	Amplitude	A	mm
	Modulation Frequency	f_m	Hz

Experimental Procedure for Modulation Assisted Machining

A three-axis computer numerical control lathe (Miyano, BNC-42CS) was used as the machining platform for this study. Particles were created using a carbide tool located in a tool holder retrofitted with modulation capabilities. The MAM tool holder (US Patent Pending # 60/667,247) [3-13] was mounted in the turret of the Miyano lathe. The cutting tool is fixed to a ball spline shaft, which is attached to a piezoelectric actuator. A waveform generator (Agilent 33220A) and signal amplifier (Physik Instrumente E-505) send electronic waveforms that drive the lead zirconate titanate linear actuator (Physik Instrumente 843.60), causing the cutting tool to displace position based on the motion of the actuator.

Aluminum 6061-T6 used in this experiment were 3–6 mm diameter round stock. The carbide tool used in the assembly had a rake angle of zero degrees, a clearance angle of 45 degrees, and edge radii of approximately 20 μm. The edge radii of the tools used in these experiments are of importance as preliminary experimental work has shown that as the edge radii values are approached with respect to particulate size, the accuracy of particulate creation becomes unpredictable. Thus, the geometrical characteristics of the tools employed in this process define the lower limit on the particulate size creation.

Continuous lubrication of the tool-chip interface is necessary in MAM. High-speed imagery of the tool-chip interface in particulate production has shown that under dry cutting conditions, particulates become bonded to each other, forming a chain of particles that resemble a serrated chip. Though the nature of the bonding mechanism has not been fully investigated, it is theorized that this phenomenon is a results of the friction present between the freshly cut particle surfaces. Experimental work has shown that isopropyl alcohol provides adequate wetting of the particulate surfaces to prevent unwanted agglomeration. Isopropyl alcohol also evaporates relatively quickly and helps to maintain the cleanliness of the particle surfaces, thus reducing any potential contamination.

The theoretical capabilities of the piezoelectric actuator and amplifier combination were considered in the experiments that were conducted. From the data provided by the manufacturer, the modulation system used in this study begins to loss the ability to achieve "instantaneous" response at full amplitude for frequencies greater than 200 Hz.

Optical microscopy equipped with digital measurement system and scanning electron microscopy (SEM, JOEL) were used to directly observe the particulate size, shape, and distribution. Electron channeling contrast imaging was utilized to analyze the deformation and the morphology of the particulates produced by the MAM. The TEM was employed to investigate the microstructural details of these machined particulates. The particulates under investigation were thinned using FIB techniques to thin the particulate to electron transparency.

Results and Discussion for Modulation Assisted Machining

Preliminary Evaluation of the MAM Process

A preliminary study in fabrication of micron-size aluminum 6061-T6 particulates with practical quantity at Sandia during the first year of this program led to the conclusion that MAM was extremely time consuming. This is particularly true for fabricating particulates with sizes less than 100 μm, which are most suitable for low temperature consolidation processes. The low output can be attributed to a low material removal rate for producing finer particulates during the machining process. As a result, particulates of different morphology used in the cold spray and shockwave consolidation evaluations were purchased from an outside vendor. Expenditure of these purchases indicates that current MAM technology in producing particulates is far less cost effective as most literature has predicted.

Controlling Particulate Size and Morphology by MAM

A preliminary evaluation of the MAM process in producing different morphology machining chips confirmed the mathematic modeling [3-13] in controlling the shape and size of particulates. Figure 3-2 demonstrates the different shapes of particulates created by the MAM process at Sandia and the respective machining and modulation parameters are given in Table 3-2. Optical images in Figure 3-2 suggest that the particulate size and shape are quite consistent with the model's prediction.

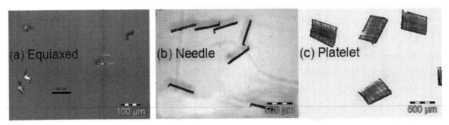

Figure 3-2. Optical microscopy images of different shape particulates created by MAM of aluminum 6061-T6 alloy. (Machining and modulation parameters are given in Table 3-2.)

The particulates used in the cold spray and the shockwave consolidation (see other parts of this SAND study) were purchased from a commercial source (M4 Sciences, West Lafayette, Indiana). The size and shape of the aluminum 6061-T6 particulate fabricated by MAM is shown in Figure 3-3, apparently these smaller size chips (<100 μm) were made with a different set of MAM parameters. The equaxied particles are slightly elongated in the long axis (the average particle size is 63.6 ± 23.55 μm), but the average size of each shape particulate, including needle and platelet shapes, are quite uniform. A dry, optical-based, particle size distribution analysis was performed for the equiaxed particulates and the result is given in Figure 3-4. The data indicates that the sizes of these machined chips varied from 20 to 110 μm, with a bimodal distribution where only a small fraction of powder (<10 Vol. %) is greater than 100 μm.

Table 3-2. Machining and modulation parameters for particulate fabrication (see Figure 3-2.)

Parameter, Symbol	Units	Equiaxed	Needle	Platelet
Feed rate, S_0	mm/rev	0.005	0.005	0.005
Particle width, W_0	mm	0.02	0.50	0.50
Rotational frequency, fw	RPM	120	180	180
Workpiece diameter, d	mm	2.96	5.10	5.10
A (Voltage to actuator)	V (peak-to peak)	5.0	3.0	3.0
Modulation frequency, f_m	Hz	375	16.5	196.5

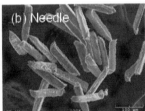

Figure 3-3. The SEM images of commercial MAM particulates made for aluminum 6061-T6 alloy (M4 Sciences, West Lafayette, IN).

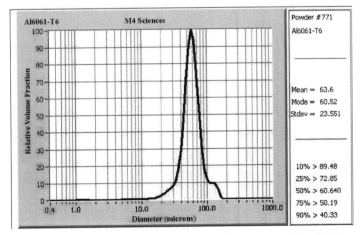

Figure 3-4. Particle size distribution of equiaxed particulates produced by MAM process (M4 Sciences, West Lafayette, IN).

Microstructure Study of MAM Particles

The SEM, electron channeling contrast imaging, electron back scatter diffraction (EBSD), and TEM were used to characterize the microstructure of commercial MAM particulates. Special emphases were placed on the microstructure refinement induced by the MAM and the possibility of texture development induced by plastic deformation. For SEM, electron channeling contrast imaging and EBSD studies, particulates were uniformly dispersed on a glass slide and then placed in an epoxy filled mold. After the resin was cured, the sample was ground and polished so that particulates of different orientations could be studied. In this section, these microstructure characterization results will be presented and discussed, based on their particle morphology. Some cautions must be taken when interpreting these micrographs since these images were taken at different locations and orientations with respect to the chip forming process. However, some generalization can be concluded based on these microstructure investigations.

(a) Equiaxed Particulate

Figure 3-5 shows the electron channeling contrast images of equiaxed particulates fabricated by MAM. The channeling technique reveals the texture development induced by the plastic deformation (in terms of image contrasts due to the differences in the electron-material interactions at various crystal orientations) during the chip fabrication process. Figures 3-5b and c suggest that there is a significant difference in the plastic deformation near the edge (Figure 3-5b) and at the center of the particulate (Figure 3-5c). The center region seems to have an elongated, fibrous structure, while the edge has a much finer structure. Participates (white spots) are uniformly distributed in the microstructure and seems to follow the direction of plastic deformation. The interface between these silicon-rich precipitates (determined by EDS, not shown here) and the aluminum alloy matrix is intact. Pits or voids (dark spots), verified by both secondary electron and backscatter electron images, are mostly submicron in size, and can be observed throughout the machined chip. Small pits (<100 nm) are concentrated at edge region (see Figure 3-6), where the microstructure of this thin layer (<2 μm) near the surface can be affected by the secondary deformation arising from tool-chip friction.

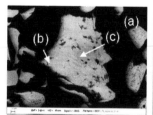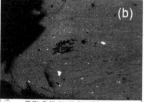

Figure 3-5. Micrographs obtained by electron channeling contrast imaging technique for equiaxed aluminum 6061-T6 particulates fabricated by MAM (a) image of a machined chip, (b) image of the edge of the chip, and (c) image at the center of the chip.

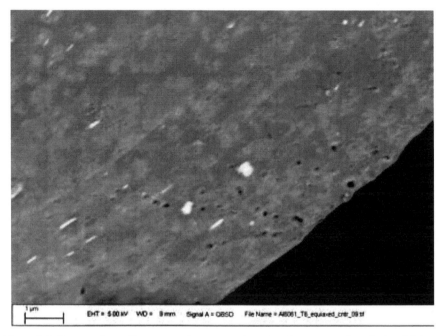

Figure 3-6. Electron channeling contrast image near the edge of an equiaxed particulate fabricated by MAM.

The observation of a different microstructure at the edge and the center of a machined chip prompted a detailed study by EBSD to map out the texture development induced by the plastic deformation, which in turn will help to discern the crystallite size and orientation for these two regions. Figure 3-7 illustrates the inverse pole figures of a MAM chip. The color figure in each image gives the orientation of the crystallite distribution with respect to sample axes (see the color of each pole in the aluminum slip systems). Results show that there is a strong gradient in grain size from the edge (UFG microstructure i.e., 100–300 nm) to the center (fine grain with grain width close to 1 μm and length more than 3 μm) of the machined chip. In addition, there is no preferred orientation of crystallites in the edge region, while the grains in the center are highly textured (as seen in the Y and Z inverse pole figures). The black regions are where patterns could not be indexed and may be attributed to heavy dislocations bands.

The graded microstructure was further investigated by TEM with a bright field (BF) and dark field (DF) image pair, using the {111} reflection (see Figures 3-8a and b). Results show there is an UFG microstructure at the surface of the sample (or the edge of the machined chip) and large elongated grains with heavy dislocations bands at the bottom (center of the machined chip). This observation is consistent with EBSD analysis. The selected area diffraction (SAD) pattern collected at the large grained region shows spotty rings suggesting that the average grain size in this region is greater than 100 nm. The strong intensities in the {220} and {111} rings indicate there is a strong texture developed in the specimen. Furthermore, detailed study indicates that the fine crystallites near the surface are approximately equiaxed in shape and between 100 and 300 nm in size (see Figure 3-9).

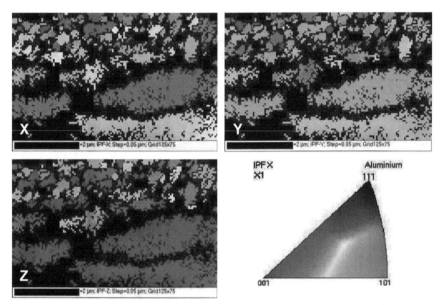

Figure 3-7. The EBSD inverse pole figures shown with respect to the sample axes of an equiaxed particulate fabricated by MAM of aluminum 6061-T6 alloy. The top region of these figures is close to the edge of a machined chip, while the large elongated grains on the bottom are near the middle of the machined chip.

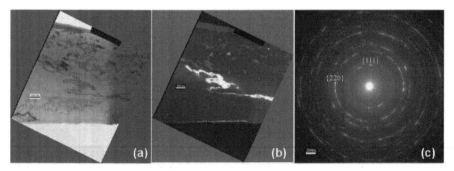

Figure 3-8. (a) The bright field. (b) dark field TEM micrographs of an equiaxed particulate, and (c) the SAD pattern (scale bar = 500nm).

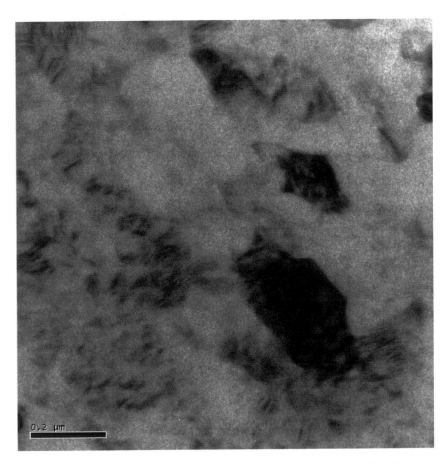

Figure 3-9. A bright field TEM micrograph near the surface of an equiaxed particulate illustrating an UFG microstructure.

(b) Needle Particulate

The needle shape particulates were first studied by an electron channeling imaging technique. Figure 3-10 gives the microstructure of these particulates. Once again, the contrast shows a core-shell structure with a fine microstructure near the surface and coarse, elongated grains stretched along the long axis of the particulate. However the overall contrast is lower in comparison to the equiaxed particles. Voids tend to aggregate and align perpendicular to the long axis. Under a back scatter electron imaging condition where these images were taken, precipitates can be seen in white and well dispersed in the alloy matrix. The long axis of some precipitates (with a high aspect ratio) tends to align perpendicular to the long axis of the particulates, and more often voids can be found near to the precipitate (to the right). For small precipitates, the interface between precipitate and matrix remains coherent. These observations suggest that the region under investigation might be close to the secondary deformation region where friction between tool and chip could modify the structure and put a compressive stress along the long axis of the chip. As a result, voids were aligned perpendicular to the compressive stress direction and less texture induced contrast was detected.

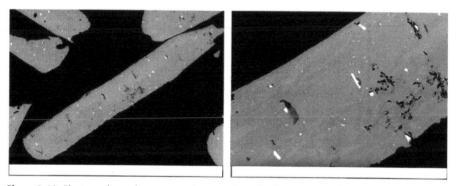

Figure 3-10. Electron channeling contrast images of needle shape particulates produced by MAM.

The TEM was used to study the microstructure of the needle shape particulates. This time a {200} reflection was used in comparison of the BF and DF images. Results of these images, together with the SAD pattern, are given in Figure 3-11. Similar to the equiaxed particulates, layers of UFG microstructure was developed near the surface, and large grains with heavy dislocation bands were observed at the inner region of the particulate. The SAD collected at the inner region shows strong texture as indicated by the strong intensity of {200} and {220} rings. Note the texture induced in these needle shape particulate is different than that of equiaxed (C.1) and platelet (C.3) particulates. Microstructure near the surface reveals layers of grains with elongated aspect ratio (see Figure 3-12). The thickness of these layers is in the 100–200 nm range.

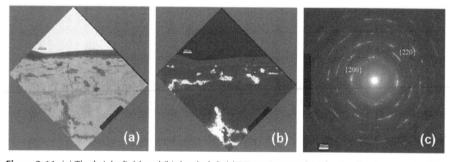

Figure 3-11. (a) The bright field and (b) the dark field TEM micrographs of a needle shape particulate, as well as (c) the SAD pattern (scale bar = 500 nm).

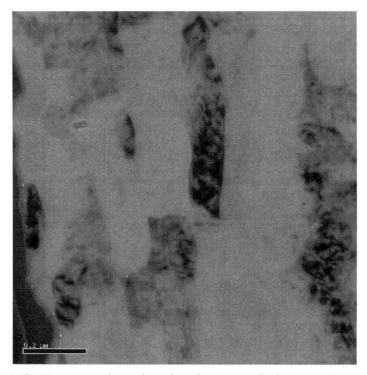

Figure 3-12. The TEM micrograph near the surface of a MAM needle shape particulate.

(c) Platelet Particulate

Electron channeling contrast images of platelet particulates are given in Figure 3-13. Similar to the equiaxed and needle shape particulates, these micrographs show there is an UFG microstructure near the edge while 2–3 µm away from the edge an elongated, fibrous structure is found in the particulate. Voids and precipitates are uniformly distributed in these machined chips. For the most part, the interface between precipitate and matrix remains coherent, and high aspect ratio precipitates are randomly oriented.

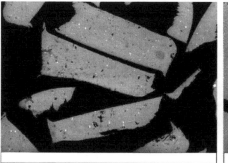

Figure 3-13. Electron channeling contrast images of platelet particles prepared by MAM.

The BF and the DF TEM micrographs, as well as the SAD pattern for a platelet particulate are given in Figure 3-14. The SAD pattern was obtained at the bottom region (see BF image) where large grains with heavy dislocations bands were observed. The SAD pattern shows stronger intensities of {111} and {220} rings, indicating that this region is highly textured. The TEM image (see Figure 3-15) reveals that there are layers of highly elongated grains (layer thickness between 50 and 200 nm) near the surface. The morphology of these grains is quite similar to the needle shape particle.

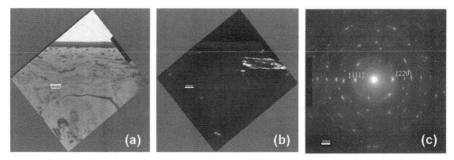

Figure 3-14. The TEM micrographs and the SAD pattern of a platelet particle created by MAM (scale bar = 500 nm).

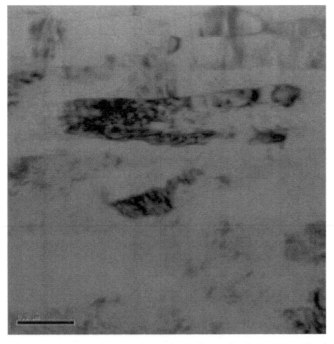

Figure 3-15. The TEM micrograph near the surface of an aluminum 6061-T6 platelet particle prepared by MAM (scale bar = 200nm).

Conclusions for Modulation Assisted Machining

This work demonstrates that modulation-assisted machining is effective to produce micron-sized particulates with tightly controlled morphology and size from aluminum 6061-T6 alloys. The resulting particulates exhibit a core-shell microstructure with UFG at the surface (100–300 nm) and coarser, elongated grains (~1–2 μm in width and 4–6 μm in length) at ~2 micron inside the particulate surface. These coarse grains show characteristics of texture and heavy dislocations bands, indicating a lesser degree of strain induced microstructure refinement. Therefore, it is expected that the surface layer will be harder, while the inner region will be soft and can accommodate more plastic deformation. This unique microstructure can play an important role on cold consolidation processes and affect their final mechanical performance.

Acknowledgments for Modulation Assisted Machining

This work was supported in part by an NSF Graduate Fellowship (to Christopher Saldana at Purdue University.). The Majority of the funding and work are support by Sandia's LDRD program. We are grateful to Professor S. Chandrasekar and C. Saldana (both from Purdue University) for their work on MAM, George R. Burns and Michael Saavedra for their assistance for equipment set-up and testing, Gilbert Benavides for encouraging development of this micro-scale manufacturing approach, and Michael Rye for the FIB sample preparation.

PREPARATION OF ALUMINUM COATINGS CONTAINING HOMOGENOUS NANOCRYSTALLINE MICROSTRUCTURES USING THE COLD SPRAY PROCESS

The thermal spray process has proven to be a successful means of not only creating bulk nanocrystalline material, but of also further refining the grain size during the high velocity impact cold working that occurs during the process. This section consists of a peer reviewed paper selected for publication by the Journal of Thermal Spray Technology. The chapter's authors are Aaron C. Hall and Luke N. Brewer of Sandia National Laboratories and Timothy J. Roemer of Ktech.

Abstract for Cold Spray

Nanostructured materials are of widespread interest because of the unique properties they offer. Well proven techniques, such as ball milling, exist for preparing powders with nanocrystalline microstructures. Nevertheless, consolidation of nanocrystalline powders is challenging and presents an obstacle to the use of nanocrystalline metals. This work demonstrates that nanocrystalline aluminum powders can be consolidated using the cold spray process. Furthermore, TEM analysis of the nanocrystalline cold spray coatings reveals that the cold spray process can cause significant grain refinement. Inert gas atomized 6061 and 5083 aluminum powders were ball milled in liquid nitrogen resulting in micron sized powder containing 250–400 nm grains. Cold spray coatings prepared using these feed stock materials exhibited homogenous microstructures with grain sizes of 30–50 nm. TEM images of the as-received powders, ball milled powders, and cold spray coatings are shown.

Introduction for Cold Spray

Nanostructured materials are of widespread interest to the scientific community because of the unique and unusual properties offered by these materials [4-1, 4-2]. The high grain boundary content of nanocrystalline materials results in grain boundary properties contributing significantly to the bulk material properties. Dislocation behavior in nanocrystals is different from larger crystals and can result in significant strengthening of nanocrystalline metals [4-3, 4-4].

Many techniques have been developed for preparing nanocrystals including inert gas condensation, precipitation from solution, ball milling, rapid solidification, and crystallization from amorphous phases [4-1, 4-2]. Many of these techniques result in free nanocrystals (precipitation and inert gas condensation) or in micron sized powders (ball milling) containing nanocrystalline microstructures. One of the most challenging problems associated with nanocrystalline materials is the consolidation of these small powders into larger shapes that can be used for practical applications [4-2]. This is a problem that Sandia National Laboratories is particularly interested in because of Sandia's interest in the manufacture of meso-scale machinery. Meso-scale machines are miniature mechanical systems. Individual components in these systems may have part dimensions on the order of a few hundred microns. The gears pictured in Figure 4-1 were manufactured using traditional meso-scale machining [4-5–4-7] approaches not LIGA [4-8, 4-9] or MEMS techniques [4-10].

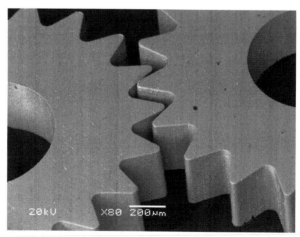

Figure 4-1. Example of a meso-scale mechanical component. Fabrication of meso-scale mechanical components requires extremely fine grained materials; otherwise component performance is determined by grain boundary locations and crystal orientations within the part.

A number of problems are encountered in meso-scale parts as part dimensions approach the grain size of the base material including unpredictable mechanical behavior and reduced component performance. These problems stem from the fact that as component dimensions approach the grain size of the base material, grain boundary properties and the orientation of individual grains dominate the mechanical behavior of the component. This problem can be solved by reducing the grain size in the base material.

One approach to preparing bulk material with nanocrystalline microstructures for the machining of meso-scale parts is to use the cold spray process to consolidate a feedstock powder containing a nanocrystalline microstructure. Preparation of micron size powder containing a nanocrystalline microstructure by ball milling under liquid nitrogen (LN_2) is well known [4-2, 4-11]. Ajdelsztajn has shown previously that cold spray can be used to successfully consolidate LN_2 ball milled powders containing a nanocrystalline microstructure (10–30 nm grain size) [4-12].

The goal of the work reported here is to demonstrate that cold spray is a feasible approach for preparing bulk metal shapes containing homogenous nanocrystalline microstructures suitable for fabrication of meso-scale mechanical components. Once prepared, the intent is that cold sprayed bulk metal shapes will be used as stock for machining meso-scale mechanical components. Near-net-shape forming followed by finish machining is not a goal of this work. Instead the intent is to prepare larger pieces of metal stock with grain sizes appropriate for meso-scale parts. The expectation is to machine multiple meso-scale components from a single cold spray nanocrystalline coating. This work only reports the feasibility of preparing metal stock with suitable microstructure, it does not address mechanical properties, aging, or machineability.

Cold spray is a well known coating process in which a metal feedstock powder is sprayed at high velocity (~1000 m/s) and low temperature (<100°C) [4-13]. A cold spray torch consists of a converging-diverging nozzle, a gas control system, a powder feed system, and a gas heater. Feedstock powder is injected into a high pressure (>200 psig) He or N_2 gas flow just upstream of a converging-diverging nozzle. The powder is entrained in the gas stream and reaches supersonic velocities as it travels through the nozzle. The propulsion gas stream is often heated to moderate temperatures (<600°C) in order to increase the sonic velocity of the gas and thus increase the particle velocity. Heating of the propulsion gas typically does not result in significant powder heating [4-14]. Upon impact with the substrate the metal powder experiences significant mechanical deformation. This deformation causes a hydrodynamic instability to form at the particle-substrate contact allowing breakup of surface oxides, shear along the particle-substrate interface, and ejection of material from the particle-substrate interface. This process results in mechanical and metallurgical bonding between the particle and substrate. It also results in significant plastic deformation of the impacting particle as well as the substrate in the immediate vicinity of the impacting particle. As subsequent particles impact and consolidate a coating or bulk shape is formed [4-15, 4-16]. Process vectors for cold spray are well understood. As particle velocity increases deposition efficiency, coating density, residual stress, and coating adhesion strength all increase [4-14].

The mechanisms responsible for nanocrystal formation during the ball milling process are also well understood. Fecht [4-11] explains that nanocrystals form during SPD in three stages. Initially high density arrays of dislocations are formed. As plastic deformation continues these dislocations annihilate and recombine forming sub-grains with low angle grain boundaries. Further deformation causes the sub-grains to rotate forming high angle grain boundaries.

Experimental Procedure for Cold Spray

Feedstock

The 5083 aluminum powder was purchased from Novemac LLC, (Dixon, CA). The 6061 aluminum powder was purchased from Valimet (Stockton, CA). Both powders were produced using inert gas atomization. All LN_2 ball milling of these powders was also conducted by Novemac LLC using a proprietary, high efficiency, LN_2 ball milling process. All powder size distribution measurements were made using a Beckman-Coulter Laser Diffraction Particle Size Analyzer. This machine is a dry-type particle size analyzer using a Fraunhaufer diffraction criterion to determine particle size distribution. A field emission SEM was used to determine the morphology of each powder.

TEM Sample Preparation and Imaging

The TEM samples were prepared using the FIB lift-out technique. This procedure used a dual electron-Ga ion beam instrument (FEI DB-235). For the starting powders, particles were sprinkled onto carbon tape on an aluminum stub. Areas on the surface of a given particle were selected and coated with electron beam and ion beam deposited platinum to protect the surface of the section during ion milling. The TEM sections approximately 10 μm by 5 μm by 200 nm were milled using a 30keV Ga ion beam. The section was then ion polished with a 5keV Ga ion beam prior to detachment and lift-out. The resultant TEM section was placed onto a thin carbon membrane on a copper TEM grid. For the sprayed coatings, the process was identical except that the sample was simply placed into the chamber in a plan view orientation, and an area for analysis was chosen.

The TEM characterization was performed using a Phillips CM30 TEM at 300keV. The BF images, DF images, and diffraction patterns were collected using both a solid state image plate system (DITABIS) and a Gatan Image Filter (GIF) camera.

Cold Spray

Cold spraying was conducted at Sandia's Thermal Spray Research Laboratory in Albuquerque, NM using a cold spray system that was designed and built by Ktech Corp. Albuquerque, NM. The cold spray nozzle used for these experiments had a 2.0 mm diameter throat, a 100 mm long supersonic region, and a 5 mm diameter exit orifice. Helium was used as the propulsion gas. All coatings were sprayed using 2410 kPa (350 psig), 350°C He flow. All samples were prepared using a raster speed of 50 mm/s, a raster step size of 1 mm, and a standoff distance of 25 mm. All coatings were prepared on 6061 aluminum substrates that were grit blasted and solvent cleaned prior to spraying.

Results for Cold Spray

Powder, Morphology, Microstructure, and Size Distribution

Powder size distributions for the as-received 5083 and 6061 powders are shown in Figure 4-2. Figure 4-3 shows the powder size distributions for the LN_2 ball milled

5083 and 6061 powders. All of the powders have gaussian powder size distributions. Table 4-1 shows the mean particle sizes and approximate mean particle velocities at the spray conditions used to prepare the coatings reported here. All particle velocities in Table 4-1 were calculated using the method described by Dykhuzien and Smith [4-17], Equation #20. The measured mean size of both the 6061 and 5083 powders increased as a result of ball milling. The SEM images of the as-received powder and the LN$_2$ ball milled powder, Figures 4-4 and 4-5, respectively, show that these powders were flattened considerably during the ball milling process and are now flake-like, as expected. It is likely that the powder size measurement was skewed by the high aspect ratio of the ball milled particles. The Beckman-Coulter instrument assumes a spherical particle.

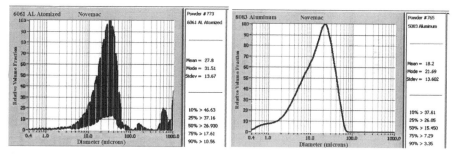

Figure 4-2. Powder size distribution measurement for the as-received 5083 atomized (left) and the 6061 atomized (right) feed stocks.

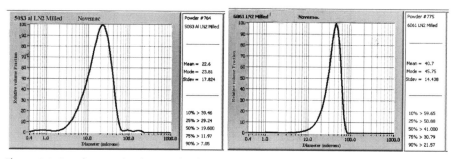

Figure 4-3. Powder size distributions for the 6061 (left) and 5083 (right) powders after LN$_2$ ball milling.

Table 4-1. Mean particle sizes and corresponding particle velocities achieved with a 2410kPa (350 psig), 350° He flow.

Powder	Mean Size (μm)	Calculated Mean Centerline Particle Velocity (m/s)
6061 As-received	27.8	1137
5083 As-received	18.2	1264
6061 LN2 Ball Milled	40.7	1022
5083 LN2 Ball Milled	22.6	1199

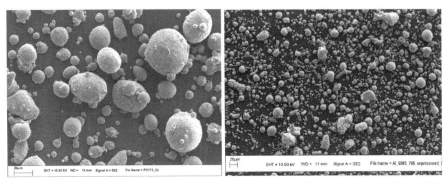

Figure 4-4. The SEM images showing the as-received morphology of the atomized 6061 (left) and the atomized 5083 (right) powders.

Figure 4-5. The SEM images showing the morphology of the LN$_2$ ball milled 6061 (left) and 5083 (right) powders.

The ball milling process causes considerable flattening of the powder. This makes the powder more difficult to feed into the cold spray process, use of a pneumatic vibrator on the hopper is necessary, but does not appear to otherwise affect its behavior.

Figures 4-6 and 4-7 show TEM images of the as received powders and the LN$_2$ ball milled powders, respectively. The LN$_2$ ball milling process has resulted in significant grain refinement creating an UFG structure in the 6061 LN$_2$ ball milled powder and an elongated nano-crystalline structure in the 5083 powder. The grain size of the as received powders is on the order of microns. After LN$_2$ ball milling the 6061 grain size is between 250 and 400 nm. The 5083 grains are 20–40 nm in width but are highly elongated. Distinct morphological texture was observed for both ball-milled powders. The microstructure consisted of elongated lamellar or pancake-shaped grains with lengths 5–15 times their thickness. These grains were also oriented parallel to the surface of the powder particle and are likely the result of deformation associated with the ball milling process. Despite the strong morphological texture in the LN$_2$ ball milled powders, there was no particularly significant crystallographic texture observed in the SAD patterns (Figure 4-7). Note: All of the grain sizes given in this chapter were determined using the TEM images shown. Thus, grain sizes are given as a range (e.g., 20–40 nm) and are somewhat approximate.

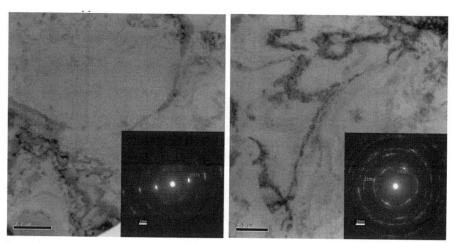

Figure 4-6. The TEM images show coatings prepared using the as-received atomized 6061(left) and 5083 (right) powders. The SADP shows 6061(left) to from a single grain with size >1μm. SADP from 5083 (right) shows heavily smeared spots from >500nm deformed grains.

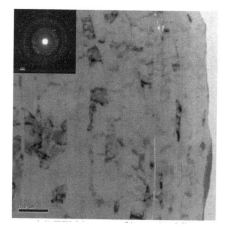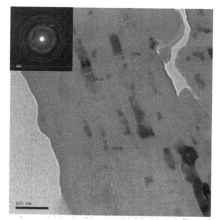

Figure 4-7. The TEM images show the microstructure of the 6061(left) and 5083 (right) powders after LN$_2$ ball milling. Notice that both powders have elongated grains.

Coating Microstructures

Nanocrystalline 5083 and 6061 coatings were prepared on 50.8 mm × 50.8 mm × 3.175 mm (2″ × 2″ × 1/8″) aluminum substrates. Figure 4-8 is a picture of the two coated test coupons. The 5083 coating is 0.058 mm (0.0023″) thick. The 6061 coating is 0.513 mm (0.0202″) thick. Figures 4-9 and 4-10 are TEM images showing microstructures of coatings prepared using the as-received and LN$_2$ ball milled powders. All coatings prepared with the as-received powders exhibited micron sized grains, as expected. However, both coatings prepared using the LN_2 ball milled powders exhibited 30–50 nm grain sizes. This was unexpected given the elongated grains in both powders and the 250–400 nm grain size in the 6061 LN_2 ball milled powder. In both cases, no crystallographic texture was observed in the diffraction patterns from the coatings. Additionally, a good deal of nano-scale porosity was observed in these coatings. The porosity was highly aligned with lath boundaries in the 5083 alloy and with much less alignment in the 6061 alloy.

Figure 4-8. As-sprayed nanocrystalline aluminum coatings of alloys 6061 (left) and 5083 (right) are shown.

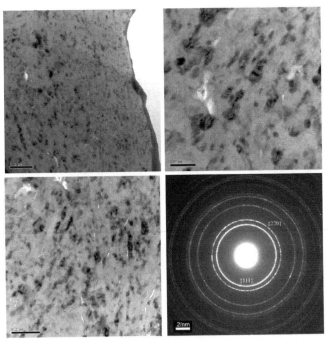

Figure 4-9. Three TEM images showing the microstructure of the cold spray coating prepared using the LN_2 ball milled 6061 powder. The selected area diffraction pattern (SADP) shows complete rings with no discernable texture.

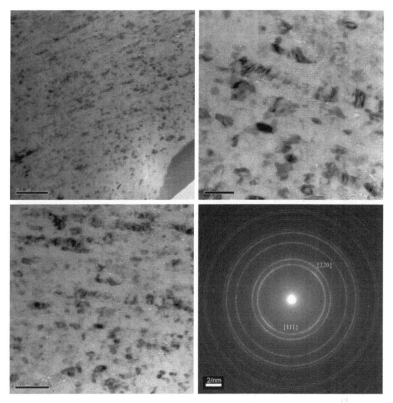

Figure 4-10. Three TEM images showing the microstructure of the cold spray coating prepared using the LN$_2$ ball milled 5083 powder. The SADP shows complete rings with no discernable texture.

Discussion for Cold Spray

As shown by the TEM images above (Figure 4-6 through Figure 4-9), the cold spray process resulted in significant grain refinement for both the 6061 and the 5083 LN$_2$ ball milled powders. The grain size reduction from 250 to 400 nm in the LN$_2$ ball milled 6061 powder to 30–50 nm in the cold sprayed coatings is approximately a factor of eight in grain size and close to a factor of 70 in grain volume. The grain refinement in the 5083 is similarly dramatic. The elongated 5083 grains are equiaxed in the cold sprayed coating. Most likely this grain refinement occurred through mechanisms similar to those proposed by Fecth [4-11]. It is reasonable to expect that the deformation associated with the cold spray process could result in sufficient dislocation generation and subgrain rotation to cause grain refinement. Grain refinement during the cold spray process has been documented before in pure aluminum of normal grain size [4-16]. The lath structure in the LN$_2$ ball milled powder appears to be retained in the 5083 cold spray coating.

Interestingly, the observation of grain refinement as a result of the cold spray process reported here appears to be in direct contrast to the results reported by Ajdelsztajn [4-11] in 2005. Ajdelsztajn also used the cold spray process to consolidate LN$_2$ ball milled 5083 aluminum powder. This study clearly showed that the cold spray process is an effective method of consolidating nanocrystalline powders. However, Ajdelsztajn reported that no grain refinement was observed during the cold spray process. The nanocrystalline grains in Ajdelsztajn's coating were of similar size to the nanocrystalline grains in the LN_2 ball milled 5083 Al powder sprayed at Sandia. While the results of these two studies may appear to be in contrast; they are not. The experiment reported here was substantially different from Ajdelsztajn's experiment in two important ways. First, Ajdelsztajn's spray conditions were significantly different from the conditions used in this experiment. This may have resulted in lower average particle velocity in Ajdelsztajn's experiment. Ajdelsztajn's experiment used a room temperature 1.7 MPa (~250 psig) helium flow to propel the particles. The experiments reported here used a 2.41 MPa (350°C), 350 psig helium flow to propel the particles. Thus, more energy and more plastic deformation were available in this experiment compared to Ajdelsztajn's experiment. Second, and more importantly, Ajdelsztajn's 5083 aluminum feedstock was ball milled to a grain size of 20–30 nm before spraying. Work by Romanov and Eckert suggests that it may become increasingly difficult or impossible to create grain sizes in aluminum smaller than 20 nm by plastic deformation mechanisms [4-3, 4-17]. Romanov explains that a critical crystal size exists below which gliding dislocations are unstable due to image forces. In aluminum Romanov estimates the critical size for dislocation stability as 18 nm. Below this size, gliding dislocations will be unstable and will rapidly move to grain boundaries or annihilate. Instability of dislocations would make it more difficult to create the dislocation networks, sub-grain boundaries, and ultimately the high angle grain boundaries required for deformation induced grain refinement. Eckert [4-17] explicitly considers the problem of minimum grain size obtainable by ball milling and explains that minimum grain size is determined by a competition between plastic deformation and the recovery behavior of the material. Minimum grain size for aluminum is shown to be between 20 and 25 nm.

Conclusions for Cold Spray

This experiment clearly demonstrates that cold spray can be used to refine the microstructure of an UFG (100's of nm grain size) powder and consolidate it to create a homogenous nanocrystalline (20–40 nm) microstructure. When compared to the work of Ajdelsztajn which shows that the cold spray process can be used to consolidate nanocrystalline aluminum withoucausing recrystallization it illustrates the flexibility of the cold spray process. This flexibility is typical of spray processes and highlights the need to further understand the process-microstructure-property relationships in nanocrystalline cold spray coatings. The colprocess is likely an extremely controllable method for preparing metal coatings and bulk metal shapes with homogenous nanocrystalline microstructures. If grain refinement in the cold spray process is occurring by the mechanism proposed by Fecth [4-11], cold spray should be capable of preparing nanocrystalline microstructures from almost any sprayable metal feedstock. The following specific conclusions were reached from the data presented in this study:

1. The cold spray process can cause the refinement of UFG powdered aluminum feed stocks creating a homogenous nanocrystalline microstructure.

2. Grain refinement observed in Al due to cold spray occurs because of the severe formation associated with the cold spray process.

Acknowledgments for Cold Spray

The authors would like to acknowledge Dr. Leonardo Ajdelsztajn for his advice on sources for LN_2 ball milled powder and for his insightful work on cold sprayed nanocrystalline aluminum. It was a foundation and significant motivator for the experiments reported here.

SHOCK CONSOLIDATION OF NANO-CRYSTALLINE 6061 ALUMINUM

The team developed methods to use shock compaction for the consolidation of nanocrystalline metal. The compacted materials were then studied to understand the effects of the process and incoming powder on the final part microstructure. The following section, written by team member D. Anthony Fredenburg, will be submitted for publication in the near future.

Abstract for Shock Consolidation

Nanostructured materials offer unique microstructure-dependent properties that have the potential to be far superior to course-grained materials; however, they also present new challenges in the formation of bulk components. In this work, both fully and partially nanocrystalline aluminum powders are consolidated into bulk form using dynamic compaction. Overall compact densities range from 98.4 to 99.0% TMD; the partially nanocrystalline compact achieved the highest density. High hardness is recorded in the fully nanocrystalline compacts, averaging, and is attributed to the fine grain structure. Fully nanocrystalline starting powders show grain refinement in the compacted state with a bimodal microstructure composed on 50–100 nm thick laths and 10–50 nm equiaxed grains. Compact hardness averages ~193 $HV_{.05}$ for fully nanocrystalline compacts. Partially nanocrystalline powders retain graded microstructures following dynamic consolidation with evidence of subgrain formation. Hardness tends to decrease with distance from the impact face for partially nanocrystalline compacts.

Introduction to Shock Consolidation

Nanostructured materials, polycrystalline materials with average grain sizes <100 nm in at least one dimension, are of engineering interest because they offer novel properties compared to their course-grained counterparts. As grain size decreases into the nanometer regime, mechanisms dominating the physical and chemical properties of materials change and have been reported to result in substantial improvements in thermal, mechanical, electrical, optical, and magnetic properties [5-1–5-6]. However, the formation of bulk materials with nanocrystalline grain structures poses significant engineering challenges.

In recent years, the production and characterization of bulk nanocrystalline aluminum alloys has garnered much attention due to their high strength to weight ratio. Generally, for these alloys research has shown a reduction in grain size is accompanied by an increase in tensile strength and hardness and a reduction in ductility and toughness [5-7]. Han et al. [5-8], investigating a cryomilled nanocrystalline 5083 Al alloy, reported an increase in yield strength of ~365% over the course-grained alloy to 713 MPa, however, this was accompanied by an elongation to failure of only ~0.3%. To make these alloys more attractive for structural applications ductility must be increased. One such method for increasing ductility is through the introduction of a bimodal grain structure composed of nano- and micron-scale grains through annealing [5-9], or the combination of nanocrystalline powders with unmilled course-grained powders [5-10]. These techniques have resulted in significant improvements in ductility with limited and controllable reductions in strength. However, special care must be undertaken during processing as Huang et al. [5-11] has found that annealing a 99.99% pure aluminum specimen at a low temperature, without the introduction of the bimodal structure, resulted in increases in flow stresses and decreases in ductility, in stark contrast to the annealing behavior of course-grained aluminum alloys. They also observed a reduction in flow stress and increase in ductility to approximate pre-anneal values through subsequent deformation of the annealed specimen. Thus through proper engineering controls, the production of high strength-moderate ductility nanocrystalline aluminum components is possible.

One of the more popular means of forming bulk nanostructured materials is through the consolidation of nanoscrystalline powders formed through cryomlling, the milling of micron-sized powders at cryogenic temperatures to form nanograined particulates. Generally, the evolution from micron to nanoscale grain structures during milling occurs in three distinct stages such that during the initial stage shear bands with high dislocation densities form as a result of large-scale plastic deformation. With further deformation, strain levels increase causing the annihilation and recombination of dislocations and result in the formation of low angle subgrains with nanometer length scales. Finally, single-crystal grains reorient to form random high-angle misorientations between grains [5-12]. In addition, when milling occurs at cryogenic temperatures, the annihilation of dislocations is suppressed allowing particles to rapidly accumulate very high dislocation densities. Another important aspect of milling is the introduction of impurities (O, N, C, Fe, etc.) during grain refinement in the form of nanoscale dispersions. These dispersions have been shown to limit grain-boundary migration and enhance thermal stability in aluminum alloys, helping to preserve nanocrystalline grain structures during consolidation [5-8].

A more recent technique of forming nanocrystallline particles for subsequent consolidation is through plain strain machining. During machining, large shear strains are introduced along a narrow deformation zone, the shear plane, transforming the original microstructure, and ejecting the chip from the bulk. This process employs a MAM technique, and through variations in undeformed chip thickness, cutting speed, frequency, and rake angle of the cutting tool chips can be produced with varying degrees of nanocrystallinity and morphology [5-13, 5-14]. However, the challenge still remains to consolidate these powders into bulk forms while retaining their original microstructures.

Some of the more prevalent consolidation methods for producing bulk nanocrystalline materials are hot- and cold-isostatic pressing, spark-plasma sintering, cold spray, and shockwave consolidation. Hot-isostatic pressing and spark-plasma sintering can be deleterious to the preservation of nanocrystalline grain structures because temperatures can reach ~0.7 Tm (solvus temperature) and can lead to grain growth [5-15, 5-16]. Cold-isostatic pressing and cold spray technology have the advantage of being lower temperature processes; however, interparticle bonding suffers as a result of low diffusion rates in cold-isostatic pressing and is highly particle size dependent for cold spray technology [5-8, 5-17]. Conversely, under the correct processing conditions, shockwave consolidation can circumvent the shortcomings of the aforementioned processes and produce fully nanocrystalline bulk materials [5-18, 5-19].

During shockwave consolidation short duration (~1 μs) stresses in excess of the materials yield strength are applied, causing particle deformation and densification of the compact. Particle deformation is highly heterogeneous with much of the deformation restricted to particle surfaces, which can result in extremely high heating and quench rates at particle surfaces while temperatures near the particle interior remain relatively constant [5-20, 5-21]. When forming components with fully nanocrystalline microstructures loading conditions must be carefully controlled such that excessive heating and cooling is minimized. At excessive stresses, temperatures can reach that of the melt and result in substantial microstructural changes as Brochu and co-workers [5-22] have observed a bimodal microstructure with larger grains localized near particle surfaces following explosive compaction of an initially equiaxed Al-Mg alloy. Stresses localization can also lead to the breakup of surface oxides which can aid in interfacial bonding [5-23]. However, as particle size decreases it becomes more difficult to break surface oxides, and Nieh et al. [5-24] have shown that for aluminum particles with average sizes between 50 and 70 nm stresses of 2–3 GPa are unable to breakup surface oxides, inhibiting the formation of metallurgical bonds between particles. Thus the formation of bulk nanocrystalline aluminum components by means of dynamic compaction is favored by the compaction of micron-scale nanocrystalline particles at moderate stresses, which leads us to our current approach.

In this work the authors investigate the use of dynamic compaction as a means of consolidating nanocrystalline aluminum alloy powders formed through SPD machining and cryomilling. First, characterization of the starting powders is presented, followed by details of the experimental approach. Results and discussion follow, covering compact density, microstructure, grain size, and microhardness for dynamically compacted samples. At this juncture no attempt is made at characterizing the elastic properties of the compacts, and reporting of said properties will be presented in a future work.

Experimental Procedure for Shock Compaction
Starting Powders
The powders used in this investigation are a liquid nitrogen ball milled aluminum 6061 alloy obtained from Novemac LLC (Dixon, CA), and an equiaxed aluminum 6061 alloy formed by a frequency MAM process obtained from M4 Sciences (West Lafayette, IN). The former will henceforth be referred to as LN_2 BM ##% and the latter M4Sci EQ ##%, where ##% indicates the initial packing density of the compacts in percent, with respect to theoretical mass density (TMD). Mean particle sizes and

standard deviations for LN_2 BM and M4Sci EQ powders were characterized using Fraunhofer diffraction as 40.7 +/– 14.4 μm and 63.6 +/– 23.5 μm, respectively. In addition to particle size, particle morphology and crystallite structures differ greatly between powders due to their highly heterogeneous processing methods. Representative particle morphologies and microstructures for the two starting powders are shown in Figures 5-1 and 5-2.

LN_2 BM powders exhibit highly deformed regions on all surfaces, are spherical/ elliptical in shape, and have a uniform lath-like grain structure with high dislocation densities. Lath thicknesses are between 200 and 400 nm. Similar elongated grain structures are observed by Han and co-workers [5-8, 5-25] in a cryomilled Al-Mg alloy and are attributed to milling times less than that required to achieve a fully equiaxed structure. The TEM also shows limited porosity located almost exclusively at the grain boundaries, which may be a result of segregation of solute atoms to the grain boundaries causing void nucleation and/or grain boundary sliding during cryomilling [5-26]. Also, SAD does not show evidence of any strong crystallographic orientation.

In contrast, M4Sci EQ powders appear physically smooth on some surfaces and jagged on others, are predominately blocky in shape, and have a graded microstructure. The TEM images show a 1–2 μm thick layer near the particle surface of approximately equiaxed crystallites between 100 and 300 nm in diameter. As distance into the particle increases the microstructure consists of larger grains (>1 μm) with high dislocation densities. Evidence of this graded microstructure is also observed in SAD patterns where discrete reflections are observed in the particle interior and spotty rings are observed near the surface.

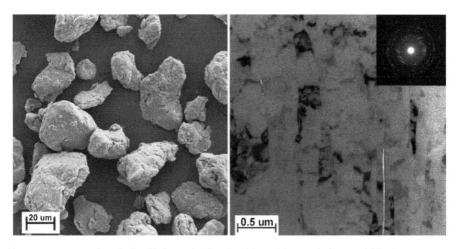

Figure 5-1. Images of LN_2 ball milled powder showing (a) particle morphology and (b) microstructure.

Figure 5-2. Images of M4Sci 6061 equiaxed showing (a) particle morphology and (b) microstructure near particle surface.

Dynamic Consolidation and Analysis

Aluminum powders in this study were dynamically consolidated within a 3-capsule recovery fixture on an 80 mm bore diameter single-stage light gas-gun at the Georgia Institute of Technology. Mass of the powders were measured using a microbalance with 10^{-4} g resolution, and powders were pressed into individual steel capsules with an inner diameter of 11.988 mm using a Carver Auto Series "M" 3890 press. Powder thickness was measured using a depth micrometer with 10^{-3} inch resolution, and initial densities of the compacts range between 68 and 80% theoretical.

Capsules were sealed with a steel plug and LOC-TITE©, and a small amount of epoxy was used to attach capsule to the surrounding fixture. This fixture is similar to the fixture reported in a previous work [5-27], with the exception that in this investigation an air-gap exists between the capsule and the surrounding ring near the impact face and the entire fixture is steel. A radial cross section showing location of the air-gap is shown in Figure 5-3. The previous fixture design was modified to accommodate for the higher impact velocities used in this investigation such that the new design increases confinement of capsules in radial direction.

Following assembly, the fixture was lapped to ensure planarity of the impact face and mounted to a set of planar stand-off blocks attached to the muzzle. The experiment tank was evacuated to ~50 mTorr, and an aluminum sabot and steel flyer plate were accelerated down the barrel of the gun using compressed helium as the driving gas. A series of four shorting pins were used to measure velocity of the incoming projectile, which was measured at ~730 m/s. Upon impact of the flyer plate with the fixture, compressive stress waves were generated causing compaction of the powder, and dimensions of the fixture and flyer are such that compaction of the powder was complete prior to arrival of release waves from the rear of the flyer.

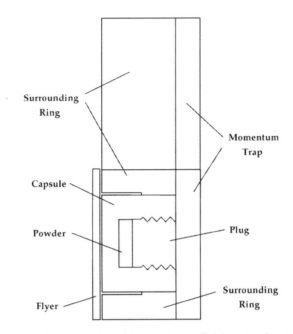

Figure 5-3. Cross-section of one capsule and fixture along radial direction showing location of air-gap in relation to powder compact.

The fixture and capsules were recovered in a soft-catch recovery tank, and compacts were machined out of the capsules with care taken to keep the temperature as low as possible. Following removal of the compact from the capsule, the rear surface of the compact was lightly sanded to remove any loose material and density determined using Archimedes displacement method with the aforementioned microbalance. Compacts were then sectioned into two halves parallel to the direction of impact for examination of compact cross-sections and TEM analysis. Optical micrographs and Vickers micro-hardness measurements of compact cross-sections were taken with a Leica DM IRM light microscope and a Leco MHT Series 200 micro-indenter. The TEM specimens were cut near the impact and rear surfaces of the compacts using a gallium FIB, where the plane of view is along the specimen cross-section. TEM imaging was carried out on an FEI Tecnai F30-ST microscope.

Results and Discussion for Shock Compaction

Dynamic Consolidation

The shock compaction stress in the powder compacts was 14.75 GPa, as determined through impedance matching techniques using the known impact velocity and equations of state of the flyer plate and fixture. However, due to large impedance mismatches between the steel fixture and the aluminum compacts, the initial compaction stresses in the powders were much lower; approximately 2.5 and 3.5 GPa for the 68 and 80% TMD compacts as inferred from continuum level simulations. Only after several stress reverberations in the compact will the specimen ring-up to the incident shock stress in the steel [5-24].

The production of a mechanically sound compact from powder constituents requires not only densification of powders, but also sufficient particle movement such that strong mechanical bonds between neighboring particles can form. The greatest amount of particle movement takes place under the influence of the lower magnitude initial stress pulse, and it is this stress which is responsible for the majority of compaction and resultant mechanical integrity of the compact. Under the experimental loading conditions the stresses present cause nearly full densification of the compacts. Compact density ranges between 98.4 and 99.0% TMD, and have approximate dimensions of diameter 12 mm and thickness 2 mm. Figure 5-4 shows compacted LN_2 6061 BM 68, and details of the initial and final compact densities for each compact are given in Table 5-1.

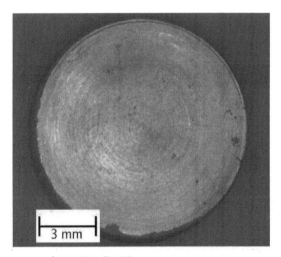

Figure 5-4. Shock compressed LN_2 6061 BM 68 compact.

Table 5-1. Density of powder compacts prior to dynamic loading ρ00 and density of shock compressed compacts ρC.

Powder	ρ_{00} (g/cm³)	ρ_C (g/cm³)
LN2 6061 BM	2.166	2.665 ± 0.004
LN2 6061 BM	1.842	2.658 ± 0.009
M4Sci 6061 EQ	1.844	2.673 ± 0.011

Microstructure of LN2 6061 Ball Milled Compacts

The TEM and STEM analysis were conducted on samples taken from specimen cross-sections at the impact and rear surfaces to determine if the compacts exhibit through thickness variations in microstructure. Grain boundaries are observable in STEM image mode as a result of liquid metal embrittlement caused by the introduction of gallium during FIB thinning of the samples [5-28]. Microstructural analysis of the compacts is presented in the following two sub-sections.

Microstructures for the LN$_2$ Ball Milled (BM) 68 and 80% compacts appear very similar, and representative TEM and STEM images are given in Figure 5-5. The crystallite structure is bimodal in character dominated by larger lath-like grains and a low volume fraction of smaller equiaxed grains having diameters between 10 and 50 nm. The lath-like structure is similar to that observed in the starting powder; however, lath thickness is significantly reduced and typical lath thicknesses are between 50 and 100 nm.

Reduction in lath thickness results from the large amount of plastic deformation imposed on the particles during consolidation, and there have been many investigations to date concerning the deformation mechanisms responsible for plastic deformation as grain size enters the nano-regime [5-29–5-33]. It is the authors' belief that the dominant deformation mechanism responsible for lath reduction is slip. This hypothesis is supported by the predominately cellular microstructure observed following compaction. Though twinning may occur at these elevated strain rates, twinning in aluminum alloys tends to be favored at lower temperatures, small crystallite sizes, and in materials without precipitates or inclusions. Following ball milling, LN$_2$ BM powders possess inclusions from the ball milling process in addition to any inherent precipitates in the alloy, all of which would serve to block twin formation. In addition, Gray [5-34] found no evidence of deformation twins in a 6061 Al alloy shocked to 13 GPa at −180°C, though grain sizes were much larger than in the present study.

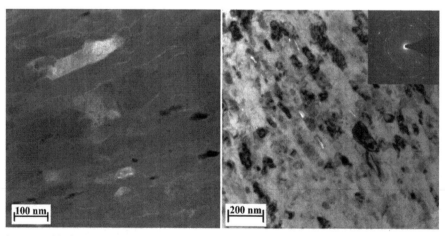

Figure 5-5. (a) The STEM image of LN$_2$ 6061 BM 68 and (b) The TEM image of LN$_2$ 6061 BM 80 near impact face of compact showing bimodal grain structure and nanoscale porosity.

Recalling the three stages of microstructure evolution that takes place during milling presented in [5-12], and that elongated grains have been attributed to insufficient milling times during nanopowder formation [5-8, 5-25], the following hypothesis regarding the reduction of lath thickness is presented. With the passage of the compaction stress a large number of dislocations are introduced through the propagation of shear bands. These dislocations interact with existing dislocations and form smaller sub-grains that are slightly reoriented with respect to the initial lath orientation. The TEM micrographs show many instances in the deformed microstructures where minimal contrast variation between neighboring laths is observed, indicating smaller laths may have originated from a similarly orientated larger lath.

Compacted samples exhibit a varied amount of microscale porosity and a fairly consistent amount of nanoscale porosity throughout the compact thickness. Nanoscale pores are located almost exclusively at grain boundaries, have typical widths of 10–20 nm, tend to be elongated in the lath direction, and can be observed in Figure 5-5. Increased amounts of micron scale porosity are observed near the rear surface of both LN_2 BM compacts, and the increase in porosity at the rear surface is thought to result from radial stress effects. To gain further insight into the stress states present in the powder compact during impact, continuum level simulations of the impact event were undertaken and revealed radial stress waves approached the rear of the compact shortly after the initial compaction front. The interaction and reflection of these stresses could cause turbulent material flow and lead to shearing of the aluminum and subsequent void formation. Simulations also revealed a late time tension region which may also contribute to the observed porosity.

Microstructure of M4Sci 6061 Equiaxed (EQ) Compact
As detailed previously, the initial M4Sci EQ powder possesses a graded bimodal microstructure with smaller equiaxed grains near the particle surface and larger (>1 μm) grains in the particle interior. Following dynamic compaction, a similar microstructure is observed in the compact. Figure 5-6a clearly shows this bimodal structure with smaller equiaxed grains extending approximately 1 μm into the particle interior. Similar to the LN_2 BM compacts, slip is believed to be the dominant deformation mechanism in the larger grains due to the T6 temper of the particles and the presence of precipitates which would prevent twinning.

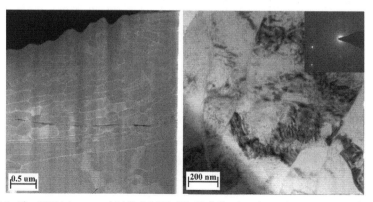

Figure 5-6. The STEM images of M4Sci 6061 EQ 68 following dynamic compaction showing (a) graded microstructure near rear surface and (b) low misorientaion of large grains.

Large grains in the compacted specimen possess high dislocation densities and show areas of dislocation tangles often observed in the form of bands spanning the length of a grain. Figure 5-6b shows diffraction contrast in a large grain region as a result of the high dislocation density within the shock compressed samples. Banded structures are observed more frequently in the larger grains, which is consistent with observations of Zhu et al. [5-32] that grains are largely dislocation free in the size range of 50–100 nm as a result of the boundaries acting as sinks for dislocations. Intra-grain shear bands are also observed which traverse several of the larger grains. The presence of these bands shows the low misorientation angles between neighboring grains, and indicates the material may be in an intermediate step toward subgrain formation, see Figure 5-7.

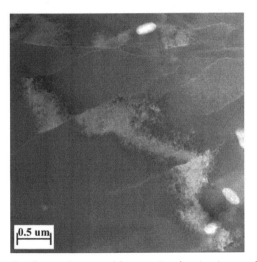

Figure 5-7. Dislocation band spanning several large grains showing intermediate step to subgrain formation as a result of plastic deformation during compaction.

Total porosity is reduced in the M4Sci EQ 68% compact compared to both LN$_2$ BM compacts, with evidence of the micron scale porosity greatly reduced. The majority of porosity is observed inside or within close proximity to the fine grained regions and is dominated by smaller 10–100 nm diameter pores at grain triple points. The M4Sci EQ particles are machined from stock in the T6 temper and inherently possess many fine-scale Mg$_2$Si precipitates. These precipitates inhibit grain boundary migration during particle formation and subsequently result in a large fraction of precipitates at particle grain boundaries. With precipitates preferentially located near grain boundaries, these could serve as void initiation sites and result in the observed void pattern. Bae and Ghosh [5-35] observed similar void patterns in an Al-Mg alloy and attribute this type of void formation to particle-matrix debonding. Larger, less frequently occurring voids are also observed and can span the length of multiple grain boundaries, reaching several microns in length, see Figure 5-7. These voids are also found predominately in the smaller grain region near particle surfaces and seem to be a result of the coalescence of multiple matrix-particle debonding instances.

Microhardness Profile

Hardness mapping of the compact cross-sections was carried out systematically over the breadth of the cross-section with 0.05″ between columns through the compact thickness and 0.06″ between rows along the compact diameter at an indentation load of 50 g. Average hardness values along the compact diameters for both LN_2 BM and M4Sci EQ compacts are given in Figure 5-8 with overall averages of 193 and 105 $HV_{.05}$, respectively. Both LN_2 BM compacts exhibit hardness values approximately 90 $HV_{.05}$ higher than the M4Sci EQ compact, and values are consistent with those reported by Brochu et al. [5-22] for a nanostructured Al-Mg alloy. High hardness values for LN_2 BM compacts are attributed to their fully nanoscale grain structure.

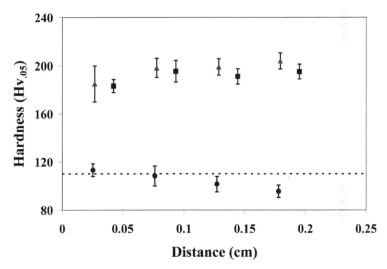

Figure 5-8. Vickers hardness values ($HV_{.05}$) averaged along specimen diameters for (■) LN_2 BM 80% (▲) LN_2 BM 68% and (●) M4Sci EQ 68%. Bulk 6061 T6 is indicated by (---).

Hardness values along the compact diameter exhibit no observable trend in hardness, and minimal scatter is observed between recorded values. Through the compact thickness the LN_2 BM compacts show a slight increase in hardness with distance from the impact face following the first series of indents; however, all values lie within their standard deviation to the overall average. Thus no definitive conclusion as to a trend in hardness for the LN_2 BM compacts can be drawn at this time. The M4Sci EQ 68% compact, however, does exhibit a clear decrease in hardness with distance from the impact face. This reduction in hardness is attributed to higher temperatures experienced by the compact rear as a result of converging radial waves.

Continuum-level simulations of the impact event show converging radial waves with magnitudes >10 GPa approaching the specimen rear less than 100 ns after the passage of the initial compaction front. The arrival of these converging radial waves leads to increased temperatures near the compact rear, and could lead to recovery of grain body dislocations and/or the reintroduction of precipitates into solution. For Al 6061 chips in the T6 temper, Swaminathan and co-workers [5-14] have shown that additional annealing results in a reduction in hardness as a result of coarsening of the originally ~80 nm grains. However, grain coarsening is not observed in the M4Sci EQ 68% compact, and the reduction in hardness to values below that of the bulk T6 alloy (dotted line in Figure 5-8) is thought to result from the reintroduction of precipitates into solution.

Shock Consolidation Conclusions

In this study, dynamic compaction was successfully used to consolidate fully and partially nanocrystalline 6061 aluminum powders, forming compacts ranging in density between 98.4 and 99.0% TMD. Microstructures of fully nanocrystalline LN_2 BM compacts exhibit grain refinement during consolidation yielding a bimodal grain structure with a large volume fraction of laths 50–100 nm in width and a smaller volume fraction of equiaxed grains 10–50 nm in diameter. Limited porosity is observed in these compacts with the level of nanovoids remaining consistent through compact thickness and a greater number of microvoids observed near the rear surface. Graded microstructural features are observed in the M4Sci EQ 68% compact, indicating preservation of initial microstructure, with larger grains showing high densities of dislocations in the form of bands which can span the length of several grains. The presence of these bands is thought to indicate the material is in an intermediate step toward subgrain formation. Compact hardness reflects microstructural features, with fully nanocrystalline compact hardness values reaching almost twice those of the partially nanocrystalline compacts. High hardness suggest components may also have high tensile strengths and work is currently underway characterize the elastic properties of the compacts.

Shock Consolidation Acknowledgments

The authors would like to thank Michael Rye and Paul Kotula for TEM sample preparation and analysis and Christopher Saldana for many useful conversations concerning the M4Sci EQ powder. Funding for this research is provided by Sandia National Laboratories, a multiprogram laboratory operated by Sandia Corporation, a Lockheed Martin Company, for the United States Department of Energy's National Nuclear Security Administration under Contract DE-AC04-94AL85000.

A COMPARISON OF NANOCRYSTALLINE ALUMINUM PREPARED USING COLD SPRAY AND DYNAMIC CONSOLIDATION METHODS

As the analysis of the different consolidation methods proceeded, it became apparent that several of the characteristics of the different methods were similar and the team felt that it was prudent to perform a more in-depth comparative analysis. The following is a draft paper that will be submitted in the future. The chapter, co-written by team members Anthony Fredenburg and Aaron Hall, is not quite complete due to the need to compare the starting powders used in each process. The materials require TEM analysis and this was not able to be completed during the project. However, knowing that the incoming materials were created by the same method, but using different parameters, the comparison is still valid and important.

Abstract for the Comparison of Consolidation Methods

Cold spray and dynamic consolidation techniques were used to successfully consolidate nanocrystalline aluminum 6061 powders to form bulk nanocrystalline components. The TEM images of bulk components showed a reduction in grain size occurs during consolidation of powders for both processes. Cold spray coatings exhibited average grain sizes between 30 and 50 nm in diameter, and dynamically consolidated compacts contained a bimodal structure with smaller equiaxed grains 20–50 nm in diameter and elongated lath-like grains having thicknesses ranging between 50 and 100 nm. Mechanical properties of the cold spray coating and dynamically consolidated compact were obtained through nano-indentation and micro-indentation hardness measurements, respectively, with the cold sprayed coating exhibiting the highest hardness.

Introduction for the Comparison of Consolidation Methods

A recent trend in design of multi-component systems has been to reduce component size while maintaining or increasing its strength. This has led to the development of new classes of materials, such as high strength-light weight nanocrystalline aluminum alloys for use in microscale components. When component size reaches 10's–100's of microns, conventionally processed large-grained components exhibit highly anisotropic mechanical properties due to directional dependence arising from a limited number of grains. Reducing grain sizes to nanometers in length eliminates this anisotropy as mechanical properties are derived from a larger aggregate of smaller crystallites. Numerous methodologies for producing bulk materials with nanocrystalline-grain structures have been explored [6-1], and in this work attention is focused on bulk nanostructured materials formed through cold spray and dynamic consolidation methods.

Cold spray is a coating process which has successfully consolidated nanocrystalline-materials including aluminum [6-2] and nickel [6-3] alloys and WC/Co [6-4] and copper alumina [6-5] composites. In this process 1–50 μ m size particles with nanocrystalline-grain structures or agglomerates of nanosized powders are deposited onto a substrate at particle velocities ranging between 500 and 1,000 m/s. Compressed gas, which can be heated up to 700°C, flows through a converging/diverging nozzle and is used to accelerate the particles toward a substrate. Upon impact with the substrate, kinetic energy of the particles is transferred into mechanical work and heat through large shear stresses at the particle/substrate interface. This results in a large amount of plastic deformation and temperature generation in the vicinity of the interface and causes solid state bonds to form. Coatings are built on a layer by layer basis with previously deposited particles serving as the substrate for incoming particles. Though the gas is heated and significant temperatures are generated at the substrate/particle interface, temperature in the bulk of the particle remain much lower and make this process ideal for the consolidation of nanostructured materials

Another relatively low temperature consolidation technique is dynamic compaction/ consolidation. Dynamic consolidation is a bulk material forming method which has been successfully used to produce both fully and partially nanostructured materials including aluminum alloys [6-6, 6-7], NiAl intermetallics [6-8], nanocrystalline-Fe [6-9], and exchange coupled magnetic materials [6-10]. Typically, micron sized particles with nanocrystalline-grain structures are quasi-statically compressed to 50–70% of their theoretical density, then consolidated to near full density by the passage of a large amplitude (in excess of a materials yield strength) short duration (~1 μs) stress pulse. Stresses are generated by the impact of a high impedance flyer/driver with the lower impedance powder, where flyers/drivers can be accelerated by means of a compressed gas (plate impact) or explosives. Under the influence of the initial stress pulse, particles rearrange and slide past one another generating friction and heat at particle interfaces. Significant temperature increases are restricted to near surface regions as a result of friction; however, the bulk temperature of the particles remains relatively constant. Similar to cold spray, the low temperatures experienced by the bulk of the powders during dynamic compaction make this process well suited for the compaction and preservation of nanostructured materials.

In this work, nanocrystalline-aluminum alloy powders compacted to near full density through cold spray and dynamic consolidation methods are examined. A brief description of the starting powders and processing methods are presented first. Microstructures and mechanical properties of the bulk materials are reported and discussed in the subsequent two sections, followed by concluding remarks.

Experimental Procedure for the Comparison of Consolidation Methods

Gas atomized 6061 aluminum powders (27.8 +/– 13.7 μm) were obtained from Valimet (Stockton, CA) and ball milled by Novemac LLC (Dixon, CA) at liquid nitrogen (LN_2) temperatures using a proprietary, high efficiency, LN_2 ball milling process. Two batches of LN_2 ball milled 6061 Al powders were acquired from Novemac LLC, henceforth referred to as Batch 1 and 2; their morphologies are shown in Figure 6-1.

Figure 6-1. Nanocrystalline starting powder morphologies for (a) Batch 1 and (b) Batch 2.

Particle sizes for Batch 1 and 2 were measured using Fraunhofer diffraction as 55.7 +/– 28.1 μm and 40.4 +/– 14.4 μm, respectively. Similar processing methods were used for both batches; however, particle morphology varies significantly as Batch 1 exhibits a plate-like morphology and Batch 2 shows an equiaxed/globular morphology. The TEM images were only available for Batch 1 powders and a representative image is given in Figure 6-2. The microstructure consists of elongated lamellar grains 250–400 nm in thickness with lengths 5–15 times their thickness. Though particle morphology differs between the two batches, it is henceforth assumed that the microstructures for both batches are similar to those shown in Figure 6-2. Furthermore, the aim of this work is to compare the microstructures of the bulk materials following cold spray and dynamic consolidation, and not the starting powders.

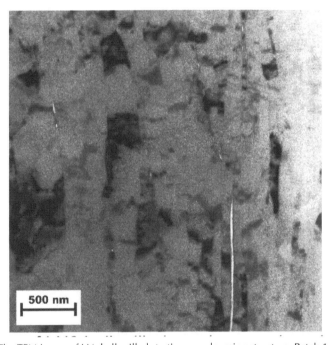

Figure 6-2. The TEM image of LN$_2$ ball milled starting powder microstructure, Batch 1.

Powders from Batch 1 were consolidated using the cold spray technique at Sandia National Laboratories Thermal Spray Research Laboratory, and Batch 2 powders were dynamically consolidated using the 80 mm gas gun at the Georgia Institute of Technology. The TEM samples of consolidated materials were prepared using the FIB lift-out technique with a Ga ion beam, and characterization was carried out on a Phillips CM30 TEM for cold sprayed specimens and an FEI Tecnai F30-ST TEM for dynamically consolidated specimens. Both microscopes operated at 300 kV. Nanoindentation Vickers hardness measurements were carried out on the cold sprayed material using an MTS Nanoindenter XP in continuous stiffness measurement mode with a standard Berkovich (T115) tip at indent depths of 250 and 1,000 nm for both loading and unloading curves; hardness values for the 250 nm loading curve are reported here. Vickers microhardness was measured for the dynamically consolidated specimens using a Leco MHT Series 200 diamond tip micro-indenter with a 50 g load and 15 second dwell time. Specific details of experimental procedures and techniques for the cold sprayed material can be found in references [6-11] and [6-12] and in reference [6-13] for the dynamically consolidated material.

Results for the Comparison of Consolidation Methods

The results presented in the following two sections compare the microstructures and mechanical properties (hardness) of bulk nanocrystalline-aluminum formed through two different processing methods, cold spray, and dynamic consolidation. Results are taken from individual studies on cold sprayed [6-11, 6-12] and dynamic consolidated [6-13] specimens and are presented here for comparison of bulk properties.

Microstructure

The TEM images of representative microstructures for the cold sprayed and dynamically consolidated aluminum powders are shown in Figure 6-3. Both cold sprayed and dynamically consolidated samples exhibit grain refinement following consolidation. Beginning with elongated grains 250–400 nm in thickness, grain sizes of cold sprayed coatings are reduced to 30–50 nm in length. Dynamically consolidated samples exhibit a bimodal grain structure with smaller equiaxed grains 20–50 nm in diameter and slightly larger elongated laths with thickness ranging between 50 and 100 nm. The SAD does not indicate any preferred crystallographic orientation for either sample.

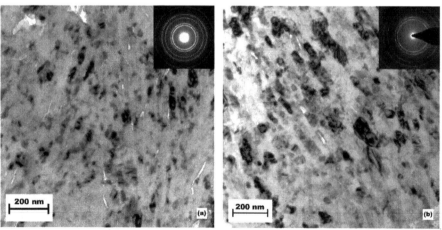

Figure 6-3. Microstructures of cold sprayed (left) and dynamically consolidated (right) nanocrystalline-6061 Al samples.

Hardness

Nano- and micro-hardness measurements for the cold sprayed and dynamically consolidated materials are reported in Table 6-1. Force-displacement nano-indentation measurements in [6-12] are reported in GPa, and the standard HV_{05} micro-hardness measurements in [6-13] are converted to GPa through the relation $GPa = 0.009807 * HV$.

Table 6-1. Nano- and micro-hardness values reported for cold sprayed and dynamically consolidated samples.

Sample ID	Nano- (GPa)	Micro- (GPa)
Bulk Al 6061 T6	1.450 +/- 0.054	1.144 +/- 0.058
Cold Spray Al 6061 Coating	2.715 +/- 0.198	X
Dynamically Consolidated Al 6061	X	1.893 +/- 0.066

Both the cold sprayed and dynamically consolidated aluminum samples have higher hardness values than bulk Al 6061 T6 measured using their respective indentation technique. Nano-indentation of the cold sprayed coating and bulk 6061 T6 indicates coating hardness is increased 1.265 GPa over that of the bulk 6061 T6. Micro-indentation of dynamically consolidated Al and bulk 6061 T6 show an increase of 0.749 over that of bulk 6061 T6. Cold spray and dynamic consolidation specimens exhibit increases in hardness of 87 and 65% respectively. In addition, hardness values measured using nano-indentation appears to be greater than those obtained through micro-indentations for the bulk Al 6061 T6 specimens.

Discussion for the Comparison of Consolidation Methods
Microstructure
Under the assumption that both initial microstructures resemble Figure 6-2, grain size reduction occurs during both cold spray consolidation and dynamic compaction of initially nanocrystalline Al powders. This reduction in grain size is thought to result from the large amount of plastic deformation that occurs during both consolidation processes. The minimum grain size in aluminum attainable through plastic deformation has been shown by Eckert [6-14] to be between 20 and 25 nm. Given that the powders start out with a lath-like structure with thicknesses between 250 and 400 nm, it is reasonable to assume plastic deformation during consolidation is responsible for the observed grain size reduction.

Plastic deformation of particles during cold spray occurs at two separate instances. As the particle initially impacts the substrate at high velocities it forms a crater in the substrate and material from the substrate and particle is jetted from the interface as the particle is flattened and bonds to the substrate [6-15]. In addition, as subsequent particles arrive at the previously deposited particles which are now acting as the substrate, additional plastic deformation occurs in the previously deposited particles as craters form to accommodate incoming particles. During dynamic consolidation, the bulk of plastic deformation occurs with the passage of the initial stress pulse with the rapid acceleration and deceleration of particles as they collide under the influence of the propagating stress pulse. This causes material to flow both parallel and perpendicular to the propagating stress pulse and can also result in turbulent (vortex) material flow at particle interfaces. The large amount of plastic deformation occurring during both consolidation processes results in substantial dislocation activity and grain refinement observed in the consolidated specimens. The extent of grain size reduction following the two consolidation techniques suggests particles undergo a greater amount of plastic deformation during the cold spray process. However, further work is needed examining the deformation behavior of particles during both consolidation techniques inclusive of the effects of particle morphology.

Hardness

Hardness results indicate that both cold sprayed and dynamically consolidated specimens show increases in hardness over bulk Al 6061 T6, with the cold sprayed coatings showing the highest overall hardness. The greater increase in hardness reported for the cold sprayed coating could be a result of the disparity in grain size between the two samples and porosity effects. Through empirical relationships like $\sigma TS \sim 3$ HV [6-16], where σ_{TS} is the tensile strength, and the Hall-Petch relationship [6-17], a decrease in grain size can generally be associated with an increase in hardness. In this work, the cold sprayed coating has average grain sizes between 30 and 50 nm, which are smaller than those observed in the dynamically consolidated specimens. This reduction in grain size may contribute to the observed increase in hardness of the cold spray coating. In addition, nano-indentation measurements do not capture the effects of micro-scale porosity. Micro- and nano-scale porosity is observed in TEM images for both cold sprayed and dynamically consolidated specimens; thus nano-indentation measurements may result in higher reported hardness values as micro-porosity is not sampled in these measurements.

Another factor which may contribute to the higher hardness value reported for the cold sprayed coating is the indentation size effect. Table 6-1 shows higher hardness values for bulk Al 6061 T6 when indents are based on nano- as compared to micro-indentation. Elmustafa and Stone [6-18] have shown that hardness tends to increase with decreasing load for f.c.c. metals, and base their results on a dislocation mechanism similar to work hardening. Their analysis predicts higher hardness values for smaller indenter tip size and shallower indents which may contribute to the higher overall hardness observed in the cold sprayed coating.

Conclusions for the Comparison of Consolidation Methods

Nanocrystalline aluminum powders have been successfully consolidated to form bulk nanocrystalline aluminum 6061 alloys using both cold spray and dynamic consolidation techniques. Both processes show grain refinement following compaction resulting in microstructures consisting of 30–50 nm diameter grains for the cold sprayed coating and a bimodal structure for dynamically consolidated compacts. Cold sprayed coatings exhibit a greater amount of grain refinement than dynamically consolidated specimens. This is attributed to the greater amount of plastic deformation that occurs during the coating process; however, further investigation into the dynamics of deformation during each consolidation process is needed. Hardness measurements indicate cold sprayed coatings have a higher hardness than dynamically consolidated specimens, which is consistent with the level of grain refinement. However, variations in hardness measurement techniques may also be responsible for some of the observed increased hardness.

Acknowledgments for the Comparison of Consolidation Methods
Funding for this research is provided by Sandia National Laboratories, a multipro-gram laboratory operated by Sandia Corporation, a Lockheed Martin Company, for the United States Department of Energy's National Nuclear Security Administration under Contract DE-AC04-94AL85000.

CONCLUSIONS

Nanocrystalline metals carry the promise of high strength, high hardness, and high wear resistance in materials that designers have used for many years. This familiarity with the materials could help nanocrystalline metals to find significant use in a short time. Standing in the way of widespread use was the difficulty in consolidating the powder created by the most accessible nanocrystalline metal production methods.

The research conducted during this project has resulted in two methods of cre-ating bulk nanocrystalline metal: cold spray and shock compaction. These methods are capable of not only creating bulk nanocrystalline material, but of further refin-ing the material during the consolidation process. The consolidated material has been evaluated for microstructure and mechanical properties and has shown great potential. Consolidation of aluminum 6061 has demonstrated a hardness increase between 65 and 87%, depending on the process. The processes have shown the ability to create relatively dense material with porosity of 1–3%.

Other processes for creating nanocrystalline material were also evaluated during the project. Two machining based processes, MAM and LSEM were evaluated for their ability to make nanocrystalline powders and bulk materials respectively. Initial testing showed the MAM powder to have a nanocrystalline shell, but to have larger nanostructured material inside the particles. The LSEM was shown to create fully nanocrystalline material and was used to make the first ever meso-scale nanocrys-talline parts. The material, however, contains a significant amount of residual stress which limited the number of applications evaluated for the material.

This research clearly demonstrated the ability to create and utilize nanocrystalline materials and developed a greater understanding of their unique properties. It will now be important to pursue opportunities to evaluate these materials in specific ap-plications to better understand best areas of opportunity for utilizing nanocrystalline material.

KEYWORDS

- **Herrmann's P-α model**
- **Hugoniot**
- **Large strain extrusion machining**
- **Modulation assisted machining**
- **Nanocrystalline**
- **Quasi-static compression**

Chapter 11

Nanoscopic Grooving on Vesicle Walls in Submarine Basaltic Glass

Jason E. French and Karlis Muehlenbachs

INTRODUCTION

Dendritic networks of nanoscopic grooves measuring 50–75 nm wide by <50 nm deep occur on the walls of vesicles in the glassy margins of mid-ocean ridge pillow basalts worldwide. Until now, their exact origin and significance have remained unclear. Here we document examples of such grooved patterns on vesicle walls in rocks from beneath the North Atlantic Ocean, and give a fluid mechanical explanation for how they formed. According to this model, individual nanogrooves represent frozen viscous fingers of magmatic fluid that were injected into a thin spheroidal shell of hot glass surrounding each vesicle. The driving mechanism for this process is provided by previous numerical predictions of tangential tensile stress around some vesicles in glassy rocks upon cooling through the glass transition. The self-assembling nature of the dendritic nanogrooves, their small size, and overall complexity in form, are interesting from the standpoint of exploring new applications in the field of nanotechnology. Replicating such structures in the laboratory would compete with state-of-the-art nanolithography techniques, both in terms of pattern complexity and size, which would be useful in the fabrication of a variety of grooved nanodevices. Dendritic nanogrooving in SiO_2 glass might be employed in the manufacturing of integrated circuits.

Vesicles in the glassy margins of submarine pillow basalts can exhibit a range of internal features that collectively record information pertaining to eruption dynamics, deformational history, the composition of magmatic fluids linked with ore deposits of economic importance, and the explosive behavior of some "popping rocks" upon dredging from the ocean floor [1-6]. Sulfide spherules commonly decorate vesicle interiors, in between which may also occur complex dendritic patterns of nanoscopic (<100 nm wide) grooves that impart a polygonal or "turtleback" appearance to some vesicle walls [3, 5]. The significance of these grooves is unclear, but it has been proposed that they might be an etching phenomenon [3] and that they resemble shrinkage cracks [5]. Evaluating their exact origin is important, because they could help elucidate a number of different geological processes, and they are similar in size and form to microbial trace fossils [7] and etch-tunnels [8] described from this geological environment.

Understanding these natural examples of nanoscopic grooving in glass may also lead to novel applications in the field of nanotechnology. Specifically, determining the physical process that leads to their formation could help to develop new methods of generating grooved patterns in synthetic materials that are useful in the fabrication of nanodevices. New technologies in sub-100 nm fine pattern formation are anticipated to replace traditional optical lithography techniques in the fabrication of integrated circuits [9], and one of the problems associated with state-of-the-art nanolithography techniques is the generation of complex patterns [10]. In seeking alternatives to nanolithography there is presently much focus on processes that are based on self-assembly, such as the myriad forms of self-assembly exhibited by biological systems [10]. From that standpoint, the complex dendritic nature of the nanogrooves makes determining their origin especially important.

In this chapter, we describe and characterize dendritic patterns of nanogrooves that occur on the interior walls of vesicles from a glassy pillow margin from the North Atlantic Ocean. The aim of this study is to determine the geological origin/meaning of the nanogrooves and to elucidate how they formed by comparison to known physical processes that lead to the formation of similar structures. In terms of their size and the morphological characteristics of their complex dendritic form, the relevance of these natural examples of nanogrooves in silicate glass to the field of nanotechnology is outlined.

Sample Descriptions and Scanning Electron Microscopy

Vesicle-bearing basaltic glass in this study originates from a sample of a glassy pillow margin from core sample 418A-75-3- [120–123], collected from Deep Sea Drilling Project (DSDP) Hole 418A. The drill hole is situated under 5511 m of water in the North Atlantic Ocean at 25°02.10′ N latitude and 68°03.44′ W longitude, and the sample originates from 785 m below the seafloor, 461 m below the top of the volcanic basement [11]. Sand to pebble-sized basaltic glass fragments were isolated from a hand crushed sample of the glassy pillow margin. These fragments were then coated using a Nanotech Semprep2 with ~150 Å of gold (Figures 1a–c and 2e) or using a Xenosput XE200 with ~10–15 Å of chromium (Figures 2a–d), for secondary electron imaging (SEI) with a JEOL 6301F field emission scanning electron microscope equipped with a PGT IMIX model X-ray analysis system.

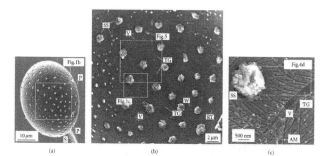

(a)   (b)   (c)

Figure 1. The SEI images of a grooved vesicle from the glassy margin of a submarine pillow basalt from DSDP-418A-75-3- [120–123]. The sample is coated with ~150 E of Au. (a) Overview of the vesicle. Note the step (S) in the fracture surface at lower right, which formed in lee of the vesicle as a propagating fracture opened it from upper left to lower right. Pits along the rim of the vesicle (P) show where sulfide spherules once were. (b) Close-up of the vesicle wall from (a) highlighting the distribution of sulfide spherules (SS) and dendritic patterns of nanoscopic grooves (viscous fingers—V) on the vesicle wall, in addition to other less commonly observed textural features including tension gashes (TG), wrinkles (W), and rill and trellis texture (RT). (c) Detail from (b) showing a Cu-bearing iron sulfide spherule (SS), dendritic nanogrooves (viscous fingers—V), a tension gash (TG), and amorphous material (AM) that postdates and partially obscures the viscous fingers and the tension gash.

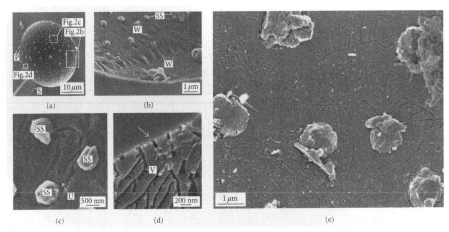

Figure 2. The SEI images of other similar vesicles. The vesicle shown in (a–d) exhibits grooved features analogous to those shown in Figure 1, but is from a different glass fragment from the same pillow margin sample, and is coated with a ~5–15 E thick layer of Cr. (a) Overview of the vesicle highlighting a step (S) in the fracture that exposed the vesicle, which formed in lee of it as the propagating fracture opened it from upper right to lower left, and pits (P) along the edge of the vesicle indicating where sulfide spherules once were. (b) Close-up of the vesicle from (a) showing numerous sulfide spherules (SS) and multiple elongate ribbon-like domains that exhibit a wrinkled surface texture (W). Such wrinkled, ribbon-like domains are interpreted as another form of pull-apart viscous deformation contemporaneous with viscous fingering and development of tension gashes (see Figure 7). (c) Close-up of the vesicle from (a) showing where a solitary viscous finger seems to have pulled a 180° U-turn (U) before stopping and spreading out into a terminal bulb as it was shielded by a sulfide spherule (SS). Some faint, more lobate viscous fingers at left are interpreted to indicate shallowing of the boundary layer to perhaps ~10 nm there. (d) Close-up of the vesicle from (a) showing where viscous fingers intersect the plane of the fracture that exposed their host vesicle, allowing for a precise estimate of finger depth at ~50 nm (arrow). (e) Overview of a portion of a vesicle that is nearby (~400 μm away on the same fragment of glass) to that shown in Figure 1, exhibiting branching grooves of the same size, but with a slightly different form. In this case the nanoscopic grooves lack the distinctive terminal bulbs and lobate forms observed in the other two vesicles (Figures 1 and 2a–d), and, instead of viscous fingers, are interpreted to represent brittle radial microfractures that formed in a thin boundary layer surrounding the vesicle. (Sample is coated with ~150 E of Au.)

Vesicles are common and typically spherical to subspherical, 10–50 μm in diameter, occasionally ranging in size up to ~100 μm in diameter. Most have smooth walls, which are embedded with numerous Cu-bearing iron sulfide spherules of varying grain size, ranging from miniscule (<200 nm) to >6 μm across, although some vesicle walls are also ornamented by complex dendritic patterns of nanoscopic grooves (Figures 1 and 2).

Sulfide Spherules on Vesicle Walls

The vesicles of interest in this study are surrounded by fresh basaltic glass and have many Cu-bearing iron sulfide grains decorating their interior walls (Figures 1 and 2). There is some debate as to whether such spherules originate from reaction of sulferous magmatic fluids with other elements diffusing into the vesicle from the host magma or from bubbles of magmatic fluids of complex composition that already contain all of the required elements dissolved within them [1]. In many cases it is clear that the sulfide grains formed along the bubble miniscus before quenching of the magma into glass, because the contacts between spherules and vesicle walls tend to form hemispherical pits [1-3, 5]. Analogous pits along the rims of the vesicles (Figures 1a and 2a) reveal where sulfide spherules that have now been popped out were once partly embedded in magma.

In both vesicles, sulfide spherules show a range of grain sizes. One vesicle (A) has large populations of equant, 0.2–0.5 and 1.0–1.5 μm wide grains, in addition to a single larger oblate spheroidal grain that measures ~4.0 μm across (Figure 1a). The other vesicle (B) also has multiple, large populations of equant sulfide spherules that have distinctive grain sizes including 0.3–0.5, 0.6–0.9, and 1.2–1.4 μm (Figure 2a). For both vesicles, spherule spacing seems to be a function of grain size, with the average distance between grains increasing with average grain size (Figure 3a) (note that the effects of foreshortening due to the spheroidal shape of the vesicle wall were considered in these measurements). This relationship could be the result of interparticle (e.g., electrostatic) repulsive forces and possibly other external forces (e.g., fluid drag and surface tension) acting to repel grains about the meniscus at the time of glass quenching. The contrasting slopes for plots of average grain size versus average spacing of sulfide spherules in the two vesicles (Figure 3a) might then reflect differences in relative accommodation space linked with varying initial fluid compositions (i.e., varying sulfide-to-vesicle volume ratios). Previous studies of such sulfide spherules also documented a regular spacing between individual grains [2, 3], with the smallest spherules spaced far from, or at maximum distance from the larger ones [1, 3, 5], and the largest spherules tending to depart from a spherical shape to a more flattened ellipsoid [2, 4].

Although the dataset is sparse (constrained by three data-points), the relationship observed for sulfide grain size versus sulfide spacing in vesicle B (Figures 2a and 3b) seems to also hint at a fractal distribution of sulfide spherules on the vesicle wall. This is because when solving for the fractal dimension (d_f) using the generalized equation:

$$d_f = \frac{\ln(N_{i+1}/N_i)}{\ln(R_i/R_{i+1})}, \tag{1}$$

where for a given order, N_i is the number of objects, and R_i is the characteristic linear dimension [15], we find that d_f is nearly identical in each case (0.91, 0.90, and 0.89), when calculating from the three possible paired sets of domains from the sulfide spacing/grain-size data (Figure 3a). We quantified the individual d_f for each paired set of data, by equating the average sulfide spherule diameter with R in (1), and considering N as the total number of sulfides that plot in a large hypothetical flat plane (e.g., 1,000 by 1,000 μm) for each domain, which is in turn controlled by the average sulfide spacing (assuming that the sulfides are arranged in hexagonal arrays). For R = 0.409, 0.751, and 1.308 μm, average spacing is 0.913, 1.203, and 1.814 μm, respectively, with corresponding N values of 1,385,246, 797,883, and 486,658. Plotting ln R) versus ln N) for each scenario (Figure 3b) also yields a composite regression line that has a slope equal to the d_f (multiplied by -1), corresponding to a value of 0.90 for d_f

(Figure 3b). The significance of this apparently fractal distribution of sulfide spherules is unknown, although it may relate to a fractal distribution of trace element concentrations in the vesicle at the time of their formation akin to that observed for some ore deposits [16], or a consequence of other combined forces acting to distribute the grains about the meniscus before glass quenching (e.g., electrostatic, fluid drag).

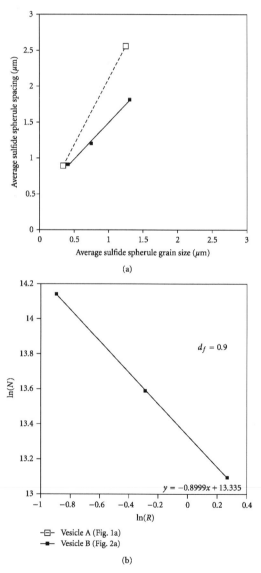

Figure 3. Grain size/spacing relationships of sulfide spherules on vesicle walls. (a) Plots of average grain size versus average spacing for five populations of sulfide spherules embedded in vesicle walls. (b) Determination of the fractal dimension d_f for the pattern of sulfide spherules embedded in the wall of vesicle B (Figure 2a).

Nanogrooves on Vesicle Walls

In between the sulfide spherules are elaborate patterns of shallow (~50 nm), branching grooves that are reminiscent of the dendritic pattern produced by the veins in a plant leaf (Figures 1b and 1c). The grooves occur in arcuate, subparallel sets that are spaced regularly and branch at locally constant angles ranging from ~30 to 90° (typically 40–60°). Subsidiary branch-sets tend to be shorter in length (although not in every case) and invariably branch off to only one preferred side of the previous parent set of grooves. Individual grooves are typically 50–75 nm wide and branch-tips typically occur as bulbous terminations that measure 150–300 nm across. In vesicle A, many of the trees seem to be rooted in wider arcuate grooves or lenses that are up to 0.5 μm wide at the center and which pinch out at the ends. Locally these grooves are postdated by thin films of amorphous material of unknown composition, which partially blankets and obscures them (Figure 1c). On the right side of the vesicle, branching grooves are rare, and the area seems to be dominated by wrinkled surfaces and regions showing a distinctive rill and trellis texture (Figure 1b). Though Fe sulfide spherules are almost ubiquitous in vesicles from the basaltic glass pillow margin studied, only a few % of all vesicles have grooved walls such as this (Figure 1), with most walls being smooth and featureless between sulfide spherules.

Origin of Dendritic Nanogrooves on Vesicle Walls

The vesicle in Figure 1 has many branching grooves that form a large number of inter-related tree-like patterns, which collectively span a large part of the total surface area of the vesicle wall (Figure 1b). Interrelationships between trees, the fine details of their appearance, and the continuity in the overall pattern have been used to place additional constraints on the origin of these grooves, through comparisons to relevant physical processes that lead to the formation of similar dendritic structures. An incredible diversity of natural growth phenomena generates dendritic patterns, many of which are scale-invariant or fractal and may be modeled by diffusion-limited aggregation [17], and these include dendritic crystals [18], vascular systems in plants and animals [19], drainage networks [20], and viscous fingering [21] to name a few.

Studies in viscous fingering involve characterizing the patterns of displacement of a viscous fluid by a relatively less viscous invading fluid, injected into it within the narrow space between two parallel transparent plates known as a Hele-Shaw cell [22]. Intricate patterns of branching viscous fingers of the invading fluid may develop as instabilities at the interface between the fluids arise and evolve [21, 23]. Typically, the mechanism of the instability is intimately linked with viscosity variations between the fluid phases, and this is why the term "viscous fingering" is commonly used [21]. As they are classically studied within the narrow gap between parallel transparent plates, images of viscous fingers are represented as quasi-two-dimensional curvilinear boundaries or traces of the interface between the two fluids [21, 24], that for complex patterns outlines the form of a highly branched dendritic structure (Figures 4a–c). The complexity of viscous fingering trees is controlled mainly by three pattern forming processes that can be summarized as tip-splitting, shielding, and spreading [21, 24].

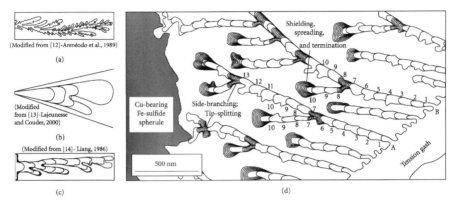

Figure 4. Interpretation of branching nanogrooves as viscous fingering trees. (a) Drawing traced from a photograph [12] of a very unstable viscous finger of air injected into silicon oil, within a linear Hele-Shaw cell of 10 cm width and 0.25 mm thickness. (b) Drawing traced from a composite image of four photographs [13] of viscous fingers at four different stages of growth, formed by air injected into silicon oil within a sector-shaped Hele-Shaw cell of angle 30°. (c) Drawing traced from a computer generated image [14] showing multiple growth stages of simulated viscous fingers that were produced using an algorithm related to diffusion-limited aggregation. (d) Drawing traced from a close-up view from Figure 1c showing 100 nm incremental growth of viscous fingers (numbers), and additional 10 nm incremental growth for all tip-splitting (side-branching) and spreading events, assuming that all viscous fingering trees started to grow simultaneously at constant rates. Examining this growth model in forward motion: finger A undergoes a tip-splitting (side-branching) event at around ~650 nm, as does finger B at around ~750 nm. The new side-branch produced by finger B stops growing and spreads out into a terminal bulb at around 1,050 nm of growth, because finger A has already passed through the region that it is approaching, effectively shielding it from further growth. At ~1,300 nm, finger A undergoes a final tip-splitting event, after which the main finger in addition to the new side-branch both stop growing and spread into terminal bulbs as they are shielded from further growth by the sulfide spherule.

Comparison of the patterns of nanoscopic branching grooves on the vesicle wall in this study to dendritic viscous fingering patterns reveals many similarities. In detail, the outline of the grooves forms a continuous curvilinear trace that is reminiscent of the rounded curvilinear traces formed by branching viscous fingers in macroscopic experiments [24] and numerical simulations (Figures 4a–c). Considering nearby individual trees to represent viscous fingers that started to grow simultaneously at identical rates, forward modeling shows that the patterns produced can be described in terms of tip-splitting (side-branching), shielding, and spreading (Figure 4d).

Two relatively long viscous fingers (fingers C and D in Figure 5) originating from two distal points seem to have arrived at the same location (star in Figure 5) at the same time and engaged in a competition to keep growing, mutually changing course in doing so. This observation was used to constrain a flat, best-fit growth model of viscous fingers across a relatively large portion of the vesicle wall where the effects of foreshortening (maximum <2.5% cumulative) could be neglected (Figure 5). Colors show the time-dependent development of the viscous fingering trees, assuming that finger velocities are constant as they increase in length, but decelerate by ~half as they spread out into terminal bulbs and then stop. This assumption is consistent with experimental viscous fingering results, which show that tip velocity is constant at constant flow rate, even during tip-splitting events and shielding of other fingers that slow to a stop [24]. One measure of the reliability of the resulting growth model is an examination of how many viscous fingers were shielded in the proper sequence of events. For most of the pattern this seems to have worked (Figure 5), but more dynamic computer models with a range of different assumptions will be required to resolve a more accurate picture.

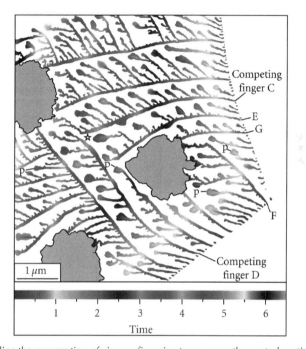

Figure 5. Modeling the propagation of viscous fingering trees across the central portion of the vesicle wall from Figure 1b. In this computer-aided interpretive drawing of dynamic viscous fingering, the colors show the time-dependent development of the branching viscous fingers according to a realistic growth model. Specifically, colors were assigned by assuming that finger velocities are constant as they increase in length but decelerate by ~half as they spread out into terminal bulbs and then stop. This assumption is consistent with experimental results for macroscopic viscous fingering, which show that tip velocity is constant at constant flow rate, even during tip-splitting events and shielding of other fingers that slow to a stop [24]. In this growth model, it is also assumed that all neighboring viscous fingers originating from the same tension gash started growing at the same time. Based on morphology, viscous fingers C and D are interpreted to have arrived at the same location at the same time (indicated on the figure with a star symbol) and consequently undergo a competition to outgrow one another in that region. This indicates that viscous fingers originating from both tension gashes were growing at the same time, constraining a delay of ~0.5 μm of growth time from one tension gash to the other as shown in the figure (i.e., competing finger D is ~0.5 μm longer than C when measuring back from the star). Examining the colors of neighboring fingers where shielding events occur is one measure of the reliability of the resulting growth model. For instance, the color of the very end of finger E shows that according to this model, it was properly shielded by finger D because it arrived approximately 1 μm of growth time after finger D had passed in front of it. Similarly, finger F was shielded by finger G in the right sequence. Some fingers seem to have been shielded prematurely however (denoted by "p"), in that their tips slowed and stopped growing before the shielding finger had already arrived. Three ~1.5 μm wide Cu-bearing Fe-sulfide spherules are also shown, which acted as obstacles that shielded viscous fingers. Note that this drawing was made using higher resolution SEI images of the area highlighted in Figure 1(b), which are not shown.

As the sulfide spherules on the vesicle wall were in place before viscous fingering began, many of them, regardless of their size, appear to have acted as obstacles that shielded viscous fingers from further growth, if the propagating tips approached within ~100–200 nm (Figure 4d). In one such shielding event, a viscous finger from another vesicle with analogous features seems to have pulled a complete 180° U-turn before stopping and spreading out into a terminal bulb (Figure 2c). The observation that sulfide spherules affected the growth patterns of propagating viscous fingers in these ways is important, because it demonstrates that the viscous fingers formed after the sulfide spherules were already in place, providing support for the model that they date to the time of pillow eruption (quenching). It is also interesting from the standpoint of understanding pattern formation during viscous fingering in general, especially because the sulfides not only caused shielding events to occur at a distance of ~100–200 nm but also caused deflection of viscous fingers to occur from a similar distance. To our knowledge, this phenomenon has not been described from classical macroscopic viscous fingering experiments—that is, the mechanisms that control the growth and pattern formation during viscous fingering are classically summarized as tip-splitting, shielding, and spreading [21, 24]. Deflection, or perhaps more effectively, steering, or redirection of viscous fingers by nearby objects, is included as another mechanism affecting pattern formation during viscous fingering in this geological environment. This also raises a question as to why sulfide spherules caused shielding events to occur in most cases and deflection/redirection in at least one isolated instance (Figure 2c).

Viscous fingers from that vesicle (Figures 2a–d) are for the most part identical in width (~50–75 nm) to those from the other vesicle (Figure 1c) and to grooves reported from vesicles in the Nazca plate [3, 5]. Where they intersect the fracture that exposed their host vesicle, they allow for measurement of their depth at ~50 nm (Figure 2d), which is interpreted to correspond to the thickness of the dehydrated boundary layer that contains them.

Quantitative Scaling Analysis of Dendritic Nanogrooves

The complex dendritic patterns commonly observed in macroscopic experiments and numerical simulations of viscous fingering can also be characterized quantitatively using the fractal approach [25, 27]. Because fractal structures are scale invariant, this allows for another method of comparing the nanoscopic viscous fingering patterns observed in this study with more classic macroscopic (cm-scale) examples.

In the experiments and numerical simulations of [25], the viscous fingering patterns produced exhibited a complex dendritic form with notably skeletal structures (i.e., narrow branches) that are similar in form to those observed here at a much smaller scale (~six orders of magnitude smaller). Consequently, in this scaling analysis of natural nanoscopic viscous fingers, we have determined the d_f of the dendritic viscous fingering patterns by the method of [25] to allow a direct comparison to the patterns observed in their study.

Scaling analysis was carried out on a representative region of the dendritic nanogrooves observed on the wall of vesicle A (Figures 1, 5, and 6). This region (Figure 6a) was selected because it is flat and exhibits a significantly large viscous fingering tree that grew from a single source finger in an unconstrained fashion (i.e., was not affected by the presence of sulfide spherules or adjacent parallel starting fingers). Growth modeling (Figure 5) was used to constrain the size of the tree, by assuming exactly 2.5 μm of growth from a single starting point on the primary stem. In this manner, a representative portion of a single, unconstrained viscous fingering tree (Figure 6a) was isolated from a larger structure (Figure 5).

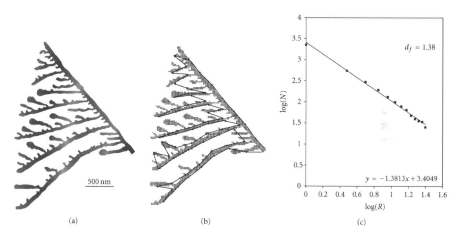

(a)   (b)   (c)

Figure 6. Quantitative scaling analysis of a representative region of dendritic nanogrooves (viscous fingering tree). (a) Overview of the selected region of dendritic nanogrooves, isolated from one of the viscous fingering trees in the growth modeling study of Figure 5. This region was selected as representative because it occurs in a flat region near the center of the vesicle and shows development of a significantly large region of dendritic nanogrooves that were not affected by nearby sulfide spherules or neighboring starting viscous fingers during pattern formation. The size of the tree was also constrained by the growth model in Figure 5 (i.e., this portion of the tree comprises a total of 2.5 μm of growth starting from the selected region of the primary stem—colors are the same as for Figure 5). (b) Pixelated image of the viscous fingering tree shown in (a). This image was used to determine the fractal dimension of the viscous fingering trees (dendritic nanogrooves) observed in this study. The method used to determine the fractal dimension is after [25], which involves measuring the apparent length of the perimeter of the pixelated image numerous times systematically with a ruler of increasing length (from one pixel to 25 pixels in length). Determination of the apparent length of the perimeter with a ruler of 15 pixels in length is shown to illustrate how these measurements were made. (c) Determination of the fractal dimension (d_f) from the apparent length measurements of the perimeter (N = apparent length in units of R; R = length of the ruler in pixels).

Measurement of the d_f was done after the method of [25]. This involved pixelating the selected viscous fingering pattern and then measuring the apparent length (N) of the perimeter numerous times systematically, with successively larger rulers of increasing length R that ranged from one pixel to 25 pixels in length (Figure 6b). If the structure is fractal, then the apparent length should decrease as R-d_f [25], and in the present study on the selected region of branching nanoscopic viscous fingers we observed this relationship to be true (Figure 6c). On a plot of log (R) versus log (N), the slope of the regression line through the data (multiplied by −1) is equal to the d_f [25], which we determined in this case to be 1.38 (Figure 6c). This value is identical to that determined (1.38) for viscous fingering of water injected into a polymer solution in a

Hele-Shaw cell [25]. This remarkable similarity indicates that even at length scales six orders of magnitude smaller, and with a very different suite of fluids (magmatic vapor injected into hot basaltic glass), the viscous fingering phenomenon yields an identical fractal structure. This supports the notion of [25], that viscous fingering yields a universal value for d_f for a wide range of fluid compositions and viscosity contrasts. Ultimately, a value of d_f determined on nanogrooves in this study (1.38) that is identical to macroscopic viscous fingering experiments provides additional quantitative evidence to support the idea that these naturally occurring nanogrooves are in fact exceptionally tiny viscous fingers.

Summary of the Model

There is enough information within the interrelated dendritic patterns of grooves on the vesicle wall in Figure 1 to conclude that they represent viscous fingering trees. The grooves themselves represent fingers of very low-viscosity ($\sim4\cdot10^{-5}$ Pa·s for CO_2 vapor at 30 MPa and 600–700°C [28]) magmatic fluid that were injected into a shell-like boundary layer of hot (~600–700°C across the glass transition for basalt [29]) dehydrating basaltic glass with much higher relative viscosity ($\sim10^9$ to $\sim10^{11}$ Pa·s across the glass transition for basalt [29]) that seems to have acted as a ~50 nm thick, natural example of a Hele-Shaw cell. The magmatic fluid was likely dominated by CO_2 vapor [30] and probably equilibrated with the ambient hydrostatic pressure (~30 MPa) in the time it took for the glassy pillow margin to form [31].

Similar, leaf-like patterns of nanoscopic branching grooves on vesicle walls in mid-ocean ridge basaltic (MORB) glass have also been recognized from DSDP site 320 in the Nazca plate [3, 5]. They have been likened to shrinkage cracks [5], which is significant because numerical modeling studies actually predict the formation of microfractures generated by elastic shrinkage on the walls of some vesicles in glassy rocks, including submarine pillow lavas [6]. According to this conceptual model (Figure 7c), when the glass transition is reached during cooling, the gases trapped in the vesicle become constrained to an isochore because the vesicle can no longer shrink by viscous relaxation of the surrounding melt [6]. As a consequence, the chemical potential of water vapor in the vesicle falls out of equilibrium with the chemical potential of water in the glass, causing water to diffuse into the vesicle from a thin boundary layer of surrounding glass [6]. Elastic shrinkage in this boundary layer occurs because the density of hydrous basaltic glass is less than that of anhydrous basaltic glass, and the resulting strain generates tangential tensile stress that causes microfractures to form in a radial array around the vesicle (Figure 7c). Branching nanogrooves on vesicle walls in this study are interpreted to represent similar pull-apart structures that formed in an analogous dehydrating boundary layer surrounding each vesicle (Figure 7).

The mechanism for generating tangential tensile stress around the vesicle during deformation is interpreted to have been the same as that proposed for generating radial microfractures around vesicles in glassy rocks [6] (Figure 7c). In this case, however, deformation in the boundary layer is interpreted to have been viscous as opposed to brittle microfracturing (Figure 7). The deformation was initiated by generation of pull-apart viscous fingers that evolved into arcuate tension gashes (Figures 7d and e). Extension on the convex-out side of the tension gashes appears to have served as a focus for the onset of continued viscous fingering (Figures 7d and e).

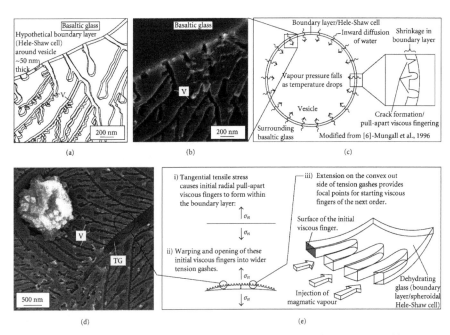

(a)   (b)   (c)

(d)   (e)

Figure 7. Summary diagram showing the conceptual model explaining the initiation and formation of nanoscopic viscous fingering patterns on vesicle walls in mid-ocean ridge basaltic glass. (a) Schematic illustration depicting a putative ~50 nm thick boundary layer around vesicle B, corresponding to the possible configuration of a previously existing, natural Hele-Shaw cell, which now contains frozen (or fossil) viscous fingers (V) of magmatic vapor. The thickness of this hypothetical boundary layer (~50 nm) is estimated from the apparent depth of the fossil viscous fingers shown in (b). Although empirical evidence for the boundary layer has not yet been determined, it is predicted to comprise basaltic glass with slightly lower water contents than ambient values in surrounding glass. Note that in this example of ultrathin film surface layer viscous fingering on the vesicle wall, the Hele-Shaw cell essentially had imaginary walls, corresponding to the inner and outer reaches of the hypothetical dehydrated boundary layer. During viscous fingering, the more viscous fluid comprised hot basaltic glass that was probably near the glass transition temperature (600–700°C)—the less viscous invading fluid being magmatic vapor that was probably dominated by CO_2. (b) Close-up SEI image (from Figure 2) showing the region of vesicle B for which a hypothetical boundary layer is illustrated in (a) (V—viscous fingers). (c) Illustration of pull-apart nanofracturing/fracturing/viscous fingering in a thin boundary layer predicted by numerical modeling of stress generation around some vesicles in glassy rocks (modified from [6]). The SEI image from (b) is located on this conceptual model. (d) Close-up SEI image of a portion of the vesicle wall from Figure 1, linking a tension gash (TG), and viscous fingers (V) to the conceptual model of their formation shown in (e). (e) Model for the stepwise initiation of viscous fingering in a ~50 nm thick boundary layer of glass around a vesicle in a glassy pillow margin. The σn denotes the components of traction normal to a surface, which define a tensile surface stress [26] that leads in this interpretation to the formation of a pull-apart viscous finger.

A different type of nanoscopic grooves observed in another vesicle (Figure 2e) from DSDP 418A-75-3- [120–123] provides additional evidence for invoking this numerical model [6] in determining the driving mechanism for viscous fingering. In this case, the branching grooves are identical in width and dendritic nature, but they do not have terminal bulbs and irregular lobate forms (as in Figures 1 and 2a–d). For that reason, these nanogrooves (Figure 2e) are interpreted instead to represent brittle deformation in the form of shallow radial microfractures, akin to those predicted by the numerical simulations [6]. The occurrence of both types of deformation in the same pillow margin may reflect the crossover from viscoelastic fracture to viscous fingering with increasing water content that is known for some geological materials [32]. Regardless, both types of deformation are interpreted to have formed as a result of the same process (Figure 7), and they result in the formation of branching grooves of similar thickness and depth surrounding the vesicles. A comprehensive summary of this conceptual model is presented in Figure 7.

The observation that vesicles with grooved walls occur locally in this pillow margin sample amongst a larger population of vesicles with smooth, featureless walls, can be explained by differences in fluid composition between vesicles. Typically, vesicles in the glassy margins of submarine pillow lavas contain predominantly (>95%) CO_2 as the vapor phase, with the remaining ~5% comprising mostly H_2O, in addition to other trace gases [4, 33]. Of interest here is that volatile concentrations in vesicles from basaltic glasses from the mid-Atlantic ridge are known to vary at the vesicle-to-vesicle scale in the same glassy pillow margin [31], and variations in the H_2O content of vesicles have also been observed in repeat measurements on glasses from the East Pacific Rise [4]. Variations in the original H_2O content from vesicle-to-vesicle should influence the degree to which dehydration (leading to nanogrooving) will occur in the surrounding glass. This is because, in the numerical modeling studies of microfracturing around vesicles in glassy rocks, the decreasing vapor pressure inside the vesicle during cooling is compensated as water is added to it from the dehydrating boundary layer [6]. Intuitively, any excess water originally present in anomalously H_2O-rich vesicles (i.e., above equilibrium values with the surrounding glass) would result in increased starting partial vapor pressure of H_2O in the vesicle, which would, in turn, decrease the amount of water entering the vesicle through dehydration of the surrounding glass, ultimately reducing the chances of microfracturing.

An alternative explanation for the occurrence of both grooved and smooth-walled vesicles in the same pillow margin is that the original water content in the glass itself might not have been perfectly homogeneous at the vesicle-to-vesicle scale in the pillow margin. For instance, vesicles originally at equilibrium with surrounding glass will undergo microfracturing at higher temperature and vapor pressure if the water content in the surrounding glasses is higher (e.g., at 500 K for ~1% and 600 K for ~2%) whereas vesicles surrounded by glass with lower water contents (<0.4%) are not predicted to undergo microfracturing at all [6]. This explanation seems more likely, because it also provides an explanation for the two contrasting styles of nanogrooving observed in this study. In particular, nanoscopic grooves around the vesicle shown in Figure 1 are the result of high-temperature viscous fingering, while the nanogrooves in a vesicle that occurs only ~400 μm away (Figure 2e) seem to have formed by brittle microfracturing at presumably lower temperature. Differences in the initial water content in the glass surrounding each of these two vesicles would have triggered deformation around each vesicle at a different temperature in each case during cooling, and therefore at different times assuming identical cooling rates for the two vesicles. This could have resulted in early viscous deformation (at high temperature) around one vesicle, and late brittle microfracturing (at low temperature) around the other. According to this model, it is inferred that water contents in the glass were highest the surrounding vesicles that show nanoscopic viscous fingering, lower around vesicles exhibiting shallow brittle microfracturing, and the lowest around the most common type of smooth-walled vesicles (i.e., those where deformation of vesicle walls simply did not ensue).

DISCUSSION

Implications for Geology

Branching nanoscopic grooves on vesicle walls in the glassy margins of mid-ocean ridge pillow basalts have now been described from the oceanic crust beneath the Pacific [3, 5] and Atlantic Oceans (this study) and are therefore predicted to be globally widespread in the oceanic crust. Their existence may now be explained by invoking the fluid mechanical process of viscous fingering, which took place at a submicroscopic scale in these rocks. Individual nanogrooves represent viscous fingers of magmatic fluid that were frozen into place after being injected into a thin (~50 nm) boundary layer of hot, dehydrating basaltic glass. Although they are not exactly microfractures, the grooves are linked with numerical predictions of microfracturing caused by elastic shrinkage during dehydration of some vesicle walls upon cooling through the glass transition [6].

Tubular dendritic microchannels are a common occurrence at the glass-palagonite alteration front in the glassy margins of mid-ocean ridge pillow basalts [7]. They are often interpreted as evidence for microbial etching [7] and in one recent study on the same rocks as the present study have been explained in terms of abiotic, preferential etch-tunneling along radiation damaged regions in the glass [8]. Branching microchannels attributed to microbial activity have actually been described from the same drill hole as rocks from the present study, and as having diameters as small as a few hundred nanometers (i.e., sample DSDP-418A-62-4- [64–70]) [7], and these might also be reinterpreted in an nonbiological context [8]. Of importance here is that some of them (see [7], Figure 5) also happen to be rooted along vesicle walls, extending outwards into fresh glass for >100 μm, and therefore the geological setting of these features [7] overlaps with that for the branching nanoscopic channels described here. Because of the similarities in size and overlap in geological setting, it is important to distinguish between these two different types of submicroscopic branching channels in basaltic glass. In contrast to those attributed to microbial activity [7] (or alternatively, abiotic etch-tunneling [8]), the branching channels (grooves) described here are only restricted to a ~50 nm thick boundary layer of glass around the vesicles (i.e., are quasi-two-dimensional channels as opposed to 3-dimensional tunnels), do not extend outwards into fresh glass, and were not produced by any kind of etching process (biotic [7] or abiotic [8]). In addition, they are not associated with palagonite alteration and are observed only within vesicles surrounded by fresh glass that were not intersected by fractures and exposed to seawater in their geological history.

Implications for Experimental Studies of Viscous Fingering

Viscous fingering experiments in macroscopic Hele-Shaw cells are widely studied because of their broad physical applications including oil recovery, chemical processing, hydrology, and filtration [21]. Several features of the natural viscous fingers observed here make them a unique and rich source of information for understanding new problems in viscous fingering. Because they formed at the glass transition in a quenching pillow margin, they are the preserved relics of a presumably high-temperature ($\sim$600–700°C) viscous fingering event. The extraordinarily small (nanoscopic) size of these viscous fingers is probably a reflection of the very high viscosity ratio ($\sim10^{15}$) of the system and the small thickness ($\sim$50 nm) of the boundary layer (natural Hele-Shaw cell), and both of these features represent new extremes in viscous fingering. Macroscopic viscous fingering experiments are classically studied in Hele-Shaw cells with relatively simple, flat geometries [21]. In contrast, the nanoscopic viscous fingering described in this study occurred within a spheroidal shell around each vesicle. Because the seemingly imposed unidirectionality of subsidiary side-branching observed in this study (e.g., Figure 5) remains entirely unexplained, this needs to be explored with viscous fingering experiments in spheroidal Hele-Shaw cells, and at high-temperature and high-viscosity ratio (e.g., with vapor/magmatic systems).

Viscous fingering trees identified in this study are notably complex skeletal structures with thin straight branches that are similar in appearance to experimental and simulated dendritic viscous fingers formed at high-viscosity ratio (e.g., 104) [34]. In the natural system studied here, the viscosity ratio was $\sim10^{15}$, which by analogy may have been an important factor in controlling the complexity and form of these skeletal viscous fingers. The complexity of dendritic viscous fingering patterns also increases with increasing flow rate [35], when driving relatively low-viscosity Newtonian fluids into complex non-Newtonian fluids such as polymeric liquids and liquid crystals [36, 37], and by narrowing the gap between plates in the Hele-Shaw cell [36]. These observations may be particularly relevant to the present study because basaltic melts exhibit non-Newtonian behavior at high strain rates [38], and the gap width of the boundary layer (Hele-Shaw cell) is only $\sim$50 nm.

Implications for Nanotechnology

Previously emplaced sulfide spherules embedded in the vesicle wall also had a very strong influence on pattern formation, acting as island walls to the Hele-Shaw cell (i.e., dehydrating boundary layer). Considered in combination with the regularity (predictability) of the viscous fingering patterns and the very small (50–75 nm) widths and depths ($\sim$50 nm) of resultant nanogrooves, this observation is interesting from the standpoint of exploring new applications in the field of nanotechnology. One problem associated with the fabrication of grooved nanodevices using conventional techniques, including soft X-ray, extreme-UV lithography, and electron beam lithography, is the generation of complex patterns [10]. Logical placement of obstacles that impede nanoscopic viscous fingers of vapor in hot glass could potentially be used to produce patterns of self-assembling branching grooves that compete with state-of-the-art nanolithography techniques, both in terms of pattern complexity and size. In addition, the practical limit of nanolithography is around 20 nm [10], which is near in width to the dendritic nanogrooves in this study.

The generation of precise patterns and linear arrays of nanogrooves in materials has a number of uses that range from the stretching, aligning, and movement of DNA molecules [39, 40] to the fabrication of nanoelectronic devices and integrated circuits [10]. As the feature size in nanoelectronic devices continues to shrink (with 32 nm chips predicted for 2009) and nanolithography techniques reach their practical size limit of ~20 nm, alternatives to such variants of optical lithography techniques are currently being sought in the fabrication of nanointegrated circuits [10, 41]. It was noted by Smith [10] that, likely, new alternatives to nanolithography will employ mechanisms of self-assembly such as the myriad forms of self-assembly exhibited by living systems. Here we propose that self-assembling dendritic patterns of nanogrooves formed by viscous fingering of vapor and silicate glass within nanoscopic (<50 nm thick) Hele-Shaw cells that have imaginary bounding plates, that is, ultra-thin film surface layer viscous fingering, could provide an important new alternative to nanolithography useful in the fabrication of nanoelectronic devices and integrated circuits. Several observations from the present study on natural glass support this claim.

Aside from the observed small width (~50–75 nm), depth (<50 nm) and complex branching form of these grooves (e.g., Figure 5), which are desirable features in modern day nanolithography, there are other advantages that potentially make this a promising technique that will be useful in fabricating integrated circuits. First is their occurrence in silicate glass. SiO_2 glass is the ideal dielectric material used in integrated circuits [42], and it may be used in passivation (protective coating) and diffusant masking [43], as gate dielectrics in Metal-Oxide-Semiconductor Field-Effect Transistors (MOSFETs) [43], or used as an insulating substrate onto which nanowire networks are formed in various nanodevices such as molecular random access memory cells (nanocells) [44] and nanowire field-effect transistors [45]. This is partly for its physical properties including stability at temperatures over 900°C, ability to be selectively removed from underlying silicon by etching, provides a barrier against chemical diffusion, and for its dielectric breakdown strength and good insulating properties [46]. But its widespread use is also because of the simplicity of growing SiO_2 glass on silicon substrates—silicon being the most commonly used semiconductor material in integrated circuits [42, 43].

In fabricating integrated circuits, the formation of a thin layer of silica glass onto the silicon substrate is most commonly done by high-temperature (900–1,100°C) oxidation in either a wet or dry atmosphere containing water vapor, or oxygen, respectively [46]. In MOSFETs, gate oxides are only a few hundred angstroms thick or less [43, 46], and by comparison the dehydrated boundary layers described in this study range downward in thickness from ~500 Å (Figures 2c and d). Forming a very thin oxide layer in gate oxides during fabrication of MOSFETs means that dielectric breakdown may occur at quite a low voltage and that a fundamental lower limit exists for oxide thickness—for instance, a 500 Å thick layer would result in breakdown around 30 V [43]. Since the 1990s, MOSFETs have become nanodevices because the thickness of the SiO_2 gate dielectric is about 3 nm for logics and 5 nm for nonvolatile memories

[46]. However, because it is horizontal sizes that limit the IC integration, the development of a new nanodevice proper presently requires that at least one of its horizontal dimensions is between 1 and 100 nm in size [46]. This is the case for the nanogrooves identified in this study, because they are ~50–75 nm wide, they have a branch spacing on the order of ~100 nm, and many of the individual branches are less than 100 nm long (Figure 1c).

Optical projection lithography has long been used in the planar fabrication process used in manufacturing integrated circuits, employing resists, patterned masks, and etching [10, 46]. The process of generating windows and grooved patterns in thin SiO_2 glass layers by this method, which exposes the underlying silicon for contact with aluminum in precise patterns that correspond to specific design portions of the integrated circuit, involves multiple steps [43]. First, a thin layer of SiO_2 glass is grown on the silicon substrate (Figure 8a). The SiO_2 layer is then coated with a radiation sensitive photoresist (Figure 8b) and a glass mask placed on top of this (Figure 8c), which is patterned with transparent and opaque areas that define a specific structural portion of the integrated circuit [43]. The wafer-mask system is then exposed to ultraviolet light, which hardens the negative-working photoresist where it is exposed, leaving the unexposed regions to be washed away during subsequent development and rinsing of the photoresist (Figure 8d) [43]. Immersing the wafer in hydrofluoric acid then etches away the regions of SiO_2 glass not covered by the remaining photoresist (Figure 8e), opening up precisely designed windows through the SiO_2 layer and exposing the underlying silicon in the desired pattern, leaving it relatively unaffected by etching [43]. Removal of the remaining photoresist leaves the final patterned SiO_2 glass layer on a silicon substrate (Figure 8f) [43].

Nanoscopic grooves in the natural glass in the present study are formed in a substrate that is not pure SiO_2 glass, but rather, basaltic glass comprised of ~51% SiO_2 in addition to numerous other oxides including ~14% Al_2O_3, ~12% CaO, ~11% total Fe-oxide, ~7% MgO, ~2% Na_2O, ~1.5% TiO_2, and other trace elements (data for adjacent glassy pillow margins at DSDP 418A: [47]). However, if complex branching nanoscopic grooves can form by ultrathin film surface layer viscous fingering between magmatic fluid and basaltic glass (Figure 7), it might be replicated in the laboratory using CO_2 (or other vapors) and pure SiO_2 glass at temperatures near the glass transition (Figures 8g–i). In this manner, generating patterns of dendritic nanogrooves useful in the fabrication of nanodevices and integrated circuits could be carried out by replicating the natural process observed on vesicle walls (Figure 7) in the laboratory, but in a thin layer of silica on a silicon substrate instead (Figures 8g–i).

If it is possible to replicate such dendritic nanogrooves formed by thin film surface viscous fingering in SiO_2 glass on a silicon substrate, it should be a relatively simple matter to design the branching networks to specified shapes. This could be done by controlling patterns of viscous finger propagation with obstacles placed at logical points in the layer. As they grew, nanoscopic viscous fingers observed in natural glass from this study were terminated (shielded) by two contrasting types of obstacles, including positive features (sulfide grains embedded in the glass) and negative features (other neighboring viscous fingers). In considering a thin layer of SiO_2 glass on silicon to represent a Hele-Shaw cell with imaginary bounding plates (i.e., the upper and lower surfaces of the SiO_2 layer), then in the lateral dimensions obstacles that could be used to shield/deflect propagating viscous fingers in certain ways could be emplaced either by embedding/growing minerals in the SiO_2 layer (positive features) or by making holes or lines (negative features) using electron beam lithography [48] or other nanolithography techniques.

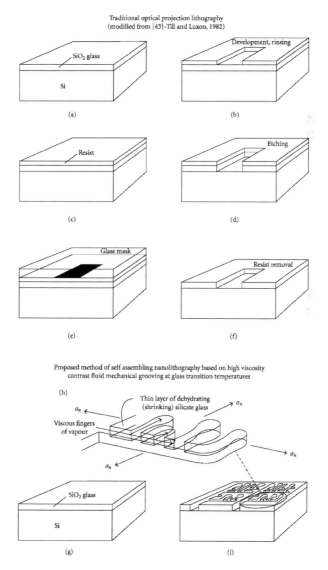

Figure 8. A comparison of traditional optical lithography techniques (a–f) used in the fabrication of integrated circuits [43] to a proposed new method of nanolithography based on high-viscosity contrast fluid mechanical grooving at glass transition temperatures in the SiO_2 layer (g–i).

There are also numerous morphological aspects of the natural grooves identified in this study that make them interesting from the standpoint of designing nanointegrated circuits. For instance, semiconducting nanowires in field-effect transistors and single-electron transistors are typically 10–100 nm in diameter [41]. Although vertical arrays of nanowires may be grown by the vapor-liquid-solid approach, the extreme difficulty of gating and circuit wiring calls for development of new methods to grow horizontal arrays of branching nanowire networks [49]. Growth of hexagonal GaAs nanowire networks has already been used to fabricate horizontal branching linear nanowire arrays to a high node density (106–108 cm^{-2}), which have applications in binary decision diagram logic architecture [49, 50]. Dendritic nanowire networks potentially have a range of applications in nanoelectronics and integrated circuits, for example, in tree structured resistor-capacitor (RC) networks used in RC delays and matched RC trees (e.g., H-tree networks) that are employed in building clocks [42]. Development of such small-scale dendritic nanowire networks could be facilitated by the fluid mechanic-based, self-assembling mechanism of nanogrooving outlined in this study (Figures 8g–i), especially because the observed node density is exceptionally high (~10^{10} cm^{-2}) in these natural examples of grooved patterns (e.g., Figure 1b).

CONCLUSIONS

We have determined the origin of dendritic patterns of nanoscopic grooves on vesicle walls, which occur globally in MORB glass. The ~50–75 nm wide grooves form as the vapor within the vesicle undergoes viscous fingering with a thin (~50 nm) shell of hot glass surrounding it, during quenching of the glassy pillow margin upon lava eruption. The pattern forming process is nearly identical to that observed in macroscopic viscous fingering experiments with Hele-Shaw cells and results in dendritic branching patterns of the same d$_f$ (1.38), despite being six orders of magnitude smaller. This viscous deformation is linked to numerical predictions of elastic shrinkage and radial microfracturing around some vesicle walls in glassy rocks, which happens upon cooling through the glass transition [6]. Because it occurred at extraordinarily high-viscosity ratio (~10^{15}) in an exceptionally thin (~50 nm), spheroidal layer, resulting in very narrow (50–75 nm wide) viscous fingers with preferred side-branch directions, we highlight a number of previously unrecognized extremes in viscous fingering.

Replicating such high-viscosity contrast fluid mechanical nanogrooving in silica glass could provide a new alternative to nanolithography that is based on self-assembly. This would provide a range of novel applications in the field of nanotechnology, particularly in the manufacturing of grooved nanodevices. For instance, if this type of ultrathin film surface layer viscous fingering is possible in SiO$_2$ glass on a silicon substrate, it might be used as a new alternative to traditional lithography techniques used in the fabrication of nanointegrated circuits.

KEYWORDS

- **Dendritic nanogrooving**
- **Fractal dimension**
- **Hele-Shaw cell**
- **Sulfide spherules**

ACKNOWLEDGMENTS

The authors would like to thank George Braybrook and Rajeev Nair from the Department of Earth and Atmospheric Sciences, University of Alberta, for assistance with scanning electron microscopy studies. In addition, they would also like to acknowledge two anonymous reviewers for their helpful comments and editorial suggestions that resulted in significant improvements to the final version of this chapter.

Chapter 12

Thermochemotherapy Effect of Nano-sized As_2O_3/ Fe_3O_4 Complex

Yiqun Du, Dongsheng Zhang, Hui Liu, and Rensheng Lai

INTRODUCTION

Both thermotherapy and arsenic have been shown to be active against a broad spectrum of cancers. To reduce the limitations of conventional thermotherapy, improve therapeutic anticancer activity, reduce the toxicity of arsenic on normal tissue, and increase tissue-specific delivery, we prepared a nano-sized As_2O_3/Fe_3O_4 complex (Fe_3O_4 magnetic nanoparticles encapsulated in As_2O_3). We assessed the thermodynamic characteristics of this complex and validated the hyperthermia effect, when combined with magnetic fluid hyperthermia (MFH), on xenograft HeLa cells (human cervical cancer cell line) in nude mice. We also measured the effect on the expression of CD44v6, VEGF-C, and MMP-9 which were related to cancer and/or metastasis.

The nano-sized As_2O_3/Fe_3O_4 particles were approximately spherical, had good dispersibility as evidenced by TEM, and an average diameter of about 50 nm. With different concentrations of the nano-sized As_2O_3/Fe_3O_4 complex, the corresponding suspension of magnetic particles could attain a steady temperature ranging from 42 to 65°C when placed in AMF for 40 min. Thermochemotherapy with the nano-sized As_2O_3/Fe_3O_4 complex showed a significant inhibitory effect on the mass (88.21%) and volume (91.57%) of xenograft cervical tumors ($p < 0.05$ for each measurement, compared with control). In addition, thermochemotherapy with the nano-sized As_2O_3/ Fe_3O_4 complex significantly inhibited the expression of CD44v6, VEGF-C, and MMP-9 mRNA ($p < 0.05$ for each).

The As_2O_3/Fe_3O_4 complex combined with MFH had is a promising technique for the minimally invasive elimination of solid tumors and may be have anticancerometastasic effect by inhibiting the expression of CD44v6, VEGF-C, and MMP-9.

Clinically, heating of certain organs or tissues to temperatures between 41 and 46°C (the procedure is termed tissue "hyperthermia") has been effective in tumor therapy. Presently, traditional thermotherapy protocols utilize radiofrequency waves, microwaves, or lasers, each of which has many limitations. With the development of nanotechnology, magnetic nanoparticles have been used not only as drug carriers but also in tumor hyperthermia, as such particles absorb energy from high frequency AMF. Nanoscaled magnetic fluid (MF) has been found to absorb much more energy than conventional materials, and this energy is further transferred to tumor cells resulting in tumor temperatures of 4245°C. This process, termed "MFH" can be used for the treatment of either non-cancer diseases or tumors [1]. Magnetic nanoparticles may be good thermoseeds for localized hyperthermia treatment of cancers [2], permitting the heating of and damage to normal tissue to be avoided, thus overcoming the limitations of conventional heat treatment. The Fe$_2$O$_3$ magnetic nanoparticles used in MFH were reported to have a significant therapeutic effect on xenograft liver cancer in nude mice [3]. Depending on the applied temperature and the duration of heating this treatment either results in direct tumor cell killing or makes the cells more susceptible to concomitant radio- or chemotherapy. Numerous groups are working in this field worldwide, but only one approach has been tested in clinical trials so far.

Metastatic spread of the solid tumor depends on a critical cascade of events, which includes tumor cell adhesion to a distant site, extracellular matrix degradation, migration, proliferation, and ultimately neovascularization. Tumors that produce higher levels of metastasis-related factors, such as proteins encoded by the "cluster of differentiation 44v6" (*CD44v6*) gene, and the genes encoding vascular endothelial growth factor-C (VEGF-C) and matrix metalloproteinase-9 (MMP-9), may show more aggressive behavior than do tumors negative for these factors. Thus, a treatment that could inhibit the expression of these tumor metastasis-related factors would be of great interest.

Arsenic is a well-documented carcinogen that also appears to be a valuable therapeutic tool in cancer treatment [4]. The first use of As$_2$O$_3$ in cancer therapy was to treat acute promyelocytic leukemia (APL) [5]. The results of *in vitro* research and clinical trials have shown that As$_2$O$_3$ is effective in inhibiting tumor growth, and in inducing the differentiation and apoptosis of APL cells. Because of its significant anti-cancer effects, As$_2$O$_3$ has been tested in patients with other tumor types, including gastric cancer [6], neuroblastoma [7], esophageal carcinoma [8], and head and neck cancers [9]. Moreover, As$_2$O$_3$ was shown to inhibit tumor metastasis by reducing the expression of metastasis-related genes [10-12].

The purpose of this study was to prepare nano-sized As$_2$O$_3$/Fe$_3$O$_4$ complex for tumor thermochemotherapy and validate its effect on xenograft tumor in nude mice as a premature treatment. We also tested the ability of a nano-sized As$_2$O$_3$/Fe$_3$O$_4$ complex combined with MFH to inhibit the expression of metastasis-associated genes.

MATERIALS AND METHODS

Reagents

The As$_2$O$_3$ was purchased from Sigma (St Louis, MO; Lot A1010). A 1 mm stock solution in RMPI 1640 medium (Gibco) was prepared, stored at 04°C, and diluted before use. Calf serum from newborn animals was obtained from Si-Ji-Qing Biotechnology Co. (Hangzhou, China). The 4-(2-hydroxyethyl)-1-piperazineethanesulfonic acid (HEPES) (the free acid) and trypsin were obtained from Amresco Corp. The RNAiso reagent, AMV retroviridase, dNTPs, Oligo(dT)18, Taq DNA polymerase, and DNA markers were purchased from Takara Biotechnology Co. (Dalian, China). The VEGF-C, CD44v6, and MMP-9 primers were obtained from Shen-neng-bo-cai Biotechnology (Shanghai, China).

Preparation of the Nano-sized As_2O_3/Fe_3O_4 Complex

The Fe_3O_4 magnetic nanoparticles were prepared by chemical coprecipitation. Briefly, solutions of $FeCl_3 \cdot 6H_2O$ and $FeCl_2 \cdot 4H_2O$ were mixed, at a molar ratio of iron (II) to iron (III) of about 0.6, under nitrogen purging and with stirring, with ammonia (1.5 mol/l) added dropwise until the pH attained pH 9. A dark precipitate (Fe_3O_4 magnetic nanoparticles) appeared rapidly, and stirring was continued for 30 min. After 30 min at 90°C, the Fe_3O_4 magnetic nanoparticles were isolated using a permanent magnet and dried in vacuo.

The nano-sized As_2O_3/Fe_3O_4 complex was prepared by an impregnation process. Briefly, Fe_3O_4 magnetic nanoparticles were added to a solution of As_2O_3 (0.01 mg/ml, pH = 5, adjusted with acetic acid) with sonication. After 30 min of thermal treatment at 80°C, the resulting nano-sized As_2O_3/Fe_3O_4 complex was centrifuged at 2,000 g/min for 10 min, rinsed twice with absolute ethanol, and dried under vacuo. The diameter of the nano-sized As_2O_3/Fe_3O_4 complex particles was measured by transmission electron microscopy (TEM; Hitachi H-600 instrument; Japan).

Heat Testing of Nano-sized As_2O_3/Fe_3O_4 Complex *in Vitro*

The nano-sized As_2O_3/Fe_3O_4 complex was dispersed in 0.9% (w/v) NaCl, at Fe_3O_4 concentrations of 0.5, 1.0, 1.5, 2.0, and 2.5 mg/ml, and 2 ml aliquots of MFs were added to flat-bottomed cuvettes and placed in a high frequency electromagnetic field. The distance from the bottom of the cuvette to the center of the source (coils) of the high frequency field was 5 mm. The output frequency was 230 kHz and the output current 20 A. Throughout 1 hr of incubation, the temperature was measured every 5 min.

Cell Culture

The HeLa cells (a human cervical cancer cell line, provided by the Institute of Biochemistry and Cell Biology, Shanghai Institute of Biological Sciences, Chinese Academy of Sciences) were maintained in RPMI-1640 medium supplemented with 10% (v/v) heat-inactivated fetal calf serum, 100 units/ml penicillin, and 100 mg/l streptomycin, at 37°C in a 5% CO_2/95% air (v/v) incubator under 95% relative humidity.

Animal Experiments

Female BALB/C nude mice, aged 6 weeks, were purchased from the Lakes Animal Experimental Center of the Institute of Biochemistry and Cell Biology, Shanghai Institute of Biological Sciences, China. The animal experiments were approved by the regional animal ethics committee and the mice treated in accordance with the international animal ethics guidelines. All of the mice were maintained in the animal facility of School of Basic Medical Sciences, Southeast University, China. Mice were housed up to eight animals per cage in individual ventilation cages and fed with specific pathogen-free mice chow ad libitum. Exponentially growing HeLa cells (density >105 cells/ml) were injected subcutaneously around the right posterior limb rump. When tumor diameters reached 0.5 cm, mice were divided into four groups of eight mice each: (1) sterile 0.9% (w/v) NaCl (control group), (2) 5 μm As_2O_3 dispersed in 0.9% (w/v) NaCl

(As_2O_3 group), (3) 1 mg/ml of Fe_3O_4 dispersed in 0.9% (w/v) NaCl (Fe_3O_4 group), and (4) 1 mg/ml of Fe_3O_4 and 5 µm As2O3 dispersed in 0.9% (w/v) NaCl (As_2O_3/Fe_3O_4 group). The solution/suspension (1, 2, 3, or 4) were directly injected into tumors with a volume equal to half the tumor volume. In our study, we applied the method of multipoint injection, following clockwise, at the 3, 6, 9, and 12 o'clock points. The tumors of the Fe_3O_4 and As_2O_3/Fe_3O_4 groups were exposed to high frequency AMF (f = 230 kHz, I = 20 A) for 30 min, three times at 24 hr intervals [13]. Tumor temperature was measured at multipoint using an infrared thermometer (ZyTemp-TN18 model; China). Mice were sacrificed after 6 weeks, and the mass and volume of each tumor were measured. Tumor growth inhibition was evaluated by measuring mass and volume inhibition proportions. Mass inhibition (IM) was calculated as (1—Relative tumor mass) × 100%, where relative tumor mass (RTM) was the mean tumor mass of the experimental group divided by the mean tumor mass of the control group. Similarly, volume inhibition (IV) was calculated as (1—Relative tumor volume) × 100%, where relative tumor volume (RTV) was the mean tumor volume of the experimental group divided by the mean tumor volume of the control group.

Expression of CD44V6, VEGF-C, and MMP-9 after Thermochemotherapy of HeLa cells

In Vitro Treatment and Sampling

The HeLa cells were seeded in 50 ml culture flasks at 6×10^5 cells per flask. After 24 hr, cells were grown in: (1) RPMI1640 medium containing 10% (v/v) fetal calf serum (positive control group); (2) 5 or 10 µm As_2O_3 (As_2O_3 groups); (3) 1 mg/ml of Fe_3O_4 magnetic nanoparticles with or without AMF exposure (Fe_3O_4 groups); or (4) 1 mg/ml of Fe_3O_4 and 5 or 10 µm As_2O_3 with AMF exposure (As_2O_3/Fe_3O_4 complex groups), with six flasks used to test each of the above conditions. The AMF exposure consisted of electromagnetic exposure for 60 min under high frequency AMF (f = 230 kHz, I = 20 A). All flasks were incubated for 48 hr and cells were thereafter isolated.

RNA Isolation

The RNA was extracted from cells using the RNAiso reagent, according to the supplier's protocol. The purity and concentration of RNA were determined by spectrophotometry at 260 nm. The quality of RNA was checked by electrophoresis of 23 µl samples in 1% (w/v) agarose gels, staining with ethidium bromide, and examining the 28S and 18S rRNA bands under UV light. No significant degradation was observed in any RNA sample.

Semi-quantitative Reverse Transcription Polymerase Chainreaction (RT-PCR)

Total RNA was denatured for 10 min at 70°C. Each RT reaction (20 µl volume) contained 2 µl of total RNA, 1 µl of Oligo dT18, 2 µl of each dNTP (10 mm), 0.5 µl RNasin, 1 µl AMV retroviridase, 4 µl 5 × AMV buffer, and 9.5 µl of DEPC water. Each reaction mixture was incubated for 1 hr at 42°C, and the reverse transcriptase was inactivated by heating at 95°C for 5 min. Each cDNA product was frozen at 20°C until use.

The PCR reactions were performed in 25 µl volumes containing 5 µl cDNA, 2 µl 10× PCR buffer, 2 µl dNTP (2.5 mm), 1 µl sense primers, and 1 µl antisense primers [14-16] (20 pmol/µl,; β-actin primers were used as an internal control), 0.5 µl Taq DNA polymerase (5 U/µl), 2.5 µl MgCl2 (25 mm), and 11 µl DEPC water. After initial denaturation at 96°C for 3 min, the mixtures were subjected to a varying number of denaturation cycles (each at 1 min at 94°C), annealing (1 min at 56°C for β-actin; 57°C for VEGF-C; and 60°C for other genes) and extension (2 min at 72°C), and a final extension at 72°C for 10 min, in a PTC-100 thermal cycler (MJ Research, Watertown, MA). Samples were stored at 20°C. For each reaction, a negative control employed distilled water instead of cDNA, and cDNA from untreated HeLa cells served as the positive control.

Table 1. PCR primers and PCR conditions.

Gene	Primers	Tm (°C)	No. of cycles	Product size (bp)	Reference
CD44v6	5'-GACACATATTGCTTCAATGCTTCAGC-3'	60	35	348	[14]
	5'-TACTAGGAGTTGCTGGATGGTAG-3'				
VEGF-C	5'-AGACTCAATGCATGCCACG-3'	57	35	435	[15]
	5'-TTGAGTCATCTCCAGCATCC-3'				
MMP-9	5'-GTGCTGGGCTGCTGCTTTGCTG-3'	60	35	303	[15]
	5'-GTCGCCCTCAAAGGTTTGGAAT-3'				
β-actin	5'-CGTCTGGACCTGGCTGGCCGGGACC-3'	56	28	600	[16]
	5'-CATGAAGCATTTGCGGTGGACGATG-3'				

Amplified PCR products were electrophoresed with DNA markers, on 2% (w/v) agarose gels containing ethidium bromide. Bands were visualized under UV light and each gel image was captured by a digital camera. Imagetool 2.0 software was used for semi-quantitative analysis of electrophoresis results.

RESULTS AND DISCUSSION

Characteristics of the Nano-sized As$_2$O$_3$/Fe$_3$O$_4$ Complex

The nano-sized As$_2$O$_3$/Fe$_3$O$_4$ complex particles were approximately spherical, uniform in size, and had good dispersibility, with an average diameter by TEM of about 50 nm (Figure 1) [17]. Upon dispersion in 0.9% (w/v) NaCl and exposure to high frequency AMF (output frequency 230 kHz, output current 20 A) for 60 min, the nano-sized As$_2$O$_3$/Fe$_3$O$_4$ complexes increased the temperature of MF from 42 to 65°C, depending on the concentration of Fe$_3$O$_4$ (Figure 2), with a permanent change seen after 40 min of electromagnetic exposure. As a 1 mg/ml concentration of Fe$_3$O$_4$ in the As$_2$O$_3$/Fe$_3$O$_4$ complex increased the MF temperature to 47°C, we chose this Fe$_3$O$_4$ concentration for further experiments.

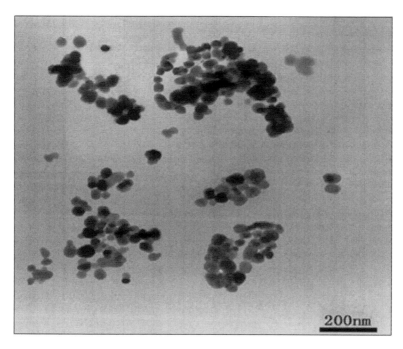

Figure 1. TEM image of the nano-sized As$_2$O$_3$/Fe$_3$O$_4$ complexes.

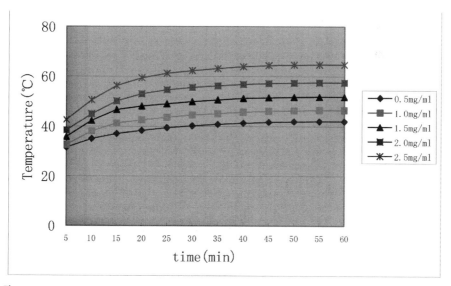

Figure 2. Heating test curve of nano-sized As$_2$O$_3$/Fe$_3$O$_4$ complex fluid in AMF *in vitro*. Output frequency = 230 kHz, output current = 20A; the concentrations of Fe$_3$O$_4$ were 0.5, 1.0, 1.5, 2.0, and 2.5 mg/ml.

In Vivo Inhibitory Effects of the As_2O_3/Fe_3O_4 Complex and AMF Exposure on Xenograft Cervical Cancer in Nude Mice

Following intratumor injection of simple Fe3O4 magnetic nanoparticles or nano-sized As_2O_3/Fe_3O_4 complexes, and exposure to high frequency AMF (f = 230 kHz, I = 20 A) for 30 min, almost the entire tumor got heated by the nanoparticles and the temperature of tumors rose to 4445°C. Compared with control tumors, the tumors of all experimental groups were smaller (Figures 3 and 4). The mass and volume inhibition rates in the As_2O_3/Fe_3O_4 group were 88.21% and 91.57%, respectively, significantly higher than observed in the control, the As_2O_3, and the Fe_3O_4 groups (p < 0.05 for each, Table 2). Histological examination revealed that, in both the As_2O_3/Fe_3O_4 and Fe_3O_4 groups, many black nanoparticles accumulated in the stroma of the tumors, with widespread tumor necrosis surrounding the nanoparticles. The necrotic areas of the As_2O_3 group were larger than those of the control group (Figure 5).

Presently, clinical thermotherapy induces substantial damage to surrounding healthy tissues. We applied a multipoint injection strategy, injecting nanoparticles into the 3, 6, 9, and 12 o'clock points of each tumor, to ensure almost whole tumor to get homogeneously heated and minimize damage to surrounding tissue. Thus, because of the targeted and localized thermogenic activity of MFH, normal tissue without magnetic particles should not be damaged. We found that, because of the dual activity (chemotherapy and hyperthermia) of the nanoparticles, the therapeutic effect of Fe_3O_4 magnetic nano-microspheres containing As_2O_3, in combination with AMF exposure, was greater than that of Fe_3O_4 magnetic nanoparticles combined with AMF exposure, and that of As_2O_3 treatment alone.

Table 2. Extent of volume and mass inhibition of xenograft cervical cancer in nude mice after treatment.

Group	Tumor volume (mm³) ($\bar{X} \pm S, n = 8$)	Tumor mass (g) ($\bar{X} \pm S, n = 8$)	Volume inhibition (%)	Mass inhibition (%)
Control group	931.51 ± 284.26	0.789 ± 0.143	0	0
Single As_2O_3 group	616.94 ± 147.96[1]	0.634 ± 0.056[1]	33.77	19.65
Single Fe_3O_4 magnetic nanoparticles group	132.79 ± 36.30[1]	0.128 ± 0.029[1]	85.74	83.78
Nanosized As_2O_3/Fe_3O_4 complex group	78.50 ± 32.73[1]	0.093 ± 0.028[1]	91.57	88.21

(1) p < 0.05 vs. control group. These measurements were obtained 6 weeks after the start of treatment.

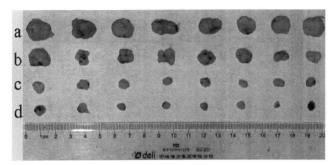

Figure 3. Morphological changes of xenograft tumors in nude mice treated by different methods. (a) control group (sterile 0.9% [w/v] NaCl); (b) As$_2$O$_3$ group (5 μm As$_2$O$_3$ dispersed in 0.9% [w/v] NaCl); (c) Fe$_3$O$_4$ group (1 mg/ml of Fe$_3$O$_4$ magnetic nanoparticles dispersed in 0.9% [w/v] NaCl plus MFH); (d) As$_2$O$_3$/Fe$_3$O$_4$ group (1 mg/ml of Fe$_3$O$_4$ and 5 μm As$_2$O$_2$ dispersed in 0.9% [w/v] NaCl plus MFH). The tumors become smaller in animals treated with protocols "a" to "d". Compared with the control, each group showed significant reductions in tumor volume and mass (p < 0.05 for both).

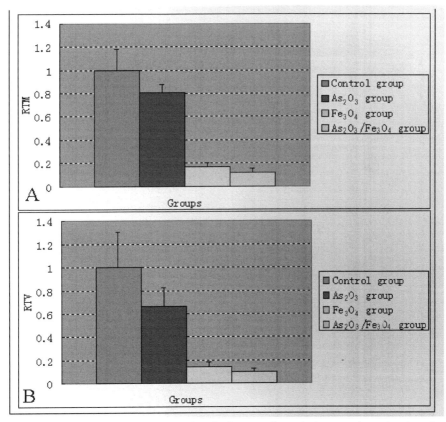

Figure 4. Part (A): Relative tumor mass (RTM) of xenograft cervical cancer in nude mice after treatment. Part (B): Relative tumor volume (RTV) of xenograft cervical cancer in nude mice after treatment. These measurements were obtained 6 weeks after the start of treatment.

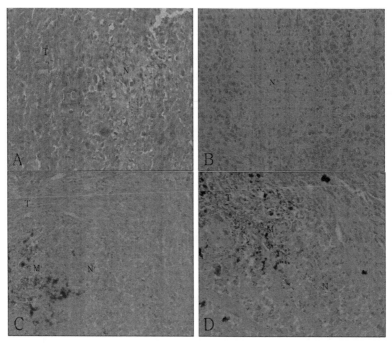

Figure 5. Histological changes in xenograft tumors (H & E) (x200). Sections from the (A) control, (B) As$_2$O$_3$, (C) Fe$_3$O$_4$, and (D) As$_2$O$_3$/Fe$_3$O$_4$ groups are shown (T, tumor tissue; N, necrosis; M, materials of Fe3O4 or As$_2$O$_3$/Fe$_3$O$_4$. Histological examination revealed that the necrotic areas in B, C, and D were much larger than the intrinsic necrosis of the tumor shown in A.

Semi-quantitative RT-PCR

The HeLa cells exposed to free As$_2$O$_3$ or to nano-sized As$_2$O$_3$/Fe$_3$O$_4$ complexes under AMF showed dose-dependent inhibition of CD44v6, VEGF-C, and MMP-9 mRNA synthesis, which indicated that As$_2$O$_3$ can inhibit CD44v6, VEGF-C, and MMP-9 mRNA expression. In contrast, exposure to Fe$_3$O$_4$ magnetic nanoparticles and AMF inhibited expression of VEGF-C, but not of CD44v6 and MMP-9 (Figure 6, Table 3).

Table 3. Results of semi-quantitative RT-PCR analysis.

	Relative gray value (gene/β-actin), ($\bar{X} \pm S$)						
Gene	P	1	2	3	4	5	6
CD44v6	0.386 ± 0.049	0.396 ± 0.052	0.438 ± 0.115	$0.210 \pm 0.075^{(1)}$	$0.244 \pm 0.120^{(1)}$	$0.142 \pm 0.064^{(1)(2)}$	$0.117 \pm 0.030^{(1)(3)}$
VEGF-C	0.605 ± 0.166	0.514 ± 0.083	$0.398 \pm 0.079^{(1)}$	$0.305 \pm 0.132^{(1)(2)}$	$0.344 \pm 0.038^{(1)}$	$0.095 \pm 0.016^{(1)(3)}$	$0.150 \pm 0.036^{(1)}$
MMP-9	0.966 ± 0.205	1.053 ± 0.222	0.997 ± 0.093	$0.461 \pm 0.170^{(1)}$	$0.466 \pm 0.121^{(1)}$	$0.244 \pm 0.107^{(1)(2)}$	$0.220 \pm 0.098^{(1)(3)}$

CD44v6: (1) p <0.05 vs. **P**, (2) p<0.05 vs. **3**, (3) p < 0.05 vs. **4**; VEGF-C: (1) p < 0.05 vs. **P**, (2) p < 0.05 vs. **2** and **4**, (3) p < 0.05 vs. **6**; MMP-9: (1) p < 0.05 vs. **P**, (2) p < 0.05 vs. **3**, (3) p < 0.05 vs. **4**.

P, positive control (untreated HeLa cells); **I**, Fe$_3$O$_4$ without MFH: **2**, Fe$_3$O$_4$ plus MFH; **3**, nanosized As$_2$O$_3$ (5μM)/Fe$_3$O$_4$ plus MFH; **4**, 5 μM As$_2$O$_3$; **5**, nanosized As$_2$O$_3$ (10μM)/Fe3O4 plus MFH group; **6**, 10 μM As$_2$O$_3$.

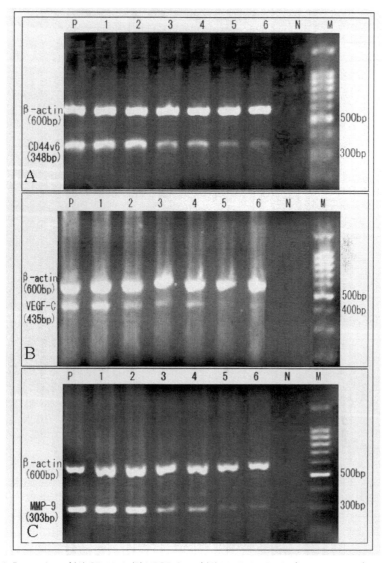

Figure 6. Expression of (A) CD44v6, (B) VEGF-C, and (C) MMP-9 mRNA after treatment of HeLa cells. Lanes: P, positive control (untreated HeLa cells); 1, Fe_3O_4 without MFH; 2, Fe_3O_4 plus MFH; 3, nano-sized As_2O_3 (5 μm)/Fe_3O_4 plus MFH; 4, 5 μm As2O3; 5, nano-sized As_2O_3 (10 μm)/Fe_3O_4 plus MFH; 6, 10 μm As_2O_3;N, negative control; M, DNA marker.

Heat therapies such as hyperthermia and thermoablation are very promising approaches in the treatment of cancer. Compared with available hyperthermia modalities, MFH yields better results in uniform heating of the deeply situated tumors. In this approach, fluid consisting of superparamagnetic nanoparticles (MF) is delivered to the tumor. An AMF is then used to heat the particles and the corresponding tumor, thereby ablating it [18]. Interstitial hyperthermia following direct injection of nanoparticles has also been proven successful in many animal models. Jordan A. et al. reported good effects of MFH on C3H mammary carcinoma *in vivo* [19]. A variety of MFs have been applied using related techniques (differing mostly in the formulation of the nanoparticles), but so far only few of these approaches has been successfully translated from research to clinical stage.

All magnetic nanoparticles used so far *in vivo* are composed of the iron oxides magnetite (Fe_3O_4) and maghemite (γ-Fe_2O_3) due to their low toxicity and their known pathways of metabolism. As demonstrated by X-ray diffraction, Fe_3O_4 nanoparticles are of the spinel type and magnetic Fe_3O_4 [17, 20] can absorb a considerable amount of energy and transform this energy into heat when exposed to AMF. We successfully prepared nano-sized As_2O_3/Fe_3O_4 complexes using co-precipitation and impregnation processes. These particles were successful in the warming of tumors upon exposure to high frequency AMF, suggesting their usefulness in hyperthermia protocols. By adjusting the concentration of Fe_3O_4, we could select a suitable temperature (4246°C) for tumor hyperthermia. At a 1 mg/ml Fe_3O_4 concentration in the As_2O_3/Fe_3O_4 complex, the temperature of the MF rose to 47°C after exposure to high frequency AMF for 30 min.

Nano-sized As_2O_3/Fe_3O_4 complexes, combined with MFH, had a significant therapeutic effect on xenograft tumor in nude mice, by inducing tumor cell apoptosis and inhibiting cell growth. We injected complexes directly into the tumor tissue, rather than into healthy tissue at the tumor boundary, thus delivering nanoparticles to the desired regions. Such nanoparticles may be internalized into tumor cells and organelles [19], and transmission electron microscopy showed that the nano-sized As_2O_3/Fe_3O_4 complex particles were indeed so internalized, both into cells and into organelles such as the nucleoli and lysosomes [13]. Although the mechanism by which nanoparticles enter tumor cells is not yet known, pinocytosis or penetration may be in play, as the nanoparticle diameter is small. Moreover, hyperthermia may increase the permeability of biological membranes. These findings suggest that the As_2O_3/Fe_3O_4 composite nanoparticles may be useful as thermoseeds for localized hyperthermia treatment of cancers, without damaging normal tissue, thus overcoming the limitations of conventional heat treatment. Moreover, by using these complexes, As_2O_3 can be localized to tumor tissue, thus reducing the As_2O_3 toxicity to normal tissue. This method may therefore allow a combination of localized thermogenic and chemotherapeutic activity, even in tumors located deep inside the body, simultaneously minimizing damage caused by heat and toxicity to normal tissue surrounding the tumor. The MFH also can act as a temperature control switch *in vivo* by controlling the concentration of Fe_3O_4, thus minimizing the risk of overheating during therapy. In comparison with traditional hyperthermia, the double function (chemotherapy and hyperthermia) and targeted activity of the combination of nano-sized As_2O_3/Fe_3O_4 complexes and MFH are most important characteristics. Based on the data in Table 2 and Figure 4, although the tumor inhibition appears to be primarily due to the heat, we can still see that the combination was more effective than either Fe_3O_4 magnetic nanoparticles plus MFH, or As_2O_3 treatment alone. In addition, the heat created during thermotherapy may directly damage tumor cells or may make cells more susceptible to accompanying chemotherapy [21].

The MFH has been undergoing clinical testing. In three clinical studies the safety and the feasibility of the technique has been proven. In March 2003, the first clinical feasibility study on magnetic nanoparticle hyperthermia was started with 14 patients suffering from glioblastoma multiforme [22]. All patients received a neuro-navigationally guided injection of the MF into the tumor. The amount of fluid and the spatial distribution of the depots were planned in advance by means of a specially developed treatment planning software following magnetic resonance imaging (MRI). The actually achieved MF distribution was measured by computed tomography (CT).

In prostate cancer, a direct injection technique of MFH was investigated in phase-I-study. The feasibility and good tolerability was shown in the trial [23, 24]. This novel approach requires specific tools for planning, quality control and thermal monitoring, based on appropriate imaging and modelling techniques. Treatment planning was carried out using CT of the prostate. Nanoparticle suspensions were injected transperineally into the prostate under transrectal ultrasound and flouroscopy guidance.

Another prospective feasibility study enrolled 22 patients with proven recurrences and residual tumors (non-resectable and pre-treated, e.g., prostate and cervix carcinoma, soft tissue sarcoma) [25]. Three implantation methods were established: Infiltration under CT fluoroscopy (group A), TRUS (transrectal ultrasound)—guided implantation with X-fluoroscopy (group B) and intra-operative infiltration under visual control (group C). In group A and B the distribution of the nanoparticles can be planned prior to implantation on the basis of three-dimensional image datasets. These approaches of MFH above were well tolerated by patients.

The MFH is suitable for treatment of superficial tumors by direct injection of nanoparticle suspension, which was first reported in 1997 [19]. For deep tumors, the feasibility and good tolerability of different injection techniques (CT, ultrasound or X-fluoroscopy guided) were shown. Endoscope such as Oesophagoscope and vaginoscope can be respectively used as guiding devices for esophageal cancer and cervical cancer. A different approach of using magnetic nanoparticles for heating tumor cells is termed ferromagnetic embolization hyperthermia. This technique, which uses a feeding artery to carry nanoparticles into the tumor, seems to be especially well suited for the treatment of hepatic malignancies due to the differences in blood supply between hepatic tumor cells and normal liver parenchyma. Several preclinical studies on arterial embolization hyperthermia of liver cancer were reported [26-28], and Granov reported this treatment on clinical use [29].

Metastasis, a unique characteristic of malignant tumors, is the most difficult issue in the clinical treatment of cancer and the major cause of death. Hyperthermia may trigger immune responses that inhibit the metastasis of graft tumors [30].

"Cluster of differentiation 44" (CD44) is a cell surface adhesion molecule that recognizes hyaluronate and mediates diverse functions, such as cell-cell and cell-matrix adhesion, lymphocyte homing, and T-cell adhesion and activation. The CD44 exists as a standard form (CD44s) and as multiple isoforms, each generated by alternative splicing of up to 10 variant exons encoding parts of the extracellular domain. The CD44, especially the CD44v6 variant, has a role in tumor progression and metastasis in human cancers [31]. Expression of CD44v6 has been found to correlate significantly with lymphatic and/or hematogenous metastasis. Following the administration of As_2O_3 to SGC9701 cells, expression of CD44 decreased, suggesting that As_2O_3 inhibited CD44 expression [32].

In the VEGF family, VEGF-C is a ligand for VEGF receptor (VEGFR)-3, a tyrosine kinase receptor that is predominantly expressed in the endothelium of lymphatic vessels. The VEGF-C plays an important role in tumor metastasis by mediating the formation of lymphatic vessels. Most cancer cells express VEGF-C, and such expression positively correlates with tumor angiogenesis and metastasis [33]. The As_2O_3 has been reported to inhibit VEGF expression, preventing the proliferation, invasion, and metastasis of solid tumors [34]. In addition, expression of VEGF mRNA and protein by fibrosarcomas was significantly inhibited after thermotherapy at 42°C [35]. Moreover, the density of tumor vessels in gliomas was reduced after thermotherapy and even more after thermochemotherapy, indicating that thermotherapy may inhibit the expression of VEGF mRNA and protein [36].

Matrix MMPs are a family of zinc- and calcium-dependent enzymes that degrade extracellular matrix proteins, such as laminin, fibronectin, and collagen. The MMPs mediate the invasive properties of most malignant cells and promote angiogenesis through their ability to degrade basement membranes and to remodel extracellular matrix architecture. In particular, MMP-9 specifically degrades collagen IV, the major component of the basement membrane. Expression of MMP-9 has been found to correlate with tumor invasiveness and metastasis, including lymph node metastasis [37]. The As_2O_3 reduced the expression of MMP-9 by nasopharyngeal carcinoma cells *in vitro*, and decreased their invasive and metastatic properties [12]. Using As_2O_3/Fe_3O_4 nanoparticles, we found that As_2O_3 dose-dependently inhibited the expression of CD44v6, VEGF-C, and MMP-9 mRNA. Moreover, thermotherapy alone inhibited the expression of VEGF-C mRNA to some extent.

CONCLUSION

The nano-sized As_2O_3/Fe_3O_4 complexes described here possess not only the chemotherapeutic activity of As_2O_3 but also the characteristic of strong magnetic responsiveness arising from the presence of Fe_3O_4. These particles are suitable for treating superficial tumors by direct injection and for deep tumors by using different injection techniques (CT, ultrasound or X-fluoroscopy guided, intraoperatively under visual control), because they absorb energy from high frequency AMF and also have chemotherapeutic effects. For early stage tumors, thermochemotherapy of nano-sized As_2O_3/Fe_3O_4 complexes could be used as preoperative treatment because of reducing tumor volume and intraoperative bleeding, creating better conditions for latter surgery. In addition, it may be inhibit tumor metastasis to some extent.

Most researches of MFH in this field applied so far have only been evaluated in preclinical studies. In spite of some Phase I trials have been carried out, it is still too early to claim therapeutic advantages for the applied method because survival benefit or time to progression were not defined endpoints of the finished feasibility studies. However, the ongoing Phase II trials will provide an initial indication whether MFH can improve survival and/or quality of life.

KEYWORDS

- **Fluid hyperthermia**
- **Hyperthermia treatment**
- **Matrix metalloproteinases**
- **Relative tumor mass**

AUTHORS' CONTRIBUTIONS

Yiqun Du was responsible for experimental design and completion of all laboratory work presented in this article. Dongsheng Zhang contributed to the conception of the study and participated in all stages of the work. Hui Liu and Rensheng Lai helped to plan and coordinate the study and helped draft the manuscript. All authors have read and approved the final manuscript.

ACKNOWLEDGMENTS

The project was supported by National Natural Science Foundation of China (Project code: 30371830, 30770584) and National 863 Plan of China (Project code: 2007AA03Z356).

Permissions

Chapter 1: Nanotechnology was originally published as "Nanotechnology: Looking As We Leap" in *Hood E 2004. Nanotechnology: Looking As We Leap. Environ Health Perspect 112:a740-a749. http://dx.doi.org/10.1289/ehp.112-a740*. Reprinted with permission under the Creative Commons Attribution License or equivalent.

Chapter 2: Cancer Nanotechnology was originally published as "Cancer Nanotechnology" by *US Department of Health and Human Services*. Reprinted with permission under the Creative Commons Attribution License or equivalent.

Chapter 3: Nanotechnology Applied to Biosensors was originally published as *"Recent Advances in Nanotechnology Applied to Biosensors" in ISSN 1424-8220 Sensors 2009, 9, 1033-1053; doi: 10.3390/s90201033*. Reprinted with permission under the Creative Commons Attribution License or equivalent.

Chapter 4: Novel Lipid-coated Magnetic Nanoparticles for *In Vivo* Imaging was originally published as "Formulation of novel lipid-coated magnetic nanoparticles as the probe for in vivo imaging" in *Journal of Biomedical Science 2009, 16:86 doi: 10.1186/1423-0127-16-86*. Reprinted with permission under the Creative Commons Attribution License or equivalent.

Chapter 5: Doxorubicin Release from PHEMA Nanoparticles was originally published as "Real time *in vitro* studies of doxorubicin release from PHEMA nanoparticles" in *Journal of Nanobiotechnology 2009, 7:5 doi: 10.1186/1477-3155-7-5*. Reprinted with permission under the Creative Commons Attribution License or equivalent.

Chapter 6: Monocyte Subset Dynamics in Human Atherosclerosis was originally published as "Monocyte Subset Dynamics in Human Atherosclerosis Can Be Profiled with Magnetic Nano-Sensors" in *PLoS ONE 5:22, 2009*. Reprinted with permission under the Creative Commons Attribution License or equivalent.

Chapter 7: Singlet-fission Sensitizers for Ultra-high Efficiency Excitonic Solar Cells was originally published as "Singlet-Fission Sensitizers for Ultra-High Efficiency Excitonic Solar Cells" in *National Renewable Energy Laboratory, Subcontract Report NREL/SR-520-44685 December 2008*. Reprinted with permission under the Creative Commons Attribution License or equivalent.

Chapter 8: Detection and Selective Destruction of Breast Cancer Cells was originally published as "Anti-HER2 IgY antibody-functionalized single-walled carbon nanotubes for detection and selective destruction of breast cancer cells" in *Biomed Central Ltd 10:2, 2009*. Reprinted with permission under the Creative Commons Attribution License or equivalent.

Chapter 9: Electrochemical Screen-printed (Bio)Sensors was originally published as "Nanotechnology: A Tool for Improved Performance on Electrochemical Screen-Printed (Bio)Sensors" in *Elena Jubete, Oscar A. Loaiza, Estibalitz Ochoteco, Jose A. Pomposo, Hans Grande, and Javier Rodríguez, "Nanotechnology: A Tool for Improved Performance on Electrochemical Screen-Printed (Bio)Sensors," Journal of Sensors, vol. 2009, Article ID 842575, 13 pages, 2009. Doi: 10.1155/2009/842575*. Reprinted with permission under the Creative Commons Attribution License or equivalent.

Chapter 10: Bulk Nanocrystalline Metal Creation was originally published as "Creating Bulk Nanocrystalline Metal" by *Sandia National Laboratories, a multiprogram laboratory operated by Sandia Corporation, a Lockheed Martin Company, for the United States Department of Energy's National Nuclear Security Administration under Contract DE-AC04-94AL85000*. Reprinted with permission under the Creative Commons Attribution License or equivalent.

Chapter 11: Nanoscopic Grooving on Vesicle Walls in Submarine Basaltic Glass was originally published as "The Origin of Nanoscopic Grooving on Vesicle Walls in Submarine Basaltic Glass:

Implications for Nanotechnology" in *Journal of Nanomaterials Volume 2009, Article ID 309208, doi: 10.1155/2009/309208*. Reprinted with permission under the Creative Commons Attribution License or equivalent.

Chapter 12: Thermochemotherapy Effect of Nanosized As_2O_3/Fe_3O_4 Complex was originally published as "Thermochemotherapy effect of nanosized As_2O_3/Fe_3O_4 complex on experimental mouse tumors and its influence on the expression of CD44v6, VEGF-C and MMP-9" in *BioMed Central Ltd. 10:5, 2009*. Reprinted with permission under the Creative Commons Attribution License or equivalent.

References

3

1. Turner, A. P. F., Karube, I., and Wilson, G. S. (1987). *BiosensorsFundamentals and Applications*. Oxford University Press, New York, NY, USA, pp. 719800.

2. Kricka, L. J. (1988). Molecular and ionic recognition by biological systems. In *Chemical Sensors*. T. E. Edmonds (Ed.). Blackie and Sons, Glasgow, UK, pp. 314.

3. Buch, R. M. and Rechnitz, G. A. (1989). Intact chemoreceptor-based biosensors: Responses and analytical limits. *Biosensors* 4, 215230.

4. Zhang, Y., Yang, M., Portney, N. G., Cui, D., Budak, G., Ozbay, E., Ozkan, M., and Ozkan, C. S. (2008). Zeta potential: A surface electrical characteristic to probe the interaction of nanoparticles with normal and cancer human breast epithelial cells. *Biomed. Microdev*. 10, 321328.

5. Cui, D. (2007). Advances and prospects on biomolecules functionalized carbon nanotubes. *J. Nanosci. Nanotechnol*. 7, 12981314.

6. Pan, B., Cui, D., Ozkan, C. S., Ozkan, M., Xu, P., Huang, T., Liu, F., Chen, H., Li, Q., He, R., and Gao, F. (2008). Effects of carbon nanotubes on photoluminescence properties of quantum dots. *J. Phys. Chem. C* 112, 939944.

7. Pan, B., Cui, D., Xu, P., Li, Q., Huang, T., He, R., and Gao, F. (2007). Study on interaction between gold nanorod and bovine serum albumin. *Colloids Surface A* 295, 217222.

8. Cui, D., Tian, F., Coyer, S. R., Wang, J., Pan, B., Gao, F., He, R., and Zhang, Y. (2007). Effects of antisense-myc-conjugated single-walled carbon nanotubes on HL-60 cells. *J. Nanosci. Nanotechnol*. 7, 16391646.

9. Pan, B., Cui, D., Sheng, Y., Ozkan, C., Gao, F., He, R., Li, Q., Xu, P., and Huang, T. (2007). Dendrimer-modified magnetic nanoparticles enhance efficiency of gene delivery system. *Cancer Res*. 67, 81568163.

10. You, X., He, R., Gao, F., Shao, J., Pan, B., and Cui, D. (2007). Hydrophilic high-luminescent magnetic nanocomposites. *Nanotechnology* 18, 035701:1035701:5.

11. Yang, D. and Cui, D. (2008). Advances and Prospects of Gold Nanorods. *Chem. Asian J*. 3, 20102022.

12. Bryant, G. W., de Abajo, F. J. G., and Aizpurua, J. (2008). Mapping the plasmon resonances of metallic nanoantennas. *Nano Lett*. 8, 631636.

13. Patolsky, F., Timko, B. P., Zheng, G., and Lieber, C. M. (2007). Nanowire-based nanoelectronic devices in the life sciences. *MRS Bull*. 32, 142149.

14. Jena, B. K. and Raj, C. R. (2008). Optical sensing of biomedically important polyionic drugs using nano-sized gold particles. *Biosens. Bioelectron*. 23, 12851290.

15. Zhao, W., Chiuman, W., Lam, J. C. F., Brook, M. A., and Li, Y. (2007). Simple and rapid colorimetric enzyme sensing assays using non-crosslinking gold nanoparticle aggregation. *Chem. Commun*. 37293731.

16. Wei, H., Li, B., Li, J., Wang, E., and Dong, S. (2007). Simple and sensitive aptamer-based colorimetric sensing of protein using unmodified gold nanoparticle probes. *Chem. Commun*. 3735–3737.

17. Liang, K. Z., Qi, J. S., Mu, W. J., and Chen, Z. G. (2008). Biomolecules/gold nanowires-doped sol-gel film for label-free electrochemical immunoassay of testosterone. *J. Biochem. Biophys. Meth*. 70, 11561162.

18. He, X., Yuan, R., Chai, Y., and Shi, Y. (2008). A sensitive amperometric immunosensor for carcinoembryonic antigen detection with porous nanogold film and nano-Au/chitosan composite as immobilization matrix. *J. Biochem. Biophys. Meth*. 70, 823829.

19. Chai, R., Yuan, R., Chai, Y., Ou, C., Cao, S., and Li, X. (2008). Amperometric immunosensors based on layer-by-layer assembly of gold nanoparticles and methylene blue on thiourea modified glassy carbon electrode

for determination of human chorionic go-nadotrophin. *Talanta* 74, 13301336.

20. Li, N. B., Park, J. H., Park, K., Kwon, S. J., Shin, H., and Kwak, J. (2008). Characterization and electrocatalytic properties of Prussian blue electrochemically deposited on nano-Au/PAMAM dendrimer-modified gold electrode. *Biosens. Bioelectron.* **23**, 15191526.

21. Gao, F., Yuan, R., Chai, Y., Chen, S., Cao, S., and Tang, M. (2007). Amperometric hydrogen peroxide biosensor based on the immobilization of HRP on nano-Au/Thi/poly (p-aminobenzene sulfonic acid)-modified glassy carbon electrode. *J. Biochem. Biophys. Meth.* **70**, 407413.

22. Shi, A. W., Qu, F. L., Yang, M. H., Shen, G. L., and Yu, R. Q. (2008). Amperometric H2O2 biosensor based on poly-thionine nanowire/HRP/nano-Au-modified glassy carbon electrode. *Sens. Actuat. B* **129**, 779783.

23. Deng, L., Wang, Y., Shang, L., Wen, D., Wang, F., and Dong, S. (2008). A sensitive NADH and glucose biosensor tuned by visible light based on thionine bridged carbon nanotubes and gold nanoparticles multilayer. *Biosens. Bioelectron.* **24**, 957963.

24. Cui, R., Huang, H., Yin, Z., Gao, D., and Zhu, J. J. (2008). Horseradish peroxidase-functionalized gold nanoparticle label for amplified immunoanalysis based on gold nanoparticles/carbon nanotubes hybrids modified biosensor. *Biosens. Bioelectron.* **23**, 16661673.

25. Wang, M. and Li, Z. (2008). Nano-composite ZrO2/Au film electrode for voltammetric detection of parathion. *Sens. Actuat. B* **133**, 607612.

26. Wasowicz, M., Viswanathan, S., Dvornyk, A., Grzelak, K., Kłudkiewicz, B., and Radecka, H. (2008). Comparison of electrochemical immunosensors based on gold nano materials and immunoblot techniques for detection of histidine-tagged proteins in culture medium. *Biosens. Bioelectron.* **24**, 284289.

27. Nusz, G. J., Marinakos, S. M., Curry, A. C., Dahlin, A., Hook, F., Wax, A., and Chilkoti, A. (2008). Label-free plasmonic detection

of biomolecular binding by a single gold nanorod. *Anal. Chem.* **80**, 984989.

28. York, J., Spetzler, D., Xiong, F., and Frasch, W. D. (2008). Single-molecule detection of DNA via sequence-specific links between F 1-ATPase motors and gold nanorod sensors. *Lab Chip* **8**, 415419.

29. Pan, B., Cui, D., He, R., Gao, F., and Zhang, Y. (2006). Covalent attachment of quantum dot on carbon nanotubes. *Chem. Phys. Lett.* **417**, 419–424.

30. Cui, D., Tian, F., Kong, Y., Titushikin, I., and Gao, H. (2004). Effects of single-walled carbon nanotubes on the polymerase chain reaction. *Nanotechnology* **15**, 154157.

31. Joshi, P. P., Merchant, S. A., Wang, Y., and Schmidtke, D. W. (2005). Amperometric biosensors based on redox polymer-carbon nanotube-enzyme composites. *Anal. Chem.* **77**, 31833188.

32. Teh, K. S. and Lin, L. (2005). MEMS sensor material based on polypyrrole-carbon nanotube nanocomposite: Film deposition and characterization. *J. Micromech. Microeng.* **15**, 20192027.

33. Qi, P., Vermesh, O., Grecu, M., Javey, A., Wang, Q., Dai, H., Peng, S., and Cho, K. J. (2003). Toward large arrays of multiplex functionalized carbon nanotube sensors for highly sensitive and selective molecular detection. *Nano Lett.* **3**, 347351.

34. Li, G., Xu, H., Huang, W., Wang, Y., Wu, Y., and Parajuli, R. (2008). A pyrrole quinoline quinone glucose dehydrogenase biosensor based on screen-printed carbon paste electrodes modified by carbon nanotubes. *Meas. Sci. Technol.* 1926.

35. Chen, X., Chen, J., Deng, C., Xiao, C., Yang, Y., Nie, Z., and Yao, S. (2008). Amperometric glucose biosensor based on boron-doped carbon nanotubes modified electrode. *Talanta* **76**, 763–767.

36. Chen, R. J., Bangsaruntip, S., Drouvalakis, K. A., Wong Shi Kam, N., Shim, M., Li, Y., Kim, W., Utz, P. J., and Dai, H. (2003). Noncovalent functionalization of carbon nanotubes for highly specific electronic biosensors. *Proc. Natl. Acad. Sci. USA* **100**, 49844989.

37. Zeng, J., Wei, W., Liu, X., Wang, Y., and Luo, G. (2008). A simple method to

fabricate a Prussian Blue nanoparticles/carbon nanotubes/poly(1,2-diaminobenzene) based glucose biosensor. *Microchim. Acta* **160**, 261267.

38. Wu, L., Lei, J., Zhang, X., and Ju, H. (2008). Biofunctional nanocomposite of carbon nanofiber with water-soluble porphyrin for highly sensitive ethanol biosensing. *Biosens. Bioelectron.* **24**, 644–649.

39. Bai, H. P., Lu, X. X., Yang, G. M., and Yang, Y. H. (2008). Hydrogen peroxide biosensor based on electrodeposition of zinc oxide nanoflowers onto carbon nanotubes film electrode. *Chin. Chem. Lett.* **19**, 314318.

40. Zhao, Y., Liu, H., Kou, Y., Li, M., Zhu, Z., and Zhuang, Q. (2007). Structural and characteristic analysis of carbon nanotubes-ionic liquid gel biosensor. *Electrochem. Commun.* **9**, 24572462.

41. Muguruma, H., Shibayama, Y., and Matsui, Y. (2008). An amperometric biosensor based on a composite of single-walled carbon nanotubes, plasma-polymerized thin film, and an enzyme. *Biosens. Bioelectron.* **23**, 827832.

42. Abe, M., Murata, K., Ataka, T., and Matsumoto, K. (2008). Calibration method for a carbon nanotube field-effect transistor biosensor. *Nanotechnology* **19**, 045505:1045505:4.

43. Weeks, M. L., Rahman, T., Frymier, P. D., Islam, S. K., and McKnight, T. E. (2008). A reagentless enzymatic amperometric biosensor using vertically aligned carbon nanofibers (VACNF). *Sens. Actuat. B* **133**, 5359.

44. Cui, D., Pan, B., Zhang, H., Gao, F., Wu, R., Wang, J., He, R., and Asahi, T. (2008). Self-assembly of quantum dots and carbon nanotubes for ultrasensitive DNA and antigen detection. *Anal. Chem.* **80**, 79968001.

45. Kim, D. H., Lee, S. H., Kim, K. N., Kim, K. M., Shim, I. B., and Lee, Y. K. (2005). Cytotoxicity of ferrite particles by MTT and agar diffusion methods for hyperthermic application. *J. Magn. Magn. Mater.* **293**, 287292.

46. Lee, H., Lee, E., Kim, D. K., Jang, N. K., Jeong, Y. Y., and Jon, S. (2006). Antibiofouling polymer-coated superparamagnetic iron oxide nanoparticles as potential magnetic resonance contrast agents for *in vivo* cancer imaging. *J. Am. Chem. Soc.* **128**, 73837389.

47. Ito, A., Ino, K., Kobayashi, T., and Honda, H. (2005). The effect of RGD peptide-conjugated magnetite cationic liposomes on cell growth and cell sheet harvesting. *Biomaterials* **26**, 61856193.

48. Sincai, M., Ganga, D., Ganga, M., Argherie, D., and Bica, D. (2005). Antitumor effect of magnetite nanoparticles in cat mammary adenocarcinoma. *J. Magn. Magn. Mater.* **293**, 438441.

49. Morishita, N., Nakagami, H., Morishita, R., Takeda, S. I., Mishima, F., Terazono, B., Nishijima, S., Kaneda, Y., and Tanaka, N. (2005). Magnetic nanoparticles with surface modification enhanced gene delivery of HVJ-E vector. *Biochem. Biophys. Res. Commun.* **334**, 11211126.

50. Guedes, M. H. A., Sadeghiani, N., Peixoto, D. L. G., Coelho, J. P., Barbosa, L. S., Azevedo, R. B., Kückelhaus, S., Da Silva, M. D. F., Morais, P. C., and Lacava, Z. G. M. (2005). Effects of AC magnetic field and carboxymethyldextran-coated magnetite nanoparticles on mice peritoneal cells. *J. Magn. Magn. Mater.* **293**, 283286.

51. Rife, J. C., Miller, M. M., Sheehan, P. E., Tamanaha, C. R., Tondra, M., and Whitman, L. J. (2003). Design and performance of GMR sensors for the detection of magnetic microbeads in biosensors. *Sens. Actuat. A* **107**, 209218.

52. Zhang, H. L., Lai, G. S., Han, D. Y., and Yu, A. M. (2008). An amperometric hydrogen peroxide biosensor based on immobilization of horseradish peroxidase on an electrode modified with magnetic dextran microspheres. *Anal. Bioanal. Chem.* **390**, 971977.

53. Lai, G. S., Zhang, H. L., and Han, D. Y. (2008). A novel hydrogen peroxide biosensor based on hemoglobin immobilized on magnetic chitosan microspheres modified electrode. *Sens. Actuat. B* **129**, 497503.

54. Janssen, X. J. A., Schellekens, A. J., van Ommering, K., van Ijzendoorn, L. J., and Prins, M. W. J. Controlled torque on superparamagnetic beads for functional biosensors. *Biosens. Bioelectron.* (in press), doi:10.1016/j.bios.2008.09.024.

55. Loh, K. S., Lee, Y. H., Musa, A., Salmah, A. A., and Zamri, I. (2008). Use of Fe3O4 nanoparticles for enhancement of biosensor response to the herbicide 2,4-dichlorophen-oxyacetic acid. *Sensors* **8**, 57755791.

56. Tamanaha, C. R., Mulvaney, S. P., Rife, J. C., and Whitman, L. J. (2008). Magnetic labeling, detection, and system integration. *Biosens. Bioelectron.* **24**, 113.

57. Park, S. H., Soh, K. S., Hwang, D. G., Rhee, J. R., and Lee, S. S. (2008). Detection of magnetic nanoparticles and Fe-hemoglobin inside red blood cells by using a highly sensitive spin valve device. *J. Magnetics* **13**, 3033.

58. Mujika, M., Arana, S., Castaño, E., Tijero, M., Vilares, R., Ruano-López, J. M., Cruz, A., Sainz, L., and Berganza, J. (2009). Magnetoresistive immunosensor for the detection of *Escherichia coli* O157:H7 including a microfluidic network. *Biosens. Bioelectron.* **24**, 12531258.

59. Jun, Y. W., Seo, J. W., and Cheon, J. (2008). Nanoscaling laws of magnetic nanoparticles and their applicabilities in biomedical sciences. *Acc. Chem. Res.* **41**, 179189.

60. Jeong, H., Chang, A. M., and Melloch, M. R. (2001). The Kondo effect in an artificial quantum dot molecule. *Science* **293**, 22212223.

61. Li, X., Wu, Y., Steel, D., Gammon, D., Stievater, T. H., Katzer, D. S., Park, D., Piermarocchi, C., and Sham, L. J. (2003). An all-optical quantum gate in a semiconductor quantum dot. *Science* **301**, 809811.

62. Lu, W., Ji, Z., Pfeiffer, L., West, K. W., and Rimberg, A. J. (2003). Real-time detection of electron tunnelling in a quantum dot. *Nature* **423**, 422425.

63. Kaul, Z., Yaguchi, T., Kaul, S. C., Hirano, T., Wadhwa, R., and Taira, K. (2003). Mortalin imaging in normal and cancer cells with quantum dot immuno-conjugates. *Cell Res.* **13**, 503507.

64. Medintz, I. L., Uyeda, H. T., Goldman, E. R., and Mattoussi, H. (2005). Quantum dot bioconjugates for imaging, labelling and sensing. *Nat. Mater.* **4**, 435446.

65. Hoshino, A., Fujioka, K., Manabe, N., Yamaya, S. I., Goto, Y., Yasuhara, M., and Yamamoto, K. (2005). Simultaneous multicolor detection system of the single-molecular microbial antigen with total internal reflection fluorescence microscopy. *Microbiol. Immunol.* **49**, 461470.

66. Han, M., Gao, X., Su, J. Z., and Nie, S. (2001). Quantum-dot-tagged microbeads for multiplexed optical coding of biomolecules. *Nat. Biotechnol.* **19**, 631635.

67. Jares-Erijman, E. A. and Jovin, T. M. (2003). FRET imaging. *Nat. Biotechnol.* **21**, 13871395.

68. Huang, X., Li, L., Qian, H., Dong, C., and Ren, J. (2006). A resonance energy transfer between chemiluminescent donors and luminescent quantum-dots as acceptors (CRET). *Angew. Chem., Int. Ed.* **45**, 51405143.

69. Du, D., Chen, S., Song, D., Li, H., and Chen, X. (2008). Development of acetylcholinesterase biosensor based on CdTe quantum dots/gold nanoparticles modified chitosan microspheres interface. *Biosens. Bioelectron.* **24**, 475479.

70. Deng, Z., Zhang, Y., Yue, J., Tang, F., and Wei, Q. (2007). Green and orange CdTe quantum dots as effective pH-sensitive fluorescent probes for dual simultaneous and independent detection of viruses. *J. Phys. Chem. B* **111**, 1202412031.

71. Ansari, A. A., Kaushik, A., Solanki, P. R., and Malhotra, B. D. (2008). Sol-gel derived nanoporous cerium oxide film for application to cholesterol biosensor. *Electrochem. Commun.* **10**, 12461249.

72. Zhang, F., Ulrich, B., Reddy, R. K., Venkatraman, V. L., Prasad, S., Vu, T.Q., and Hsu, S.T. (2008). Fabrication of submicron IrO2 nanowire array biosensor platform by conventional complementary metal-oxide-semiconductor process. *Jpn. J. Appl. Phys., Pt. 1* **47**, 11471151.

73. Ghanbari, K., Bathaie, S. Z., and Mousavi, M. F. (2008). Electrochemically fabricated polypyrrole nanofiber-modified electrode as a new electrochemical DNA biosensor. *Biosens. Bioelectron.* **23**, 18251831.

74. Shan, D., Zhu, M., Han, E., Xue, H., and Cosnier, S. (2007). Calcium carbonate nanoparticles: A host matrix for the construction of highly sensitive amperometric

phenol biosensor. *Biosens. Bioelectron.* **23**, 648654.

75. Ekanayake, E. M. I. M., Preethichandra, D. M. G., and Kaneto, K. (2007). Polypyrrole nanotube array sensor for enhanced adsorption of glucose oxidase in glucose biosensors. *Biosens. Bioelectron.* **23**, 107113.

76. Dai, Z., Bai, H., Hong, M., Zhu, Y., Bao, J., and Shen, J. (2008). A novel nitrite biosensor based on the direct electron transfer of hemoglobin immobilized on CdS hollow nanospheres. *Biosens. Bioelectron.* **23**, 18691873.

77. Li, J. and Lin, X. (2007). Simultaneous determination of dopamine and serotonin on gold nanocluster/overoxidized-polypyrrole composite modified glassy carbon electrode. *Sens. Actuat. B* **124**, 486493.

78. Miao, X. M., Yuan, R., Chai, Y. Q., Shi, Y. T., and Yuan, Y. Y. (2008). Direct electrocatalytic reduction of hydrogen peroxide based on Nafion and copper oxide nanoparticles modified Pt electrode. *J. Electroanal. Chem.* **612**, 157163.

79. Liu, A. (2008). Towards development of chemosensors and biosensors with metal-oxide-based nanowires or nanotubes. *Biosens. Bioelectron.* **24**, 167177.

80. Cheng, J., Di, J., Hong, J., Yao, K., Sun, Y., Zhuang, J., Xu, Q., Zheng, H., and Bi, S. (2008). The promotion effect of titania nanoparticles on the direct electrochemistry of lactate dehydrogenase sol-gel modified gold electrode. *Talanta* **76**, 10651069.

81. Xia, C., Wang, N., Lidong, L., and Lin, G. (2008). Synthesis and characterization of waxberry-like microstructures ZnO for biosensors. *Sens. Actuat. B* **129**, 268273.

82. Huang, Z. B., Yan, D. H., Yang, M., Liao, X. M., Kang, Y. Q., Yin, G. F., Yao, Y. D. and Hao, B. Q. (2008). Preparation and characterization of the biomineralized zinc oxide particles in spider silk peptides. *J. Colloid Interface Sci.* **325**, 356362.

83. Lu, X., Zhang, H., Ni, Y., Zhang, Q., and Chen, J. (2008). Porous nanosheet-based ZnO microspheres for the construction of direct electrochemical biosensors. *Biosens. Bioelectron.* **24**, 9398.

84. Chen, Y., Wang, X., Hong, M., Erramilli, S., and Mohanty, P. (2008). Surface-modified

silicon nano-channel for urea sensing. *Sens. Actuat. B* **133**, 593598.

85. Zhang, Z. L., Asano, T., Uno, H., Tero, R., Suzui, M., Nakao, S., Kaito, T., Shibasaki, K., Tominaga, M., Utsumi, Y., Gao, Y. L., and Urisu, T. (2008). Fabrication of Si-based planar type patch clamp biosensor using silicon on insulator substrate. *Thin Solid Films* **516**, 28132815.

86. Wang, D., Sun, G., Xiang, B., and Chiou, B. S. (2008). Controllable biotinylated poly(ethylene-co-glycidyl methacrylate) (PE-co-GMA) nanofibers to bind streptavidin-horseradish peroxidase (HRP) for potential biosensor applications. *Eur. Polym. J.* **44**, 20322039.

87. Kusakari, A., Izumi, M., and Ohnuki, H. (2008). Preparation of an enzymatic glucose sensor based on hybrid organic-inorganic Langmuir-Blodgett films: Adsorption of glucose oxidase into positively charged molecular layers. *Colloids Surf. A* **321**, 4751.

88. Kim, G. Y., Kang, M. S., Shim, J., and Moon, S. H. (2008). Substrate-bound tyrosinase electrode using gold nanoparticles anchored to pyrroloquinoline quinone for a pesticide biosensor. *Sens. Actuat. B* **133**, 14.

89. Lu, B. W. and Chen, W. C. (2006). A disposable glucose biosensor based on drop-coating of screen-printed carbon electrodes with magnetic nanoparticles. *J. Magn. Magn. Mater.* **304**, e400402.

90. Yu, J., Yu, D., Zhao, T., and Zeng, B. (2008). Development of amperometric glucose biosensor through immobilizing enzyme in a Pt nanoparticles/mesoporous carbon matrix. *Talanta* **74**, 15861591.

91. Manesh, K. M., Kim, H. T., Santhosh, P., Gopalan, A. I., and Lee, K. P. (2008). A novel glucose biosensor based on immobilization of glucose oxidase into multiwall carbon nanotubes-polyelectrolyte-loaded electrospun nanofibrous membrane. *Biosens. Bioelectron.* **23**, 771779.

92. Zou, Y., Xiang, C., Sun, L. X., and Xu, F. (2008). Glucose biosensor based on electrodeposition of platinum nanoparticles onto carbon nanotubes and immobilizing enzyme with chitosan-SiO2 sol-gel. *Biosens. Bioelectron.* **23**, 10101016.

93. Cao, C., Kim, J. H., Yoon, D., Hwang, E. S., Kim, Y. J., and Baik, S. (2008). Optical detection of DNA hybridization using absorption spectra of single-walled carbon nanotubes. *Mater. Chem. Phys.* **112**, 738741.

94. Zhang, W., Yang, T., Huang, D., Jiao, K., and Li, G. (2008). Synergistic effects of nano-ZnO/multi-walled carbon nanotubes/chitosan nanocomposite membrane for the sensitive detection of sequence-specific of PAT gene and PCR amplification of NOS gene. *J. Membr. Sci.* **325**, 245251.

95. Zhang, W., Yang, T., Huang, D. M., and Jiao, K. (2008). Electrochemical sensing of DNA immobilization and hybridization based on carbon nanotubes/nano zinc oxide/chitosan composite film. *Chin. Chem. Lett.* **19**, 589591.

96. Galandova, J., Ziyatdinova, G., and Labuda, J. (2008). Disposable electrochemical biosensor with multiwalled carbon nanotubesChitosan composite layer for the detection of deep DNA damage. *Anal. Sci.* **24**, 711716.

97. García, T., Casero, E., Revenga-Parra, M., Martín-Benito, J., Pariente, F., Vázquez, L., and Lorenzo, E. (2008). Architectures based on the use of gold nanoparticles and ruthenium complexes as a new route to improve genosensor sensitivity. *Biosens. Bioelectron.* **24**, 184190.

98. McKenzie, F., Faulds, K., and Graham, D. (2007). Sequence-specific DNA detection using high-affinity LNA-functionalized gold nanoparticles. *Small* **3**, 18661868.

99. Ma, Y., Jiao, K., Yang, T., and Sun, D. (2008). Sensitive PAT gene sequence detection by nano-SiO2/p-aminothiophenol self-assembled films DNA electrochemical biosensor based on impedance measurement. *Sens. Actuat. B* **131**, 565571.

100. Maki, W. C., Mishra, N. N., Cameron, E. G., Filanoski, B., Rastogi, S. K., and Maki, G. K. (2008). Nanowire-transistor based ultra-sensitive DNA methylation detection. *Biosens. Bioelectron.* **23**, 780–787.

101. Vamvakaki, V. and Chaniotakis, N. A. (2007). Pesticide detection with a liposome-based nano-biosensor. *Biosens. Bioelectron.* **22**, 28482853.

102. Zhang, W., Tang, H., Geng, P., Wang, Q., Jin, L., and Wu, Z. (2007). Amperometric method for rapid detection of *Escherichia coli* by flow injection analysis using a bismuth nano-film modified glassy carbon electrode. *Electrochem. Commun.* **9**, 833838.

103. Seo, S., Dobozi-King, M., Young, R. F., Kish, L. B., and Cheng, M. (2008). Patterning a nanowell sensor biochip for specific and rapid detection of bacteria. *Microelectron. Eng.* **85**, 14841489.

104. Medley, C. D., Smith, J. E., Tang, Z., Wu, Y., Bamrungsap, S., and Tan, W. (2008). Gold nanoparticle-based colorimetric assay for the direct detection of cancerous cells. *Anal. Chem.* **80**, 10671072.

105. Prow, T.W., Bhutto, I., Grebe, R., Uno, K., Merges, C., McLeod, D. S., and Lutty, G. A. (2008). Nanoparticle-delivered biosensor for reactive oxygen species in diabetes. *Vision Res.* **48**, 478485.

106. Li, B., Du, Y., Wei, H., and Dong, S. (2007). Reusable, label-free electrochemical aptasensor for sensitive detection of small molecules. *Chem. Commun.* 37803782.

107. Kerman, K., Saito, M., Tamiya, E., Yamamura, S., and Takamura, Y. (2008). Nanomaterial-based electrochemical biosensors for medical applications. *TrAC-Trend. Anal. Chem.* **27**, 585–592.

4

1. Grodzinski, P., Silver, M., and Molnar, L. K. (2006). Nanotechnology for cancer diagnostics: Promises and challenges. *Expert Rev. Mol. Diagn.* **6**, 307–318.

2. Gupta, A. K. and Gupta, M. (2005). Synthesis and surface engineering of iron oxide nanoparticles for biomedical applications. *Biomaterials* **26**, 3995–4021.

3. Norman, A. B., Thomas, S. R., Pratt, R. G., Lu, S. Y., and Norgren, R. B. (1992). Magnetic resonance imaging of neural transplants in rat brain using a superparamagnetic contrast agent. *Brain Res.* **594**, 279–283.

4. Hawrylak, N., Ghosh, P., Broadus, J., Schlueter, C., Greenough, W. T., and Lauterbur, P. C. (1993). Nuclear magnetic resonance

(NMR) imaging of iron oxide-labeled neural transplants. *Exp. Neurol.* **121**, 181–92.

5. Flexman, J. A., Minoshima, S., Kim, Y., and Cross, D. J. (2006). Magneto-optical labeling of fetal neural stem cells for *in vivo* MRI tracking. *Conf. Proc. IEEE Eng. Med. Biol. Soc.* **1**, 5631–5634.

6. Lewin, M., Carlesso, N., Tung, C. H., Tang, X. W., Cory, D., Scadden, D. T., and Weissleder, R. (2000). Tat peptide-derivatized magnetic nanoparticles allow *in vivo* tracking and recovery of progenitor cells. *Nat. Biotechnol.* **18**, 410–414.

7. Bulte, J. W., Douglas, T., Witwer, B., Zhang, S. C., Strable, E., Lewis, B. K., Zywicke, H., Miller, B., van Gelderen, P., Moskowitz, B. M., Duncan, I. D., and Frank, J. A. (2001). Magnetodendrimers allow endosomal magnetic labeling and *in vivo* tracking of stem cells. *Nat. Biotechnol.* **19**, 1141–1147.

8. Hoehn, M., Küstermann, E., Blunk, J., Wiedermann, D., Trapp, T., Wecker, S., Föcking, M., Arnold, H., Hescheler, J., Fleischmann, B. K., Schwindt, W., and Bührle, C. (2002). Monitoring of implanted stem cell migration *in vivo*: A highly resolved *in vivo* magnetic resonance imaging investigation of experimental stroke in rat. *Proc. Natl. Acad. Sci. USA* **99**, 16267–16272.

9. Kircher, M. F., Allport, J. R., Graves, E. E., Love, V., Josephson, L., Lichtman, A. H., and Weissleder, R. (2003). *In vivo* high resolution three-dimensional imaging of antigen-specific cytotoxic T-lymphocyte trafficking to tumors. *Cancer Res.* **63**, 6838–6846.

10. Arbab, A. S., Yocum, G. T., Wilson, L. B., Parwana, A., Jordan, E. K., Kalish, H., and Frank, J. A. (2004). Comparison of transfection agents in forming complexes with ferumoxides, cell labeling efficiency, and cellular viability. *Mol. Imaging* **3**, 24–32.

11. Berger, C., Rausch, M., Schmidt, P., and Rudin, M. (2006). Feasibility and limits of magnetically labeling primary cultured rat T cells with ferumoxides coupled with commonly used transfection agents. *Mol. Imaging* **5**, 93–104.

12. Matuszewski, L., Persigehl, T., Wall, A., Schwindt, W., Tombach, B., Fobker, M., Poremba, C., Ebert, W., Heindel, W., and Bremer, C. (2005). Cell tagging with clinically approved iron oxides: Feasibility and effect of lipofection, particle size, and surface coating on labeling efficiency. *Radiology* **235**, 155–161.

13. Miyoshi, S., Flexman, J. A., Cross, D. J., Maravilla, K. R., Kim, Y., Anzai, Y., Oshima, J., and Minoshima, S. (2005). Transfection of neuroprogenitor cells with iron nanoparticles for magnetic resonance imaging tracking: Cell viability, differentiation, and intracellular localization. *Mol. Imaging Biol.* **7**, 286–295.

14. Montet-Abou, K., Montet, X., Weissleder, R., and Josephson, L. (2007). Cell internalization of magnetic nanoparticles using transfection agents. *Mol. Imaging* **6**, 1–9.

15. Hong, R. L., Huang, C. J., Tseng, Y. L., Pang, V. F., Chen, S. T., Liu, J. J., and Chang, F. H. (1999). Direct comparison of liposomal doxorubicin with or without polyethylene glycol coating in C-26 tumor-bearing mice: Is surface coating with polyethylene glycol beneficial? *Clin. Cancer Res.* **5**, 3645–52.

16. Torchilin, V. (2007). Micellar nanocarriers: Pharmaceutical perspectives. *Pharm. Res.* **24**, 1–16.

17. Simberg, D., Danino, D., Talmon, Y., Minsky, A., Ferrari, M. E., Wheeler, C. J., and Barenholz, Y. (2001). Phase behavior, DNA ordering, and size instability of cationic lipoplexes. *J. Biol. Chem.* **276**, 47453–47459.

18. Lukyanov, A. N., Gao, Z., Mazzola, L., and Torchilin, V. P. (2002). Polyethylene glycol-diacyllipid micelles demonstrate increased acculumation in subcutaneous tumors in mice. *Pharm. Res.* **19**, 1424–1429.

19. Riboni, L., Prinetti, A., Bassi, R., Caminiti, A., and Tettamant, G. (1995). A mediator role of ceramide in the regulation of neuroblastoma neuro2a cell differentiation. *J. Biol. Chem.* **270**, 26868–26875.

20. Sato, C., Matsuda, T., and Kitajima, K. (2002). Neuronal differentiation-dependent expression of the disialic acid epitope on CD166 and its involvement in neurite formation in neuro2A cells. *J. Biol. Chem.* **277**, 45299–45305.

21. Weissleder, R. and Pittet, M. J. (2008). Imaging in the era of molecular oncology. *Nature* **452**, 580–589.

22. Dandamudi, S. and Campbell, R. B. (2007). Development and characterization of magnetic cationic liposomes for targeting tumor microvasculature. *Biochim. Biophys. Acta.* **1768**, 427–438.

23. Kawai, N., Ito, A., Nakahara, Y., Futakuchi, M., Shirai, T, Honda, H., Kobayashi, T., and Kohri, K. (2005). Anticancer effect of hyperthermia on prostate cancer mediated by magnetite cationic liposomes and immune-response induction in transplanted syngeneic rats. *Prostate* **64**, 373–381.

24. Martina, M. S., Fortin, J. P., Menager, C., Clement, O., Barratt, G., Grabielle-Madelmont, C., Gazeau, F., Cabuil, V., and Lesieur, S. (2005). Generation of superparamagnetic liposomes revealed as highly efficient MRI contrast agents for *in vivo* imaging. *J. Am. Chem. Soc.* **127**, 10676–10685.

25. Moghimi, S. M., Hunter, A. C., and Murray, J. C. (2001). Long-circulating and target-specific nanoparticles: Theory to practice. *Pharmacol. Rev.* **53**, 283–318.

26. Immordino, M. L., Dosio, F., and Cattel, L. (2006). Stealth liposomes: Review of the basic science, rationale, and clinical applications, existing and potential. *Int. J. Nanomedicine* **1**, 297–315.

27. Simberg, D., Weisman, S., Talmon, Y., and Barenholz, Y. (2004). DOTAP (and other cationic lipids): Chemistry, biophysics, and transfection. *Crit. Rev. Ther. Drug. Carrier. Syst.* **21**, 257–317.

28. Lee, C. H., Ni, Y. H., Chen, C. C., Chou, C., and Chang, F. H. (2003). Synergistic effect of polyethylenimine and cationic liposomes in nucleic acid delivery to human cancer cells. *Biochim. Biophys. Acta.* **1611**, 55–62.

29. Cinteza, L. O., Ohulchanskyy, T. Y., Sahoo, Y., Bergey, E. J., Pandey, R. K., and Prasad, P. N. (2006). Diacyllipid micelle-based nanocarrier for magnetically guided delivery of drugs in photodynamic therapy. *Mol. Pharm.* **3**, 415–23.

30. Chang, C. F., Chen, C.Y., Chang, F. H., Tai, S. P., Chen, C. Y., Yu, C. H., Tseng, Y. B., Tsai, T. H., Liu, I. S., Su, W. F., and Sun, C. K. (2008). Cell tracking and detection of molecular expression in live cells using lipid-enclosed CdSe quantum dots as contrast agents for epi-third harmonic generation microscopy. *Opt. Express.* **16**, 9534–48.

31. Krötz, F., Sohn, H. Y., Gloe, T., Plank, C., and Pohl, U. (2003). Magnetofection potentiates gene delivery to cultured endothelial cells. *J. Vasc. Res.* **40**, 425–434.

32. Pisanic, T. R. II., Blackwell, J. D., Shubayev, V. I., Fiñones, R. R., and Jin, S. (2007). Nanotoxicity of iron oxide nanoparticle internalization in growing neurons. *Biomaterials* **28**, 2572–2581.

33. Kraitchman, D. L. and Bulte, J. W. (2008). Imaging of stem cells using MRI. *Basic Res. Cardiol.* **103**, 105–113.

34. Anderson, S. A., Shukaliak-Quandt, J., Jordan, E. K., Arbab, A. S., Martin, R., McFarland, H., and Frank, J. A. (2004). Magnetic resonance imaging of labeled T-cells in a mouse model of multiple sclerosis. *Ann. Neurol.* **55**, 654–659.

35. Kraitchman, D. L., Heldman, A. W., Atalar, E., Amado, L. C., Martin, B. J., Pittenger, M. F., Hare, J. M., and Bulte, J. W. (2003). *In vivo* magnetic resonance imaging of mesenchymal stem cells in myocardial infarction. *Circulation* **107**, 2290–2293.

36. Zhang, Z., Jiang, Q., Jiang, F., Ding, G., Zhang, R., Wang, L., Zhang, L., Robin, A. M., Katakowski, M., and Chopp, M. (2004). *In vivo* magnetic resonance imaging tracks adult neural progenitor cell targeting of brain tumor. *Neuroimage* **23**, 281–287.

37. Guzman, R., Uchida, N., Bliss, T. M., He, D., Christopherson, K. K., Stellwagen, D., Capela, A., Greve, J., Malenka, R. C., Moseley, M. E., Palmer, T. D., and Steinberg, G. K. (2007). Long-term monitoring of transplanted human neural stem cells in developmental and pathological contexts with MRI. *Proc. Natl. Acad. Sci. USA* **104**, 10211–10216.

38. Nelson, G. N., Roh, J. D., Mirensky, T. L., Wang, Y., Yi, T., Tellides, G., Pober, J. S., Shkarin, P., Shapiro, E. M., Saltzman, W. M., Papademetris, X., Fahmy, T. M., and Breuer, C. K. (2008). Initial evaluation of the use of USPIO cell labeling and noninvasive MR monitoring of human tissue-engineered vascular grafts *in vivo*. *FASEB J.* **22**, 3888–95.

5

1. Stewart, B. W. and Weihues, P. (2003). *World Cancer Report: World Health Organizaton.*

2. Sinha, R., Kim, G. J., Nie, S., and Shin, D. M. (2006). Nanotechnology in cancer therapeutics: Bioconjugated nanoparticles for drug delivery. *Mol. Cancer Ther.* **5**(8), 1909.

3. Nie, S., Xing, Y., Kim, G. J., and Simons, J. W. (2007). Nanotechnology applications in cancer. *Annu. Rev. Biomed. Eng.* **9**, 257.

4. Soloviev, M. (2007). Nanobiotechnology today: Focus on nanoparticles. *J. Nanobiotechnol.* **5**, 1113.

5. Jain, R. K. (2001). Delivery of molecular and cellular medicine to solid tumour's. *Adv. Drug del. Rev.* **46**, 149.

6. Priya, P. A. and Katiyar, V. K. (2007). Multifunctional nanoparticles and their role in cancer drug deliveryA review. *Azonano.* **3**, 1.

7. Cho, K., Wang, X., Nie, S., Chen, Z. G., and Shin, D. M. (2008). Therapeutic nanoparticles for drug delivery in cancer. *Clin. Cancer Res.* **14**(5), 1310.

8. Frank, G., Langer, R., and Farokhzad, O. C. (2008). Precise engineering of targeted nanoparticles by using self-assembled biointegrated block copolymers. *PNAS* **105**(7), 2586.

9. Vashir, J. K., Reddy, M. K., and Labhasetwar, V. V. (2005). Nanosystems in drug targeting: Opportunities and challenges. *Current Nanosci.* **1**, 45.

10. Vieira, D. B. and Carmona-Ribeiro, A. M. (2008). Cationic nanoparticles for delivery of amphotericin B: Preparation, characterization and activity *in vitro. J. Nanobiotechnol.* **6**, 6.

11. Maitra, A., et al. (2007). Polymeric nanoparticle-encapsulated curcumin ("nanocurcumin"): Anovel strategy for human cancer therapy. *J. Nanobiotechnol.* **5**, 3.

12. Michalek, J., Pradny, M., Arthyukhov, A., and Sloufm, M. (2005). Macroporous hydrogels based on 2-hydroxyethyl methacrylate part III hydrogels as carriers for immobilization of proteins. *J. Mater. Sci. Med.* **16**, 783.

13. Manneti, C., Casciani, L., and Pescosolido, N. (2002). Diffusive contribution to permeation of hydrogel contact lenses: Theoretical model and experimental evaluation by nuclear magnetic resonance techniques. *Polymer* **43**(1), 87.

14. Ruiz, J., Mantecon, A., and Cadiz, V. (2002). Investigation of loading and release in PVA-based hydrogels. *J. Appl. Polym. Sci.* **85**, 1644.

15. Ostrovidova, G. U., Makeev, A. V., and Shamtsiam, M. M. (2003). Polyfunctional film coatings for medical use. *Mater. Sci. Eng. C* **23**, 545.

16. Mao, J., Zhao, L., Deyao, K., Shang, Q., Yang, G., and Cao, Y. (2003). Study of novel chitosan-gelatin artificial skin *in vitro. J. Biomed. Mater. Res.* **64**, 301.

17. Dalton, P. D., Flynn, L., and Schoichet, M. S. (2002). Fibre templating of poly (2-hydroxyethyl methacrylate) for neural tissue engineering. *Biomaterials* **28**, 3843.

18. Chou, K. F., Han, C. C., and Lee, S. (2000). Water transport in crosslinked 2-hydroxyethyl methacrylate. *Polym. Eng. Sci.* **40**(4), 1004.

19. Reddy, L. H. and Murthy, R. S. R. (2004). Pharmacokinetics and biodistribution studies of doxorubicin loaded poly (butyl cyanoacrylate) nanoparticles synthesized by two different techniques. *Biomed. Papers* **148**(2), 161.

20. Hede, S. and Huilgol, N. (2006). Nano: The new nemesis of cancer. *J. Cancer Res. Ther.* **2**(4), 186.

21. Moghimi, S. M., Hunter, A. C., and Murray, J. C. (2005). Nanomedicine: Current status and future prospects. *FASEB J.* **19**(3), 311.

22. Hoet, P. H. M., Brüske-Hohlfeld, I., and Salata, O. V. (2004). NanoparticlesKnown and unknown health risks. *J. Nanobiotechnol.* **2**, 12.

23. Lu, W., Zhang, Y., Tan, Y-. Z., Hu, K-. L., Jiang, X-. G., and Fu, S-. K. (2005). Cationic albumin-conjugated PEGlyted nanoparticles as novel drug carrier for brain delivery. *J. Cont. Rel.* **107**, 428.

24. Jahanshahi, M. and Babaei, Z. (2008). Protein nanoparticle: A unique system as drug delivery vehicles. *African J. Biotech.* **7**(25), 4926.

25. Higuchi, T. (1961). Rate of release of me-dicaments from ointments bases containing drugs in suspension. *Pharmaceut. Sci.* **56**, 874.

26. Budhian, A., Seigel, S. J., and Winey, K. I. (2008). Controlling the *in-vitro* release profiles for a system of haloperidol-loaded PLGA nanoparticles. *Int. J. Pharm.* **346**, 151.

27. Bajpai, A. K. and Rajpoot, M. (2000). Re-lease and diffusion of sulphamethoxazole through acrylamide-based hydrogel. *J. Appl. Polym. Sci.* **81**, 12381247.

28. Risbud, M. V., Hardikar, A. A., Bhat, S. V., and Bhonde, R. R. (2000). pH-sensitive freeze-dried Chitosan-polyvinyl hydrogels as controlled release system for antibiotic delivery. *J. Control Release* **68**, 23.

29. Singh, A. A., Narvi, S. S., Dutta, P. K., and Pandey, N. D. (2006). External stimuli re-sponse on a novel chitosan hydrogel cross-linked with formaldehyde. *Bull. Mater. Sci.* **29**(3), 233.

30. Bajpai, A. K. (2004). Adsorption of a blood protein on to hydrophillic sponges based on poly(2-hydroxyethyl methacrylate). *J. Mater. Sci. Mater. Med.* **15**, 583.

31. Bajpai, A. K. and Mishra, A. (2005). Prepa-ration and characterization of tetracycline-loaded interpenetrating polymer networks of carboxymethyl cellulose and poly(acrylic acid): water sorption and drug release study. *Polym. Int.* **54**, 1347.

32. Ganta, S., Devalpally, H., Shahiwala, A., and Amiji, M. (2008). A review of stimuli-responsive nanocarriers for drug and gene delivery. *J. Cont. Rel.* **126**, 187.

33. Devalpally, H., Shenoy, D., Little, S., Langer, R., and Amiji, M. (2007). Poly (eth-ylene oxide)-modified poly (epsiloncapro-lactone) nanoparticles for targeted delivery of tamoxifen in breast cancer. *Cancer Che-mother. Pharmacol.* **59**(4), 477.

34. Wang, X., Yang, L., Chen, Z., and Shin, D. M. (2008). Application of nanotechnology in cancer therapy and imaging. *CA Cancer J. Clin.* **58**, 97.

35. Bajpai, A. K. and Saini, R. (2005). Prepa-tion and characterization of spongy cryogels of poly(vinyl alcohol)-casein system: Water

sorption and blood compatibility study. *Polym. Int.* **54**(5), 796.

36. El-Shabouri, M. H. (2002). Positively charged nanoparticles for improving the oral bioavailability of cyclosporine. *Int. J. Pharm.* **249**, 101.

37. Zhang, Q., Zhen, Z., and Nagai, T. (2001). Prolonged hypoglycemic effect of insulin-loaded polybutyl cynoacrylate nanopar-ticles after pulmonary administration to normal rats. *Int. J. Pharm.* **218**, 75.

38. Kaparissides, C., Alexandridou, S., Kam-mona, O., and Dini, E. (2002). Polymeric nano- and microparticles for controlled re-lease applications. *Workshop of CPERI*.

39. Peppas, N. A. and Langer, R. (2002). Ori-gins and development of biomedical en-gineering within chemical engineering. *AICHE J.* **50**, 536.

40. Budhian, A., Seigel, S. J., and Winey, K. I. (2005). Production of haloperidol loaded PLGA nanoparticles for extended controlled drug release of haloperidol. *J. Microencap-sul.* **22**, 773.

41. Siepmann, J. and Peppas, N. A. (2001). Modeling of drug release from delivery sys-tems based on hydroxypropyl methyl cellu-lose. *Adv. Drug Deliv. Rev.* **48**, 139.

42. Wang, L. F., Chen, W. B., Chen, Y. B., and Lu, S. C. (2003). Effects of preparation methods of hydroxyl propylmethyl cel-lulose polyacrylic acid blended films on drug release. *J. Biomater. Sci. Polymer Edn.* **14**(1), 27.

6

1. Gordon, S. and Taylor, P. R. (2005). Mono-cyte and macrophage heterogeneity. *Nat. Rev. Immunol.* **5**, 953–964.

2. Hansson, G. K. and Libby, P. (2006). The immune response in atherosclerosis: Adou-ble-edged sword. *Nat. Rev. Immunol.* **6**, 508–519.

3. Weber, C., Zernecke, A., and Libby, P. (2008). The multifaceted contributions of leukocyte subsets to atherosclerosis: Les-sons from mouse models. *Nat. Rev. Immu-nol.* **8**, 802–815.

4. Auffray, C., Sieweke, M. H., and Geissmann, F. (2009). Blood monocytes: Development,

heterogeneity, and relationship with dendritic cells. *Annu. Rev. Immunol.*

5. Libby, P. (2002). Inflammation in atherosclerosis. *Nature* **420**, 868–874.

6. Swirski, F. K., Pittet, M. J., Kircher, M. F., Aikawa, E., Jaffer, F. A., et al. (2006). Monocyte accumulation in mouse atherogenesis is progressive and proportional to extent of disease. *Proc. Natl. Acad. Sci. USA* **103**, 10340–10345.

7. Schlitt, A., Heine, G. H., Blankenberg, S., Espinola-Klein, C., Dopheide, J. F., et al. (2004). CD14+CD16+ monocytes in coronary artery disease and their relationship to serum TNF-alpha levels. *Thromb. Haemost.* **92**, 419–424.

8. Nahrendorf, M., Swirski, F. K., Aikawa, E., Stangenberg, L., Wurdinger, T., et al. (2007). The healing myocardium sequentially mobilizes two monocyte subsets with divergent and complementary functions. *J. Exp. Med.* **204**, 3037–3047.

9. Swirski, F. K., Libby, P., Aikawa, E., Alcaide, P., Luscinskas, F. W., et al. (2007). Ly-6Chi monocytes dominate hypercholesterolemia-associated monocytosis and give rise to macrophages in atheromata. *J. Clin. Invest.* **117**, 195–205.

10. Tacke, F., Alvarez, D., Kaplan, T. J., Jakubzick, C., Spanbroek, R., et al. (2007). Monocyte subsets differentially employ CCR2, CCR5, and CX3CR1 to accumulate within atherosclerotic plaques. *J. Clin. Invest.* **117**, 185–194.

11. Dragu, R., Huri, S., Zuckerman, R., Suleiman, M., Mutlak, D., et al. (2008). Predictive value of white blood cell subtypes for long-term outcome following myocardial infarction. *Atherosclerosis* **196**, 405–412.

12. Waterhouse, D. F., Cahill, R. A., Sheehan, F., and McCreery, C. (2008). Prediction of calculated future cardiovascular disease by monocyte count in an asymptomatic population. *Vasc. Health Risk Manag.* **4**, 177–187.

13. Lee, H., Sun, E., Ham, D., and Weissleder, R. (2008). Chip-NMR biosensor for detection and molecular analysis of cells. *Nat. Med.* **14**, 869–874.

14. Stroh, A., Zimmer, C., Werner, N., Gertz, K., Weir, K., et al. (2006). Tracking of systemically administered mononuclear cells in the ischemic brain by high-field magnetic resonance imaging. *Neuroimage* **33**, 886–897.

15. Passlick, B., Flieger, D, and Ziegler-Heitbrock, H. W. (1989). Identification and characterization of a novel monocyte subpopulation in human peripheral blood. *Blood* **74**, 2527–2534.

16. Ziegler-Heitbrock, H. W., Fingerle, G., Strobel, M., Schraut, W., Stelter, F., et al. (1993). The novel subset of CD14+/CD16+ blood monocytes exhibits features of tissue macrophages. *Eur. J. Immunol.* **23**, 2053–2058.

17. Wunderbaldinger, P., Josephson, L., and Weissleder R (2002). Crosslinked iron oxides (CLIO): A new platform for the development of targeted MR contrast agents. *Acad. Radiol.* **9**(Suppl 2), S304–306.

18. Metz, S., Bonaterra, G., Rudelius, M., Settles, M., Rummeny, E. J., et al. (2004) Capacity of human monocytes to phagocytose approved iron oxide MR contrast agents *in vitro. Eur. Radiol.* **14**, 1851–1858.

19. Weissleder, R., Stark, D. D., Engelstad, B. L., Bacon, B. R., Compton, C. C., et al. (1989). Superparamagnetic iron oxide: Pharmacokinetics and toxicity. *AJR Am. J. Roentgenol.* **152**, 167–173.

20. Harisinghani, M. G., Barentsz, J., Hahn, P. F., Deserno, W. M., Tabatabaei, S., et al. (2003). Noninvasive detection of clinically occult lymph-node metastases in prostate cancer. *N. Engl. J. Med.* **348**, 2491–2499.

21. Libby, P., Nahrendorf, M., Pittet, M. J., and Swirski, F. K. (2008). Diversity of denizens of the atherosclerotic plaque: Not all monocytes are created equal. *Circulation* **117**, 3168–3170.

22. An, G., Wang, H., Tang, R., Yago, T., McDaniel, J. M., et al. (2008). P-selectin glycoprotein ligand-1 is highly expressed on Ly-6Chi monocytes and a major determinant for Ly-6Chi monocyte recruitment to sites of atherosclerosis in mice. *Circulation* **117**, 3227–3237.

23. Waldo, S. W., Li, Y., Buono, C., Zhao, B., Billings, E. M., et al. (2008). Heterogeneity of human macrophages in culture and in atherosclerotic plaques. *Am. J. Pathol.* **172**, 1112–1126.

24. Draude, G., von Hundelshausen, P., Frankenberger, M., Ziegler-Heitbrock, H. W., and Weber, C. (1999). Distinct scavenger receptor expression and function in the human CD14(+)/CD16(+) monocyte subset. *Am. J. Physiol.* **276**, H1144–1149.

25. Lee, J. H., Huh, Y. M., Jun, Y. W., Seo, J. W., Jang, J. T., et al. (2007) Artificially engineered magnetic nanoparticles for ultrasensitive molecular imaging. *Nat. Med.* **13**, 95–99.

26. Fingerle, G., Pforte, A., Passlick, B., Blumenstein, M., Strobel, M., et al. (1993) The novel subset of CD14+/CD16+ blood monocytes is expanded in sepsis patients. *Blood* **82**, 3170–3176.

27. Pulliam, L., Gascon, R., Stubblebine, M., McGuire, D., and McGrath, M. S. (1997). Unique monocyte subset in patients with AIDS dementia. *Lancet* **349**, 692–695.

28. Schinkel, C., Sendtner, R., Zimmer, S, and Faist, E. (1998) Functional analysis of monocyte subsets in surgical sepsis. *J. Trauma.* **44**, 743–749.

29. Schmidt, H., Bastholt, L., Geertsen, P., Christensen, I. J., Larsen, S., et al. (2005). Elevated neutrophil and monocyte counts in peripheral blood are associated with poor survival in patients with metastatic melanoma: A prognostic model. *Br. J. Cancer.* **93**, 273–278.

30. Murdoch, C., Tazzyman, S., Webster, S., and Lewis, C. E. (2007). Expression of Tie-2 by human monocytes and their responses to angiopoietin-2. *J. Immunol.* **178**, 7405–7411.

31. Ulrich, C., Heine, G. H., Garcia, P., Reichart, B., Georg, T., et al. (2006) Increased expression of monocytic angiotensin-converting enzyme in dialysis patients with cardiovascular disease. *Nephrol. Dial. Transplant.* **21**, 1596–1602.

32. Szaflarska, A., Baj-Krzyworzeka, M., Siedlar, M., Weglarczyk, K., Ruggiero, I., et al. (2004). Antitumor response of CD14+/CD16+ monocyte subpopulation. *Exp. Hematol.* **32**, 748–755.

33. Josephson, L., Kircher, M. F., Mahmood, U., Tang, Y., and Weissleder, R. (2002). Near-infrared fluorescent nanoparticles as combined MR/optical imaging probes. *Bioconjug. Chem.* **13**, 554–560.

34. Simon, G. H., Bauer, J., Saborovski, O., Fu, Y., Corot, C., et al. (2006). T1 and T2 relaxivity of intracellular and extracellular US-PIO at 1.5T and 3T clinical MR scanning. *Eur. Radiol.* **16**, 738–745.

7

1. Paci, I., Johnson, J. C., Chen, X., Rana, G., Popović, D., David, D. E., Nozik, A. J., Ratner, M. A., and Michl, J. (2006). Singlet fission for dye-sensitized solar cells: Can a suitable sensitizer be found? *J. Am. Chem. Soc.* **128**, 1654616553.

2. Hanna, M. and Nozik, A. J. (2006). Solar conversion efficiency of photovoltaic and photoelectrolysis cells with carrier multiplication absorbers. *J. Appl. Phys.* **100**, 074510/1–074510/8.

3. Müller, A. M., Avlasevich, Y. S., Müllen, K., and Bardeen, C. J. (2006). Evidence for exciton fission and fusion in a covalently linked tetracene dimer. *Chem. Phys. Lett.* **421**, 518522.

4. Michl, J. and Bonačić-Koutecký, V. (1990). *Electronic Aspects of Organic Photochemistry.* John Wiley and Sons, Inc., New York.

5. Pariser, R. (1956). Theory of the electronic spectra and structure of the polyacenes and of alternant hydrocarbons. *J. Chem. Phys.* **24**, 250268.

6. Koutecky, J. (1967). Contribution to the theory of alternant systems. *J. Chem. Phys.* **44**, 37023706. Some properties of semiempirical Hamiltonians. *J. Chem. Phys.* **47**, 15011511.

7. Bonačić-Koutecký, V., Koutecký, J., and Michl, J. (1987). Neutral and charged biradicals, zwitterions, funnels in S1, and proton translocation: Their role in photochemistry, photophysics, and vision. *Angew. Chem. Internat. Ed. Engl.* **26**, 170189.

8. Michl, J. (1992). Singlet and triplet states of an electron pair in a molecule—A simple model. *J. Mol. Struct. (THEOCHEM.)* **260**, 299311.

9. Michl, J., Nozik, A. J., Chen, X., Johnson, J. C., Rana, G., Akdag, A., and Schwerin, A. F.

(2007). Toward singlet fission for excitonic solar cells. In *Organic Photovoltaics* VIII. Z. H. Kafafi and P. A. Lane (Eds.). *Proc. of SPIE* **6656**, 66560E166560E1.

10. Flynn, C. R. and Michl, J. (1974). π,π-Biradicaloid hydrocarbons: o-Xylylene. photochemical preparation from 1,4-Dihydrophthalazine in rigid glass, electronic spectroscopy, and calculations. *J. Am. Chem. Soc.* **96**, 32803288.

11. Herkstroeter, W. G. and Merkel, P. B. (1981). The triplet state energies of rubrene and diphenylisobenzofuran. *J. Photochem.* **16**, 331342.

12. Schütz, M. and Schmidt, R. R. (1996). Deactivation of 9,9'-Bianthryl in Solution Studied by Photoacoustic Calorimetry and Fluorescence. *J. Phys. Chem.* **100**, 20122018.

8

1. Peer, D., Karp, J. M., Hong, S., Farokhzad, O. C., Margalit, R., and Langer, R. (2007). Nanocarriers as an emerging platform for cancer therapy. *Nat. Nanotechnol.* **2**, 751–760.

2. Sapra, P. and Allen, T. M. (2003). Ligand-targeted liposomal anticancer drugs. *Prog. Lipid. Res.* **42**, 439–462.

3. Hirsch, L. R., Gobin, A. M., Lowery, A. R., Tam, F., Drezek, R. A., Halas, N. J., and West, J. L. (2006). Metal nanoshells. *Ann. Biomed. Eng.* **34**, 15–22.

4. Yang, W., Thordarson, P., Gooding, J. J., Ringer, S. P., and Braet, F. (2007). Carbon nanotubes for biological and biomedical applications. *Nanotechnology* **18**, 412001.

5. Chen, R. J., Bangsaruntip, S., Drouvalakis, K. A., Kam, N. W., Shim, M., Li, Y., Kim, W., Utz, P. J., and Dai, H. (2003). Noncovalent functionalization of carbon nanotubes for highly specific electronic biosensors. *Proc. Natl. Acad. Sci. USA* **100**, 4984–4989.

6. Narayan, R. J., Jin, C., Menegazzo, N., Mizaikoff, B., Gerhardt, R. A., Andara, M., Agarwal, A., Shih, C. C., Shih, C. M., Lin, S. J., and Su, Y. Y. (2007). Nanoporous hard carbon membranes for medical applications. *J. Nanosci. Nanotechnol.* **7**, 1486–1493.

7. Kam, N. W., O'Connell, M., Wisdom, J. A., and Dai, H. (2005). Carbon nanotubes as multifunctional biological transporters and near-infrared agents for selective cancer cell destruction. *Proc. Natl. Acad. Sci. USA* **102**, 11600–11605.

8. Radomski, A., Jurasz, P., Alonso-Escolano, D., Drews, M., Morandi, M., Malinski, T., and Radomski, M. W. (2005). Nanoparticle-induced platelet aggregation and vascular thrombosis. *Br. J. Pharmacol.* **146**, 882–893.

9. Li, X., Peng, Y., Ren, J., and Qu, X. (2006). Carboxyl-modified single-walled carbon nanotubes selectively induce human telomeric i-motif formation. *Proc. Natl. Acad. Sci. USA* **103**, 19658–19663.

10. Zhang, L. W., Zeng, L., Barron, A. R., and Monteiro-Riviere, N. A. (2007). Biological interactions of functionalized single-wall carbon nanotubes in human epidermal keratinocytes. *Int. J. Toxicol.* **26**, 103–113.

11. Asuri, P., Bale, S. S., Pangule, R. C., Shah, D. A., Kane, R. S., and Dordick, J. S. (2007). Structure, function, and stability of enzymes covalently attached to single-walled carbon nanotubes. *Langmuir* **23**, 12318–12321.

12. Kam, N. W. S. and Dai, H. (2005). Carbon nanotubes as intracellular protein transporters: generality and biological functionality. *J. Am. Chem. Soc.* **127**, 6021–6026.

13. Singh, R., Pantarotto, D., McCarthy, D., Chaloin, O., Hoebeke, J., Partidos, C. D., Briand, J. P., Prato, M., Bianco, A., and Kostarelos, K. (2005). Binding and condensation of plasmid DNA onto functionalized carbon nanotubes: Toward the construction of nanotube-based gene delivery vectors. *J. Am. Chem. Soc.* **127**, 4388–4396.

14. Liu, Y., Wu, D. C., Zhang, W. D., Jiang, X., He, C. B., Chung, T. S., Goh, S. H., and Leong, K. W. (2005). Polyethylenimine-grafted multiwalled carbon nanotubes for secure noncovalent immobilization and efficient delivery of DNA. *Angew. Chem. Int. Edn. Engl.* **44**, 4782.

15. Kam, N. W., Liu, Z., and Dai, H. (2005). Functionalization of carbon nanotubes via cleavable disulfide bonds for efficient intracellular delivery of siRNA and potent gene

silencing. *J. Am. Chem. Soc.* **127**, 12492–12493.

16. Jorio, A., Saito, R., Dresselhaus, G, and Dresselhaus, M. S. (2004). Determination of nanotubes properties by Raman spectroscopy. *Philos. Transact. A. Math. Phys. Eng. Sci.* **362**, 2311–2336.

17. Liu, Z., Cai, W., He, L., Nakayama, N., Chen, K., Sun, X., Chen, X., and Dai, H. (2007). *In vivo* biodistribution and highly efficient tumor targeting of carbon nanotubes in mice. *Nat. Nanotechnol.* **2**, 47–52.

18. Keren, S., Zavaleta, C., Cheng, Z., de la Zerda, A., Gheysens, O., and Gambhir, S. S. (2008). Noninvasive molecular imaging of small living subjects using Raman spectroscopy. *Proc. Natl. Acad. Sci. USA* **105**, 5844–5849.

19. Liu, Z., Li, X., Tabakman, S. M., Jiang, K., Fan, S., and Dai, H. (2008). Multiplexed multicolor Raman imaging of live cells with isotopically modified single walled carbon nanotubes. *J. Am. Chem. Soc.* **130**, 13540–13541.

20. Shao, N., Lu, S., Wickstrom, E., and Panchapakesan, B. (2007). Integrated molecular targeting of IGF1R and HER2 surface receptors and destruction of breast cancer cells using single wall carbon nanotubes. *Nanotechnology* **18**, 315101.

21. Chakravarty, P., Marches, R., Zimmerman, N. S., Swafford, A. D., Bajaj, P., Musselman, I. H., Pantano, P., Draper, R. K., and Vitetta, E. S. (2008). Thermal ablation of tumor cells with antibody-functionalized single-walled carbon nanotubes. *Proc. Natl. Acad. Sci USA* **105**, 8697–8702.

22. Zhou, F., Xing, D., Ou, Z., Wu, B., Resasco, D. E., and Chen, W. R. (2009). Cancer photothermal therapy in the near-infrared region by using single-walled carbon nanotubes. *J. Biomed. Opt.* **14**, 021009.

23. McDevitt, M. R., Chattopadhyay, D., Kappel, B. J., Jaggi, J. S., Schiffman, S. R., Antczak, C., Njardarson, J. T., Brentjens, R., and Scheinberg, D. A. (2007). Tumor targeting with antibody-functionalized, radiolabeled carbon nanotubes. *J. Nucl. Med.* **48**, 1180–1189.

24. Gabizon, A. A. (2001). Pegylated liposomal doxorubicin: metamorphosis of an old drug into a new form of chemotherapy. *Cancer Invest.* **19**, 424–436.

25. Zhang, W. W. (2003). The use of gene-specific IgY antibodies for drug target discovery. *Drug Discov. Today* **8**, 364–371.

26. Xiao, Y., Gao, X., Gannot, G., Emmert-Buck, M. R., Srivastava, S., Wagner, P. D., Amos, M. D., and Barker, P. E. (2008). Quantitation of HER2 and telomerase biomarkers in solid tumors with IgY antibodies and nanocrystal detection. *Int. J. Cancer.* **122**, 2178–2186.

27. Wang, Y., Iqbal, Z., and Mitra, S. (2006). Rapidly functionalized, water-dispersed carbon nanotubes at high concentration. *J. Am. Chem. Soc.* **128**, 95–99.

28. Papadopoulos, N. G., Dedoussis, G. V., Spanakos, G., Gritzapis, A. D., Baxevanis, C. N., and Papamichail, M. (1994). An improved fluorescence assay for the determination of lymphocyte-mediated cytotoxicity using flow cytometry. *J. Immunol. Methods* **177**, 101–111.

29. Bankfalvi, A., Boecker, W., and Reiner, A. (2004). Comparison of automated and manual determination of HER2 status in breast cancer for diagnostic use: A comparative methodological study using the Ventana BenchMark automated staining system and manual tests. *Int. J. Oncol.* **25**, 929–935.

30. Xiao, Y., Telford, W. G., Ball, J. C., Locascio, L. E., and Barker, P. E. (2005). Semiconductor nanocrystal conjugates, FISH and pH. *Nat. Methods* **2**, 723.

31. Xiao, Y. and Barker, P. E. (2004). Semiconductor nanocrystal probes for human metaphase chromosomes. *Nucleic. Acids. Res.* **32**, e28.

32. Huang, W., Taylor, S., Fu, K., Lin, Y., Zhang, D., Hanks, T. W., Rao, A. M., and Sun, Y. P. (2002). Attaching proteins to carbon nanotubes via diimide-activated amidation. *Nano. Lett.* **2**, 311–314.

33. O'Connell, M. J., Bachilo, S. M., Huffman, C. B., Moore, V. C., Strano, M. S., Haroz, E. H., Rialon, K. L., Boul, P. J., Noon, W. H., Kittrell, C., Ma, J., Hauge, R. H., Weisman, R. B., and Smalley, R. E. (2002). Band gap fluorescence from individual single-walled carbon nanotubes. *Science* **297**, 593–596.

34. Bachilo, S. M., Strano, M. S., Kittrell, C., Hauge, R. H., Smalley, R. E., and Weisman, R. B. (2002). Structure-assigned optical spectra of single-walled carbon nanotubes. *Science* **298**, 2361–2366.

35. Dyke, C. A. and Tour, J. M. (2003). Solvent-free functionalization of carbon nanotubes. *J. Am. Chem. Soc.* **125**, 1156–1157.

36. Hudson, J. L., Casavant, M. J., and Tour, J. M. (2004). Water-soluble, exfoliated, non-roping single-wall carbon nanotubes. *J. Am. Chem. Soc.* **126**, 11158–11159.

37. Choo-Smith, L. P., Edwards, H. G., Endtz, H. P., Kros, J. M., Heule, F., Barr, H., Robinson, Jr. J. S., Bruining, H. A., and Puppels, G. J. (2002). Medical applications of Raman spectroscopy: from proof of principle to clinical implementation. *Biopolymers* **67**, 1–9.

38. Nijssen, A., Koljenovi, S., Bakker Schut, T. C., Caspers, P. J., and Puppels, G. J. (2009). Towards oncological application of Raman spectroscopy. *J. Biophotonics* **2**, 29–36.

39. Matousek, P. and Stone, N. (2009). Emerging concepts in deep Raman spectroscopy of biological tissue. *Analyst* **134**, 1058–1066.

40. Schulmerich, M. V., Cole, J. H., Dooley, K. A., Morris, M. D., Kreider, J. M., Goldstein, S. A., Srinivasan, S., and Pogue, B. W. (2008). Noninvasive Raman tomographic imaging of canine bone tissue. *J. Biomed. Opt.* **13**, 020506.

41. Srinivasan, S., Schulmerich, M., Cole, J. H., Dooley, K. A., Kreider, J. M., Pogue, B. W., Morris, M. D., and Goldstein, S. A. (2008). Image-guided Raman spectroscopic recovery of canine cortical bone contrast in situ. *Opt. Express.* 16, 12190–12200.

42. Tini, M., Jewell, U. R., Camenisch, G., Chilov, D., and Gassmann, M. (2002). Generation and application of chicken egg-yolk antibodies. *Comp. Biochem. Physiol. A. Mol. Integr. Physiol.* **131**, 569–574.

43. Carlander, D., Stålberg, J., and Larsson, A. (1999). Chicken antibodies: a clinical chemistry perspective. *Ups. J. Med. Sci.* **104**, 179–89.

44. Carlander, D., Kollberg, H., Wejåker, P. E., and Larsson, A. (2000). Peroral immunotherapy with yolk antibodies for the prevention and treatment of enteric infections. *Immunol. Res.* **21**, 1–6.

45. Schade, R., Calzado, E. G., Sarmiento, R., Chacana, P. A., Porankiewicz-Asplund, J., and Terzolo, H. R. (2005). Chicken egg yolk antibodies (IgY-technology): A review of progress in production and use in research and human and veterinary medicine. *Altern. Lab. Anim.* **33**, 129–154.

46. Larsson, A. and Carlander, D. (2003). Oral immunotherapy with yolk antibodies to prevent infections in humans and animals. *Ups. J. Med. Sci.* **108**, 129–140.

47. Mine, Y. and Kovacs-Nolan, J. (2002). Chicken egg yolk antibodies as therapeutics in enteric infectious disease: A review. *J. Med. Food.* **5**, 159–169.

48. Avedisian, C. T., Cavicchi, R. E., McEuen, P. L., and Zhou, X. (2009). Nanoparticles for cancer treatment: Role of heat transfer. *Ann. N. Y. Acad. Sci.* **1161**, 62–73.

49. Akiyama, T., Sudo, C., Ogawara, H., Toyoshima, K., and Yamamoto, T. (1986). The product of the human c-erbB-2 gene: A 185-kilodalton glycoprotein with tyrosine kinase activity. *Science* **232**, 1644–1646.

50. Clift, M. J., Rothen-Rutishauser, B., Brown, D. M., Duffin, R., Donaldson, K., Proudfoot, L., Guy, K., and Stone, V. (2008). The impact of different nanoparticle surface chemistry and size on uptake and toxicity in a murine macrophage cell line. *Toxicol. Appl. Pharmacol.* **232**, 418–427.

51. Chavanpatil, M. D., Khdair, A., and Panyam, J. (2006). Nanoparticles for cellular drug delivery: Mechanisms and factors influencing delivery. *J. Nanosci. Nanotechnol.* **6**, 2651–2663.

9

1. Baumbach, P. L. (1981). Electrochemical sensor construction. UK patent no. 2073891.

2. Lambrechts, M., Suls, J., and Sansen, W. (1987). A thick film glucose sensor. In *Proceedings of the 9th Annual Conference of the IEEE Engineering in Medicine and Biology Society* (EMBS '87), pp. 789–799, Boston, Mass, USA, November.

3. Matthews, D. R., Holman, R. R., and Bown, E. et al. (1987). Pen-sized digital 30-second blood glucose meter. *The Lancet* **1**(8536), 778–779.

4. Liu, C. C. (1987). Apparatus and method for sensing species, substances and substrates using oxidase. US Patent 4655880.

5. Zhang, X.-E. (2004). Screen printing methods for biosensor production. In *Biosensors,* 2nd edition. J. Cooper and T. Cass (Eds.). Oxford University Press, Oxford, UK, pp. 41–58.

6. Andreescu, S., Barthelmebs, L., and Marty, J. -L. (2002). Immobilization of acetylcholinesterase on screen-printed electrodes: Comparative study between three immobilization methods and applications to the detection of organophosphorus insecticides. *Analytica. Chimica. Acta*, **464**(2), 171–180.

7. Sánchez, S., Pumera, M., Cabruja, E., and Fàbregas, E. (2007). Carbon nanotube/polysulfone composite screen-printed electrochemical enzyme biosensors. *Analyst* **132**, (2), 142–147.

8. Arduini, F., Ricci, F., Tuta, C. S., Moscone, D., Amine, A., and Palleschi, G. (2006). Detection of carbamic and organophosphorous pesticides in water samples using a cholinesterase biosensor based on Prussian Blue-modified screen-printed electrode. *Analytica Chimica Acta.* **580**(2), 155–162.

9. Sántha, H., Dobay, R., and Harsányi, G. (2003). Amperometric uric acid biosensors fabricated of various types of uricase enzymes. *IEEE Sensors J.* **3**(3), 282–287.

10. Cui, G., Kim, S. J., Choi, S. H., Nam, H., Cha, G. S., and Paeng, K.-J. (2000). A disposable amperometric sensor screen printed on a nitrocellulose strip: A glucose biosensor employing lead oxide as an interference-removing agent. *Anal. Chem.***72**(8), 1925–1929.

11. Galán-Vidal, C. A., Muñoz, J., Domínguez, C., and Alegret, S. (1998). Glucose biosensor strip in a three electrode configuration based on composite and biocomposite materials applied by planar thick film technology. *Sensors Actuators B.* **52**(3), 257–263, 1998.

12. Rawson, F. J., Purcell, W. M., Xu, J., et al. (2009). A microband lactate biosensor fabricated using a water-based screen-printed carbon ink. *Talanta*, **77**(3), 1149–1154.

13. Pemberton, R. M., Pittson, R., Biddle, N., and Hart, J. P. (2009). Fabrication of microband glucose biosensors using a screen-printing water-based carbon ink and their application in serum analysis. *Biosens. Bioelec.* **24**(5), 1246–1252.

14. Crouch, E., Cowell, D. C., Hoskins, S., Pittson, R. W., and Hart, J. P. (2005). Amperometric, screen-printed, glucose biosensor for analysis of human plasma samples using a biocomposite water-based carbon ink incorporating glucose oxidase. *Anal. Biochem.* **347**(1), 17–23.

15. Crouch, E., Cowell, D. C., Hoskins, S. Pittson, R. W., and Hart, J. P. (2005). A novel, disposable, screen-printed amperometric biosensor for glucose in serum fabricated using a water-based carbon ink. Biosens. Bioelec. **21**(5), 712–718.

16. Tudorache, M. and Bala, C. (2007).Biosensors based on screen-printing technology, and their applications in environmental and food analysis. *Anal. Bioanal. Chem.***388**(3), 565–578.

17. Albareda-Sirvent, M., Merkoçi, A., and Alegret, S. (2000). Configurations used in the design of screen-printed enzymatic biosensors. A review. *Sens. Actuat. B.* **69**(1), 153–163.

18. Solná, R., Sapelnikova, S., Skládal, P., et al. (2005). Multi-enzyme electrochemical array sensor for determination of phenols and pesticides. *Talanta.* **65**(2), 349–357.

19. Solná, R., Dock, E., Christenson, A., et al. (2005). Amperometric screen-printed biosensor arrays with co-immobilised oxidoreductases and cholinesterases. *Analytica Chimica Acta*, **528**(1), 9–19.

20. Crew, A., Hart, J. P., Wedge R., Marty, J. L., and Fournier, D. (2004). A screen-printed, amperometric, biosensor array for the detection of organophosphate pesticides based on inhibition of wild type, and mutant acetylcholinesterases, from Drosophila melanogaster. *Anal. Lett.***37**(8), 1601–1610.

21. Laschi, S., Palchetti, I. Marrazza, G., and Mascini, M. (2006). Development of disposable low density screen-printed electrode arrays for simultaneous electrochemical

measurements of the hybridisation reaction. *J. Electroanal.l Chem.***593**(1–2), 211–218.

22. Willner, I., Baron, R., and Willner, B. (2007). Integrated nanoparticle-biomolecule systems for biosensing and bioelectronics. *Biosens.Bioelectron.***22**(9–10), 1841–1852.

23. Pedrosa, V. A., Luo, X., Burdick, J., and Wang, J. (2008). 'Nanofingers' based on binary gold-polypyrrole nanowires. *Small* **4**(6), 738–741.

24. Morrow, T. J., Li, M., Kim, J., Mayer, T. S., and Keating, C. D. (2009). Programmed assembly of DNA-coated nanowire devices *Science.* **323**(5912), 352.

25. He, B., Morrow, T. J., and Keating, C. D. (2008). Nanowire sensors for multiplexed detection of biomolecules. *Curr. Opin. Chem. Biol.* **12**(5), 522–528.

26. Piccin, E., Laocharoensuk, R., Burdick, J., Carrilho, E., and Wang, J. (2007). Adaptive nanowires for switchable microchip devices. *Anal. Chem.* **79**(12), 4720–4723.

27. Wang, J., Scampicchio, M., Laocharoensuk, R., Valentini, F., González-García, O., and Burdick, J. (2006). Magnetic tuning of the electrochemical reactivity through controlled surface orientation of catalytic nanowires. *J Am. Chem. Soc.***128**, (14), 4562–4563.

28. Laocharoensuk, R., Bulbarello, A., Hocevar, S. B., Mannino, S., Ogorevc, B., and Wang, J. (2007). On-demand protection of electrochemical sensors based on adaptive nanowires. *J. Am. Chem. Soc.***129**, (25), 7774–7775.

29. Laocharoensuk, R., Bulbarello, A., Mannino, S., and Wang, J. (2007). Adaptive nanowire-nanotube bioelectronic system for on-demand bioelectrocatalytic transformations. *Chem. Commun.* **32**, 3362–3364.

30. Yogeswaran, U. and Chen, S.-M. (2008). A review on the electrochemical sensors and biosensors composed of nanowires as sensing material. *Sensors.* **8**(1), 290–313.

31. Loaiza, Ó. A., Laocharoensuk, R., Burdick, J., et al. (2007). Adaptive orientation of multifunctional nanowires for magnetic control of bioelectrocatalytic processes. *Angewandte Chemie International Edition* **46**(9), 1508–1511.

32. Shen, J., Dudik, L., and Liu, C.-C. (2007). An iridium nanoparticles dispersed carbon based thick film electrochemical biosensor and its application for a single use, disposable glucose biosensor. *Sens. Actuat. B.* **125**(1), 106–113.

33. Fanjul-Bolado, P. Queipo, P., Lamas-Ardisana, P. J., and Costa-García, A. (2007). Manufacture and evaluation of carbon nanotube modified screen-printed electrodes as electrochemical tools. *Talanta,* **74**(3), 427–433.

34. Wang, J. Musameh, M., and Lin, Y. (2003). Solubilization of carbon nanotubes by Nafion toward the preparation of amperometric biosensors. *J. Am. Chem. Soc.***125**(9), 2408–2409.

35. Laschi, S., Bulukin, E., Palchetti, I., Cristea, C., and Mascini, M. (2008). Disposable electrodes modified with multi-wall carbon nanotubes for biosensor applications. *IRBM.* **29**(2–3), 202–207.

36. Rivas, G. A., Rubianes, M. D., Rodríguez, M. C., et al. (2007). Carbon nanotubes for electrochemical biosensing. *Talanta.* **74**(3), 291–307.

37. Domínguez-Renedo, O., Ruiz-Espelt, L., García-Astorgano, N., and Arcos-Martínez, M. J. (2008). Electrochemical determination of chromium(VI) using metallic nanoparticle-modified carbon screen-printed electrodes. *Talanta.***76**(4), 854–858.

38. Wang, J. (2005). Nanomaterial-based electrochemical biosensors. *Analyst.* **130**(4), 421–426.

39. Sergeev, G. B. (2006). Nanoparticles in science and technology. In *Nanochemistry.* Elsevier, Amsterdam, The Netherlands, pp. 175–208.

40. Welch, C. M. and Compton, R. G. (2006). The use of nanoparticles in electroanalysis: A review. *Anal. Bioanal. Chem.* **384**(3), 601–619.

41. Alivisatos, P. (2004). The use of nanocrystals in biological detection. *Nature Biotechnol.* **22**(1), 47–52.

42. Bakker, E. (2004). Electrochemical sensors. *Anal. Chem.***76**, (12), 3285–3298.

43. Ambrosi, A., Merkoçi, A., and Escosura-Muñiz, A. de la (2008). Electrochemical analysis with nanoparticle-based biosys-

tems. *TrAC Trends Anal. Chem.***27**(7), 568–584.

44. Erdem, A. (2007). Nanomaterial-based electrochemical DNA sensing strategies. *Talanta*, **74**(3), 318–325.

45. Wang, J. (2005). Carbon-nanotube based electrochemical biosensors: A review. *Electroanalysis.* **17**(1), 7–14.

46. Pumera, M., Sánchez, S., Ichinose, I., and Tang, J. (2007). Electrochemical nanobiosensors. *Sens. Actuat. B* **123**(2), 1195–1205.

47. Huang, X.-J. and Choi, Y.-K. (2007). Chemical sensors based on nanostructured materials. *Sens. Actuat. B* **122**(2), 659–671.

48. Willner, I., Baron, R., and Willner, B. (2007). Integrated nanoparticle-biomolecule systems for biosensing and bioelectronics. *Biosens. Bioelec.* **22**(9–10), 1841–1852.

49. Yun, Y., Dong, Z., Shanov, V., et al. (2007). Nanotube electrodes and biosensors. *Nano Today.* **2**(6), 30–37.

50. Vamvakaki, V. and Chaniotakis, N. A. (2007). Carbon nanostructures as transducers in biosensors. *Sens. Actuat. B* **126**(1), 193–197.

51. Zhang, X.-E. (2004). Screen-printing methods for biosensor production. In *Biosensors.* J. Cooper and T. Cass (Eds.). *Practical Approach Series*, Oxford University Press, Oxford, UK, pp. 41–58.

52. Domínguez-Renedo, O., Alonso-Lomillo, M. A., and Arcos-Martínez, M. J. (2007). Recent developments in the field of screen-printed electrodes and their related applications. *Talanta.* **73**(2), 202–219.

53. Hart, J. P., Crew, A., Crouch, E., Honeychurch, K. C., and Pemberton, R. M. (2007). Screen-printed electrochemical (bio)sensors in biomedical, environmental and industrial applications. In *Electrochemical Sensor Analysis,* Vol. 49. S. Alegret and A. Merkoçi (Eds.). Elsevier, Amsterdam, The Netherlands, pp. 497–557.

54. Laschi, S. and Mascini, M. (2006). Planar electrochemical sensors for biomedical applications. *Med. Eng.Phys.* **28**(10), 934–943, 2006.

55. Hart, J. P., Crew, A., Crouch, E., Honeychurch, K. C., and Pemberton, R. M. (2004). Some recent designs and developments of screen-printed carbon electrochemical sensors/biosensors for biomedical, environmental, and industrial analyses. *Anal. Lett.* **37**(5), 789–830.

56. Avramescu, A., Andreescu, S., Noguer, T., Bala, C., Andreescu, D., and Marty, J.-L. (2002). Biosensors designed for environmental and food quality control based on screen-printed graphite electrodes with different configurations. *Anal. Bioanal. Chem.* **374**(1), 25–32.

57. Nascimento, V. B. and Angnes, L. (1998). Eletrodos fabricados por 'silk-screen'. *Química Nova.* **21**(5), 614–629.

58. Newman, J. D. and Turner, A. P. F. (2005). Home blood glucose biosensors: A commercial perspective. *Biosens. Bioelectron.* **20**(12), 2435–2453.

59. Hecht, H. J., Kalisz, H. M., Hendle, J., Schmid, R. D., and Schomburg, D. (1993). Crystal structure of glucose oxidase from Aspergillus niger refined at 2.3 Å resolution. *J. Mol. Biol.* **229**(1), 153–172.

60. de Mattos, I. L., Gorton, L., and Ruzgas, T. (2002). Sensor and biosensor based on Prussian Blue modified gold and platinum screen printed electrodes. *Biosens. Bioelectron.* **18**(2–3), 193–200.

61. Derwinska, K., Miecznikowski, K., Koncki, R., Kulesza, P. J., Glab, S., and Malik, M. A. (2003). Application of Prussian Blue based composite film with functionalized organic polymer to construction of enzymatic glucose biosensor. *Electroanalysis.* **15**(23–24), 1843–1849.

62. Zhao, W., Xu, J.-J., Shi, C.-G., and Chen, H.-Y. (2005). Multilayer membranes via layer-by-layer deposition of organic polymer protected Prussian Blue nanoparticles and glucose oxidase for glucose biosensing. *Langmuir.* **21**(21), 9630–9634.

63. Li, T., Yao, Z., and Ding, L. (2004). Development of an amperometric biosensor based on glucose oxidase immobilized through silica sol-gel film onto Prussian Blue modified electrode. *Sens. Actuat. B.* **101**(1–2), 155–160.

64. Lupu, A., Compagnone, D., and Palleschi, G. (2004). Screen-printed enzyme electrodes for the detection of marker analytes during winemaking. *Anal. Chimi. Acta.* **513**(1), 67–72.

65. Zuo, S., Teng, Y., Yuan, H., and Lan, M. (2008). Development of a novel silver nanoparticles-enhanced screen-printed amperometric glucose biosensor. *Analy. Lett.* **41**(7), 1158–1172.

66. Fiorito, P. A., Gonçales, V. R., Ponzio, E. A., and Córdoba de Torresi, S. I. (2005). Synthesis, characterization and immobilization of Prussian Blue nanoparticles. A potential tool for biosensing devices. *Chem. Comm.* **3**, 366–368.

67. Guan, W.-J., Li, Y., Chen, Y.-Q., Zhang, X.-B, and Hu, G.-Q. (2005). Glucose biosensor based on multi-wall carbon nanotubes and screen printed carbon electrodes. *Biosens. Bioelec.* **21**(3), 508–512.

68. Rubianes, M. D. and Rivas, G. A. (2003). Carbon nanotubes paste electrode. *Electrochem. Comm.* **5**(8), 689–694.

69. Lu, B.-W. and Chen, W.-C. (2006). A disposable glucose biosensor based on drop-coating of screen-printed carbon electrodes with magnetic nanoparticles. *J. Magnetism and Magnetic Mater.***304**(1), e400–402.

70. Rossi, L. M., Quach, A. D., and Rosenzweig, Z. (2004). Glucose oxidase-magnetite nanoparticle bioconjugate for glucose sensing. *Anal. Bioanal. Chem.* **380**(4), 606–613.

71. Gao, Z., Xie, F., Shariff, M., Arshad, M., and Ying, J. Y. (2005). A disposable glucose biosensor based on diffusional mediator dispersed in nanoparticulate membrane on screen-printed carbon electrode. *Sens. Actuat. B* **111-112**, 339–346.

72. Gao, Z., Xu, G., Ying, Y. -R. J., Shariff, M., Arshad, M., and Xie, F. (2005). *Biosensor.* WO/2005/040404.

73. Bragagnolo, N. and Rodriguez-Amaya, D. B. (2002). Simultaneous determination of total lipid, cholesterol and fatty acids in meat and backfat of suckling and adult pigs. *Food Chem.* **79**(2), 255–260.

74. Hwang, B. -S. Wang, J. -T. and Choong, Y. -M. (2003). A simplified method for the quantification of total cholesterol in lipids using gas chromatography. *J. Food Comp. Anal.* **16**(2), 169–178.

75. Gilmartin, M. A. T. and Hart, J. P. (1994). Fabrication and characterization of a screen-printed, disposable, amperometric cholesterol biosensor. *Analyst* **119**(11), 2331–2336.

76. Charpentier, L. and El Murr, N. (1995). Amperometric determination of cholesterol in serum with use of a renewable surface peroxidase electrode. *Anal. Chim. Acta*, **318**(1), 89–93.

77. Nakaminami, T., Kuwabata, S., and Yoneyama, H. (1997). Electrochemical oxidation of cholesterol catalyzed by cholesterol oxidase with use of an artificial electron mediator. *Anal. Chem.* **69**(13), 2367–2372.

78. Ricci, F. and Palleschi, G. (2005). Sensor and biosensor preparation, optimisation and applications of Prussian Blue modified electrodes. *Biosens. Bioelec.* **21**(3), 389–407.

79. Cosnier, S., Senillou, A., Grätzel, M., et al. (1999). Glucose biosensor based on enzyme entrapment within polypyrrole films electrodeposited on mesoporous titanium dioxide. *J. Electroanalytical Chem.* **469**(2), 176–181.

80. Wang, J., Zhang, X., and Prakash, M. (1999). Glucose microsensors based on carbon paste enzyme electrodes modified with cupric hexacyanoferrate. *Analytica Chimica Acta* **395**(1–2), 11–16.

81. Arya, S. K., Datta, M., and Malhotra, B. D. (2008). Recent advances in cholesterol biosensor. *Biosens. Bioelec.* **23**(7), 1083–1100.

82. Shumyantseva, V. V., Carrara, S., Bavastrello, V., et al. (2005). Direct electron transfer between cytochrome P450scc and gold nanoparticles on screen-printed rhodium-graphite electrodes. *Biosens. Bioelec.* **21**(1), 217–222.

83. Carrara, S., Shumyantseva, V. V., Archakov, A. I., and Samorì, B. (2008). Screen-printed electrodes based on carbon nanotubes and cytochrome P450scc for highly sensitive cholesterol biosensors. *Biosens. Bioelec.* **24**(1), 148–150.

84. Bistolas, N., Wollenberger, U., Jung, C., and Scheller, F. W. (2005). Cytochrome P450 biosensors—A review. *Biosens. Bioelec.* **20**(12), 2408–2423.

85. Shumyantseva, V., Deluca, G., Bulko, T., et al. (2004). Cholesterol amperometric biosensor based on cytochrome P450scc. *Biosens. Bioelec.* **19**(9), 971–976.

86. Li, G., Liao, J. M., Hu, G. Q., Ma, N. Z., and Wu, P. J. (2005). Study of carbon nanotube modified biosensor for monitoring

total cholesterol in blood. *Biosens. Bioelec.* **20**(10), 2140–2144.

87. Jelen, F., Yosypchuk, B., Kourilová, A., Novotný, L., and Paleček, E. (2002). Label-free determination of picogram quantities of DNA by stripping voltammetry with solid copper amalgam or mercury electrodes in the presence of copper. *Analy. Chem.* **74**(18), 4788–4793.

88. Kara, P., Kerman, K., Ozkan, D., et al. (2002). Electrochemical genosensor for the detection of interaction between methylene blue and DNA. *Electrochem. Comm.* **4**(9), 705–709.

89. Caruana, D. J. and Heller A. (1999). Enzyme-amplified amperometric detection of hybridization and of a single base pair mutation in an 18-base oligonucleotide on a 7-μm-diameter microelectrode. *J. Am. Chem. Soc.* **121**(4), 769–774.

90. Wang, J., Polsky, R., Merkoçi, A., and Turner, K. L. (2003). 'Electroactive beads' for ultrasensitive DNA detection. *Langmuir* **19**(4), 989–991.

91. Merkoçi, A. (2007). Electrochemical biosensing with nanoparticles. *FEBS Journal* **274**(2), 310–316.

92. Du, D., Ding, J., Tao, Y., Li, H., and Chen, X. (2008). CdTe nanocrystal-based electrochemical biosensor for the recognition of neutravidin by anodic stripping voltammetry at electrodeposited bismuth film. *Biosens. Bioelec.* **24**(4), 869–874.

93. Niu, S., Zhao, M., Hu, L., and Zhang, S. (2008). Carbon nanotube-enhanced DNA biosensor for DNA hybridization detection using rutin-Mn as electrochemical indicator. *Sens. Actuat. B.* **135**(1), 200–205.

94. Castañeda, M. T., Merkoçi, A., Pumera, M., and Alegret, S. (2007). Electrochemical genosensors for biomedical applications based on gold nanoparticles. *Biosens. Bioelec.* **22**(9–10), 1961–1967.

95. Authier, L., Grossiord, C., Brossier, P., and Limoges, B. (2001). Gold nanoparticle-based quantitative electrochemical detection of amplified human cytomegalovirus DNA using disposable microband electrodes. *Analy. Chem.* **73**(18), 4450–4456.

96. Wang, J., Rincón, O., Polsky, R., and Dominguez, E. (2003). Electrochemical detection of DNA hybridization based on DNA-templated assembly of silver cluster. *Electrochem. Comm.* **5**(1), 83–86.

97. Wang, J., Xu, D., and Polsky, R. (2002). Magnetically-induced solid-state electrochemical detection of DNA hybridization. *J. Am. Chem. Soc.* **124**(16), 4208–4209.

98. Suprun, E., Shumyantseva, V., Bulko, T., et al. (2008). Au-nanoparticles as an electrochemical sensing platform for aptamer-thrombin interaction. *Biosens. Bioelec.* **24**(4), 831–836.

99. Kerman, K. and Tamiya, E. (2008). Aptamer-functionalized Au nanoparticles for the electrochemical detection of thrombin. *J. Biomed. Nanotech.* **4**(2), 159–164.

100. Shih, W.-C., Yang, M.-C., and Lin, M. S. (2009). Development of disposable lipid biosensor for the determination of total cholesterol. *Biosens. Bioelec.* **24**(6), 1679–1684.

101. Burgoa Calvo, M. E., Domínguez-Renedo, O., and Arcos-Martínez, M. J. (2007). Determination of lamotrigine by adsorptive stripping voltammetry using silver nanoparticle-modified carbon screen-printed electrodes. *Talanta* **74**(1), 59–64.

102. Alonso-Lomillo, M. A., Yardimci, C., Domínguez-Renedo, O., and Arcos-Martínez, M. J. (2009). CYP450 2B4 covalently attached to carbon and gold screen printed electrodes by diazonium salt and thiols monolayers. *Analytica Chimica Acta* **633**(1), 51–56.

103. Alonso-Lomillo, M. A., Domínguez-Renedo, O., Matos, P., and Arcos-Martínez, M. J. (2009). Electrochemical determination of levetiracetam by screen-printed based biosensors. *Bioelectrochemistry* **74**(2), 306–309.

104. Martinez, N. A., Messina, G. A., Bertolino, F. A., Salinas, E., and Raba, J. (2008). Screen-printed enzymatic biosensor modified with carbon nanotube for the methimazole determination in pharmaceuticals formulations. *Sens. Actuat. B* **133**(1), 256–262.

105. Boujtita, M., Hart, J. P., and Pittson, R. (2000). Development of a disposable ethanol biosensor based on a chemically modified screen-printed electrode coated with

alcohol oxidase for the analysis of beer. *Biosens. Bioelec.* **15**(5–6), 257–263.

106. Wang, J., Timchalk, C., and Lin, Y. (2008). Carbon nanotube-based electrochemical sensor for assay of salivary cholinesterase enzyme activity: An exposure biomarker of organophosphate pesticides and nerve agents. *Envi. Sci. Tech.* **42**(7), 2688–2693.

107. Liao, M.-H., Guo, J.-C., and Chen, W.-C. (2006). A disposable amperometric ethanol biosensor based on screen-printed carbon electrodes mediated with ferricyanide-magnetic nanoparticle mixture. *J. Magn. Mag. Mat.* **304**(1), e421–423.

108. Andreescu, S. and Marty, J.-L. (2006). Twenty years research in cholinesterase biosensors: From basic research to practical applications. *Biomol. Engineer.* **23**(1), 1–15.

109. Lin, Y., Lu, F., and Wang, J. (2004). Disposable carbon nanotube modified screen-printed biosensor for amperometric detection of organophosphorus pesticides and nerve agents. *Electroanalysis* **16**(1–2), 145–149.

110. Cai, J. and Du, D. (2008). A disposable sensor based on immobilization of acetylcholinesterase to multi-wall carbon nanotube modified screen-printed electrode for determination of carbaryl. *J. Appl. Electrochem.* **38**(9), 1217–1222.

111. Bontidean, I., Mortari, A., Leth, S., et al. (2004). Biosensors for detection of mercury in contaminated soils. *Envi. Pollution* **131**(2), 255–262.

112. Honeychurch, K. C. and Hart, J. P. (2003). Screen-printed electrochemical sensors for monitoring metal pollutants. *TrAC Trends in Analy. Chem.* **22**(7–8), 456–469.

113. Ogończyk, D., Tymecki, L., Wyżkiewicz, I., Koncki, R., and Głąb, S. (2005). Screen-printed disposable urease-based biosensors for inhibitive detection of heavy metal ions. *Sens. Actuat. B* **106**(1), 450–454.

114. Rodriguez, B. B., Bolbot, J. A., and Tothill, I. E. (2004). Development of urease and glutamic dehydrogenase amperometric assay for heavy metals screening in polluted samples. *Biosens. Bioelec.* **19**(10), 1157–1167.

115. Rodriguez, B. B., Bolbot, J. A., and Tothill, I. E. (2004). Urease-glutamic dehydrogenase biosensor for screening heavy metals in water and soil samples. *Analy. Bioanaly. Chem.* **380**(2), 284–292.

116. Domínguez-Renedo, O. and Arcos-Martínez, M. J. (2007). A novel method for the anodic stripping voltammetry determination of Sb(III) using silver nanoparticle-modified screen-printed electrodes. *Electrochem. Comm.* **9**(4), 820–826.

117. Domínguez-Renedo, O. and Arcos-Martínez, M. J. (2007). Anodic stripping voltammetry of antimony using gold nanoparticle-modified carbon screen-printed electrodes. *Analytica Chimica Acta* **589**(2), 255–260.

118. Song, Y.-S., Muthuraman, G., Chen, Y.-Z., Lin, C.-C., and Zen, J.-M. (2006). Screen printed carbon electrode modified with poly(L-lactide) stabilized gold nanoparticles for sensitive as(III) detection. *Electroanalysis* **18**(18), 1763–1770.

119. Song, Y.-S., Muthuraman, G., and Zen, J.-M. (2006). Trace analysis of hydrogen sulfide by monitoring As(III) at a poly(L-lactide) stabilized gold nanoparticles modified electrode. *Electrochem. Comm.* **8**(8), 1369–1374.

120. Santos, W. J. R., Lima, P. R., Tanaka, A. A., Tanaka, S. M. C. N., and Kubota, L. T. (2009). Determination of nitrite in food samples by anodic voltammetry using a modified electrode. *Food Chem.* **113**(4), 1206–1211.

121. Sohail, M. and Adeloju, S. B. (2008). Electroimmobilization of nitrate reductase and nicotinamide adenine dinucleotide into polypyrrole films for potentiometric detection of nitrate. *Sens. Actuat. B* **133**(1), 333–339.

122. Almeida, M. G., Silveira, C. M., and Moura, J. J. G. (2007). Biosensing nitrite using the system nitrite redutase/Nafion/methyl viologen—A voltammetric study. *Biosens. Bioelec.* **22**(11), 2485–2492.

123. Chen, H., Mousty, C., Chen, L., and Cosnier, S. (2008). A new approach for nitrite determination based on a HRP/catalase biosensor. *Mat. Sci. Engineer. C* **28**(5–6), 726–730.

124. Chen, H., Mousty, C., Cosnier, S., Silveira, C., Moura, J. J. G., and Almeida, M. G. (2007). Highly sensitive nitrite biosensor based on the electrical wiring of nitrite reductase by [ZnCr-AQS] LDH. *Electrochem. Comm.* **9**(9), 2240–2245.

125. Cosnier, S., Da Silva, S., Shan, D., and Gorgy, K. (2008). Electrochemical nitrate biosensor based on poly(pyrrole-viologen) film-nitrate reductase-clay composite. *Bioelectrochemistry* **74**(1), 47–51.

126. Dhaoui, W., Bouzitoun, M., Zarrouk, H., Ouada, H. B., and Pron, A. (2008). Electrochemical sensor for nitrite determination based on thin films of sulfamic acid doped polyaniline deposited on Si/SiO2 structures in electrolyte/insulator/semiconductor (E.I.S.) configuration. *Synthetic Metals* **158**(17–18), 722–726.

127. Zhang, Z., Xia, S., Leonard, D., et al. (2009). A novel nitrite biosensor based on conductometric electrode modified with cytochrome c nitrite reductase composite membrane. *Biosens. Bioelec.* **24**(6), 1574–1579.

128. Dai, Z., Bai, H., Hong, M., Zhu, Y., Bao, J., and Shen, J. (2008). A novel nitrite biosensor based on the direct electron transfer of hemoglobin immobilized on CdS hollow nanospheres. *Biosens. Bioelec.* **23**(12), 1869–1873.

129. Huang, X., Li, Y., Chen, Y., and Wang, L. (2008). Electrochemical determination of nitrite and iodate by use of gold nanoparticles/poly(3-methylthiophene) composites coated glassy carbon electrode. *Sens. Actuat. B* **134**(2), 780–786.

130. Salimi, A., Hallaj, R., Mamkhezri, H., and Hosaini, S. M. T. (2008). Electrochemical properties and electrocatalytic activity of FAD immobilized onto cobalt oxide nanoparticles: Application to nitrite detection. *J. Electroanalytical Chem.* **619–620**(1–2), 31–38.

131. Xu, X., Liu, S., Li, B., and Ju, H. (2003). Disposable nitrite sensor based on hemoglobin-colloidal gold nanoparticle modified screen-printed electrode. *Analy. Lett.* **36**(11), 2427–2442.

132. Wang, J. and Pamidi, P. V. A. (1995). Disposable screen-printed electrodes for monitoring hydrazines. *Talanta* **42**(3), 463–467.

133. Yang, C.-C., Kumar, A. S., Kuo, M.-C., Chien, S.-H., and Zen, J.-M. (2005). Copper-palladium alloy nanoparticle plated electrodes for the electrocatalytic determination of hydrazine. *Analytica Chimica Acta* **554**(1–2), 66–73.

10

Please note that there is some repetition in the references listed here. This is due to many of the chapters being self-contained papers that came as a result of this project. The duplication of references was permitted in order to retain the wholeness of the individual chapters, especially since some of the papers are still in draft format and have not yet been submitted, and therefore cannot be sited at the time of writing this study.

1-1. Barker, L. M. and Hollenbach, R. E. (November 1972). Laser interferometer for measuring high velocities of any reflecting surface. *J. Appl. Phy.* **43**(11), 4669–4675.

1-2. Vandersall, K. S. and Thadhani, N. N. (January 2003). Investigation of 'shock-induced' and "shock-assisted" chemical reactions in Mo + 2Si powder mixtures. *Met. Mat. Trans. A* **34**A(1), 15–23.

1-3. Xu, X. and Thadhani, N. N. (August 2004). Investigation of shock-induced reaction behavior of as-blended and ball-milled Ni+Ti powder mixtures using time-resolved stress measurements. *J. Appl. Phy.* **96**(4), 2000–2009.

1-4. Ferranti, L. and Thadhani, N. N. (November 2005). Dynamic impact characterization of Al + Fe2O3 + 30% epoxy composites using time synchronized high-speed camera and VISAR measurements. *Mat. Res. Soc. Sym. Proc.* **896**, 197–202.

1-5. Martin, M. (April 2005). *Processing and Characterization of Energetic and Structural Behavior of Nickel Aluminum with Polymer Binders*. PhD Thesis, Georgia Institute of Technology.

1-6. Thadhani, N. N. (1993). Shock-induced chemical reactions and synthesis of materials. *Progr. Mat. Sci.* 37, 117–226.

1-7. Graham, R. A. (1993). *Solids Under High-Pressure Shock Compression.* Springer-Verlag, New York.

1-8. Rice, M. H., McQueen, R. G., and Walsh, J. M. (1958). Compression of solids by strong shock waves. In *Solid State Physics*, Vol. VI. F. Seitz and D. Turnbull (Eds.). Academic Press, New York, pp. 1–63.

1-9. Herrmann, W. (May 1969). Constitutive equation for the dynamic compaction of ductile porous materials. *J. Appl. Phy.* 40(6), 2490–2499.

1-10. Carroll, M. M. and Holt, A. C. (February 1972). Suggested modification of the P-α model for porous materials. *J. Appl. Phy.* 43(2), 759–761.

1-11. Mackenzie, J. K. (January 1950). The elastic constants of a solid containing spherical holes. *Proc. Phy. Soc. Sec. B* 63, 1–11.

1-12. Carroll, M. M. and Holt, A. C. (April 1972). Static and dynamic pore-collapse relations for ductile porous materials. *J. Appl. Phy.* 43(4), 1626–1636.

1-13. Kenkre, V. M. and Endicott, M. R. (December 1996). A theoretical model for compaction of granular materials. *J. Am. Ceramic Soc.* 79(12), 3045–3054.

1-14. Raybould, D., Morris, D. G., and Cooper, G. A. (October 1979). A new powder metallurgy method. J. Mat. Sci. **14**(10), 2523–2526.

1-15. Raybould, D. (March 1981). The properties of stainless steel compacted dynamically to produce cold interparticle welding. J. Mat. Sci. **16**(3), 589–998.

1-16. Morris, D. G. (March 1981). Melting and solidification during dynamic powder compaction of tool steel. *Met. Sci.* 15(3), 116–124.

1-17. Kasiraj, P., Vreeland Jr., T., Schwarz, R. B., and Ahrens, T. J. (August 1984). Shock consolidation of a rapidly solidified steel powder. *Acta Metall.* 32(8), 1235–1241.

1-18. Schwarz, R. B., Kasiraj, P., Vreeland Jr., T., and Ahrens, T. J. (August 1984). A theory for the shock-wave consolidation of powders. *Acta Metall.* 32(8), 1243–1252.

1-19. Raybould, D. (January 1980). The cold welding of powders by dynamic compaction. *Int. J. Powder Met. Powder Tech.* 16(1), 9–14, 16–19.

1-20. Gourdin, W. H. (January 1984). Energy deposition and microstructural modification in dynamically consolidated metal powders. *J. Appl. Phy.* 55(1), 172–181.

1-21. Gourdin, W. H. (1984). Prediction of microstructural modifications in dynamically consolidated metal powders. In *Shock Waves in Condensed Matter – 1983*. Proceedings of the American Physical Society Topical Conference. J. R. Asay, R. A. Graham, and G. K. Straub, (Eds.). Elsevier, New York, pp. 379–382.

1-22. Gourdin W. H. (1986). Dynamic consolidation of metal powders. *Prog. Mat. Sci.* 30(1), 39–80.

1-23. Gourdin, W. H. (September 1984). Local microstructural modification in dynamically consolidated metal powders. *Metall. Trans. A* 15A(9), 1653–1664.

1-24. Roman, O. V., Nesterenko, V. F., and Pikus, I. M. (September–October 1979). Influence of the powder particle size on the explosive pressing process. *Combustion, Explostion, and Shock Waves* (English Translation of Fizika Goreniya i Vzryva) 15(5), 644–649.

1-25. Santiso, E. and Müller, E. A. (August 2002). Dense packing of binary and polydisperse hard spheres. *Mol. Phy.* 100(15), 2461–2469.

1-26. Cumberland, D. J. and Crawford, R. J. (1987). *The Packing of Particles*, Chap. 4. Elsevier, Amsterdam.

1-27. Cumberland, D. J. and Crawford, R. J. (1987). *The Packing of Particles*, Chap. 8. Elsevier, Amsterdam.

1-28. Yang, R. Y., Zou, R. P., and Yu, A. B. (September 2003). Effect of material properties on the packing of fine particles. *J. Appl. Phy.* 94(5), 3025–3034.

2-1. Benavides, G. L., Adams, D. P., and Yang, P. (2001). *Meso-scale Machining Capabilities*, SAND2001-1708, Sandia National Laboratory, Albuquerque, NM.

2-2. Geiger, M., Kleine, M., Eckstein, R., Tiesler, N., and Engel, U. (2001). Microforming.

CIRP Annals-Manufacturing Tech. 50(2), 445.

2-3. Schuster, R., Kirchner, V., Allongue, P., and Ertl, G. (2000). *Electrochemcial micromachining, Science* 289(5476), 98.

2-4. Valiev, R. Z., Isamgaliev, R. K., and Alexandrov, I. V. (2000). Bulk nanostructured materials from severe plastic deformation. *Prog. Mater. Sci.* 45, 103.

2-5. Valiev, R. Z., Estrin, Y., Horita, Z., Langdon, T. G., Zehetbauer, M. J., and Zhu, Y. T. (2003). Producing bulk ultrafine-grained materials by severe plastic deformation. *JOM* 58(4), 43.

2-6. Stolyarov, V. V., Zhu, Y. T., Alexandarov, I. V., Lowe, T. C., and Valiev, Z. Z. (2003). Grain refinement and properties of pure Ti processed by warm ECAP and cold rolling. *Mater. Sci. Eng. A* 343(1–2), 43.

2-7. Shankar, M. R., Verma, R., Rao, B. C., Chandrasekar, S., Compton, W. D., King, A. H., and Trumble, K. P. (2007). Severe plastic deformation of difficult-to-deform materials at near-ambient temperatures. *Metal. Mater. Trans. A* 38, 1899.

2-8. Brown, T. L., Swaminathan, S., Chandrasekar, S., Compton, W. D., King, A. H., and Trumble, K. P. (2002). Low-cost manufacturing process for nanostructured metals and alloys. *J. Mater. Res.* 17, 2484.

2-9. Shankar, M. R., Rao, B. C., Lee, S., Chandrasekar, S., King, A. H., and Compton, W. D. (2006). Severe plastic deformation (SPD) of titanium at near-ambient temperature. *Acta. Mater.* 54(14), 3691.

2-10. Moscoso, W., Shankar, M. R., Mann, J. B., Compton, W. D., and Chandrasekar, S. (2007). Bulk nanostructured materials by large strain extrusion machining. *J. Mater. Res.* 22, 201.

2-11. Embury, J. D. and Fisher, R. M. (1966). The structure and properties of drawn pearlite. *Acta Metall.* 14, 147.

2-12. Segal, V. M., Reznikov, V. I., Drobyshevskiy, A. E., and V. I. Kopylov (1981). Plastic working on metals by simple shear. *Russian Metall.* 1, 99 and Segal, V. M. (1995). Materials processing by simple shear. *Mater. Sci. Eng. A* 197, 157.

2-13. Horita, Z., Fujinami, T., Nemoto, N., and Langdon, T. G. (2000). Equal-channel angular pressing of commercial aluminum alloys: Grain refinement, thermal stability and tensile properties. *Metall. Mater. Trans. A* 31, 691.

2-14. Zhiyaev, A. P., Lee, S., Nurislamova, G. V., Valiev, R. Z., and Langdon, T. G. (2001). Microhardness and microstructural evolution in pure nickel during high-pressure torsion. *Scripta Mater.* 44, 2753.

2-15. Hughes, D. A. and Hansen, N. (2000). Microstructure and strength of nickel at large strains. *Acta Mater.* 48, 2985.

2-16. De Chiffre, L. (1983). Extrusion cutting of brass strips. *Int. J. Mach. Tool Des. Res.* 23, 141.

2-17. Oliver, W. C. and Pharr, G. M. (1992). An improved technique for determining hardness and modulus using loading and displacement sensing indentation technique. *J. Mater. Res.* 17(6), 1564.

2-18. Benavides, G. L., Bieg, L. F., Saavedra, M. P., and Bryce, E. A. (2002). High aspect ratio meso-scale parts enabled by wire micro-EDM. *Microsyst. Technol.* 8(6), 295.

2-19. Wynick, G. L. and Boehlert, C. J. (2005). Use of electropolishing for enhanced metallic specimen preparation for electron backscatter diffraction analysis. *Mater. Charact.* 55(3), 190.

2-20. Shankar, M. R., Rao, B. C., Chandrasekar, S., Compton, W. D., and King, A. H. (2007). *Scr. Mater.* Published online, doi:10.1016/j.scriptamat.2007.11.040

2-21. Seaminathan, S., Brown, T. L., Chandrasekar, S., McNelly, T. R., and Compton, W. D. (2007). Severe plastic deformation of copper by machining: Microstructure refinement and nanostructure evaluation with strain. *Scr. Mater.* 56(12), 1047.

2-22. Shankar, M. R., Chandrasekar, S., King, A. H., and Compton, W. D. (2005). Microstructure and stability of nanocrystalline aluminum created by large strain machining. *Acta Mater.* 53, 4781.

2-23. Liu, W. C., Chen, Z., and Yao, M. (1999). Effect of cold rolling on the precipitation behavior of delta phase in Inconel 718. *Metall. Mater. Trans. A*, 30A(1), 31 (1999).

2-24. Liu, W. C., Xiao, F. R., Yao, M., Chen, Z. L., Wang, S. G., and Li, W. H. (1997). Quantitative phase analysis of Inconel 718 by X-ray diffraction. *J. Mater. Sci. Lett.* 16(9), 769.

2-25. Cai, D. Y., Liu, W. C., Li, R. B., Zhang, W. H., and Yao, M. (2004). On the accuracy of X-ray diffraction quantitative phase analysis method in Inconel 718. *J. Mater. Sci.* 39(2), 719.

2-26. Slama, C., Servant, C., and Cizeron, G. (1997). Aging of Inconel 718 alloy between 500 and 750 degrees C. *J. Mater. Res.* 12(9), 2298.

2-27. Chapman, L. A. (2004). Application of high temperature DSC technique to nickel based superalloys. *J. Mater. Sci.* 39(24), 7229.

3-1. Kumar, K. S., Van Swygenhoven, H., et al. (2003). Mechanical behavior of nanocrystalline metals and alloys. *Acta materialia* 51(19), 5743.

3-2. Lu, L., Sui, M. L., et al. (2000). Superplastic extensibility of nanocrystalline copper at room temperature. *Science* 287(5457), 1463.

3-3. Meyers, M. A., Mishra, A., et al. (2006). Mechanical properties of nanocrystalline materials. *Prog. Mat. Sci.* 51(4), 427.

3-4. Briesen, H. (1998). Electrically assisted aerosol reactors using ring electrodes. *MRS Symposium Proceeding* 520(3).

3-5. Embury, J. D. and Fisher, R. M. (1966). Structure and properties of drawn pearlite. *Acta Metall.* 14(2), 147.

3-6. Langford, G. and Cohen, M. (1969). Strain hardening of iron by server plastic deformation. *ASM Transactions Quarterly* 62(3), 623.

3-7. Brown, T. L., Swaminathan, S., et al. (2002). Low-cost manufacturing process for nanostructured metals and alloys. *J. Mater. Res.* 17(10), 2484.

3-8. Shankar, M. R., Chandrasekar, S., et al. (2004). Characteristics of aluminum 6061-T6 deformed to large strain by machining. *JOM* 56(11), 224.

3-9. Swaminathan, S., Ravi Shankar, M., et al. (2005). Large strain deformation and ultrafine grained materials by machining. *Mater. Sci. Eng. A* 410–411, 358.

3-10. Shankar, M. R., Rao, B. C., et al. (2006). Severe plastic deformation (SPD) of titanium at near-ambient temperature. *Acta materialia*, 54(14), 3691.

3-11. Hall, E. O. (1951). Deformation and aging of mild steel. *Physical Society – Proceeding* 64(381B), 747 and Hall, E. O. (1954). Variation of hardness of metals with grain size. *Nature* 173(4411), 948.

3-12. Petch, N. J. (1953). Cleavage strength of polycrystals. *Iron and Steel Institute—J.* 174(Part 1), 25.

3-13. Saldana, C. J. (2006). *Nanostructured Particulate by Modulation-assisted Machining*. Master thesis, Purdue University, West Lafayette, Indiana.

3-14. Valiev, R. Z. and Langdon, T. G. (2006). Principle of equal-channel angular processing as a processing tool for grain refinement. *Prog. Mat. Sci.* 51(7), 881 and Dalla Toore, F. H., Pereloma, E.V., et al. (2006). Strain hardening behavior and deformation kinetics of Cu deformed by equal channel angular extrusion from 1 to 16 passes. *Acta Materialia* 54(4), 1135.

3-15. Valiev, R. Z., Islamgaliev, R. K., et al. (2000). Bulk nanostructured materials from severe plastic deformation. *Prog. Mat. Sci.* 45(2), 103.

4-1. Gleiter, H. (1995). Nanostructured materials: State of the art and perspectives. *Nanostructured Materials* 6(1–4), 3–14.

4-2. Koch, C. C. (1997). Synthesis of nanostructured materials by mechanical milling: Problems and Opportunities. *Nanostructured Materials* 9(1), 13–22.

4-3. Romanov, A. E. (1995). Continuum theory of defects in nanoscaled materials. *Nanostructured Materials* 6(1–4) 125–134.

4-4. Trelewicz, J. R. and Schuh, C. A. (2007). The Hall-Petch breakdown in nanocrystalline metals: A crossover to glass-like deformation. *Acta Materialia* 55, 5948–5958.

4-5. Frazier, A. B., Warrington, R. O., and Friedrich, C. (1995). The miniaturization technologies: Past; present; and future. *IEEE Transactions on Industrial Electronics* 42(5), 423–430.

4-6. Bhat, K. N. (1998). Micromachining for microeletromechanical systems. *Defense Sci. J.* 48(1), 5–19.

4-7. Adams, D. P., Vasile, M. J., and Krishnan, A. S. M. (2000). Microgrooving and microthreading tools for fabricating curvilinear features. *Precision Engineering-Journal of the Int. Soc. Precision Engineer. Nanotech.* 24(4), 347–356.

4-8. Mohr, J. (1998). LIGA: A technology for fabricating microstructures and Microsystems. *Sens. Mat.* 10(6), 363–373.

4-9. Bacher, W., Menz, W., and Mohr, J. (1995). The LIGA technique and its potential for Microsystems—A survey. *IEEE Transactions on Industrial Electronics* 42(5), 431–441.

4-10. Spearing, S. M. (2000). Materials issues in microelectromechanical systems. *Acta Materialia* 48, 179–196.

4-11. Fecht, H. J. (1995). Nanostructure formation by mechanical attrition. *Nanostructured Mat.* 6(1–4), 33–42.

4-12. Ajdelsztajn, L., Jodoin, B., Kim, G. E., Schoenung, J. M., and Mondoux, J. (2005). Cold spray deposition of nanocrystalline aluminum alloys. *Metall. Mat.—Trans. A* 36A, 657–666.

4-13. Alkimov, P., Kosarev, V. F., and Papyrin, A. N. (1990). A method of cold gas-dynamic deposition. *Sov. Phys. Dokl* 35(12), 1047–1049.

4-14. Gilmore, D. L., Dykhuizen, R. C., Neiser, R. A., Roemer, T. J., and Smith, M. F. (1999). Particle velocity and deposition efficiency in the cold spray process. *J. Ther. Spray Tech.* 8(4), 576–582.

4-15. Dykhuizen, R. C., Smith, M. F., Gilmore, D. L., Neiser, R. A., Jiang, X., and Sampath, S. (1999). Impact of high velocity cold spray particles. *J. Ther. Spray Tech.* 8(4), 559–564.

4-16. Hall, A. C., Cook, D. J., Neiser, R. A., Roemer, T. J., and Hirschfeld, D. A. (2006). The effect of a simple annealing heat treatment on the mechanical properties of cold-sprayed aluminum. *J. Ther. Spray Tech.* 15(2), 233–238.

4-17. Dykhuizen, R. C. and Smith M. F. (1998). Gas dynamic principles of cold spray. *J. Ther. Spray Tech.* 7(2), 205–212.

4-18. Eckert, J. (1995). Relationships governing the grain size of nanocrystalline metals and alloys. *Nanostructured Materials* 6(1–4), 431–416.

5-1. Welch, C. M. and Compton, R. G. (2006). The use of nanoparticles in electroanalysis: A review. *Anal. Bioanal. Chem.* 384(3), 601–619.

5-2. Dresselhaus, M. S., et al. (2007). New directions for low-dimensional thermoelectric materials. *Adv. Mat.* 19(8), 1043–1053.

5-3. Dao, M., et al. (2007). Toward a quantitative understanding of mechanical behavior of nanocrystalline metals. *Acta Materialia* 55(12) 4041–4065.

5-4. Das, S. K., Chow, S., and Patel, H. (2006). Heat transfer in nanofluids—A review. *Heat Transfer Engineering* 27(10), 3–19.

5-5. Koch, C. C. (2007). *Nanostructured Materials: Processing, Properties, and Applications,* 2nd edition. William Andrews Publishing, New York, NY.

5-6. Oelhafen, P. and Schüler, A. (2005). Nanostructured materials for solar energy conversion. *Solar Energy* 79(2), 110–121.

5-7. Suryanarayana, C. (2005). Recent developments in nanostructured materials. *Adv. Engineer. Mat.* 7(11), 983–992.

5-8. Han, B. Q., et al. (2007). Processing and behavior of nanostructured metallic alloys and composites by cryomilling. *J. Mat. Sci.* 42(5), 1660–1672.

5-9. Han, B. Q., Mohamed, F. A., Bampton, C. C., and Lavernia, E. J. (2005). Improvement of toughness and ductility of a cryomilled Al-Mg alloy via microstructural modification. *Metall. Mat. Trans. A* 36A(8), 2081–2091.

5-10. Lee, Z, et al. (2005). Bimodal microstructure and deformation of cryomilled bulk nanocrystalline Al-7.5Mg alloy. *Mat. Sci. Engineer. A* 410–411, 462–467.

5-11. Huang, X., Hansen, N., and Tsuji, N. (2006). Hardening by annealing and softening by deformation in nanostructured metals. *Science* 312(5771), 249–251.

5-12. Fecht, H. J. (1995). Nanostructure formation by mechanical attrition. *Nanostructured Materials* 6, 33–42.

5-13. Swaminathan, S., et al. (2005). Nanostructured materials by machining. *Am. Soc. Mech. Engineers, MED* **16**(2), 981–985.

5-14. Swaminathan, S., et al. (2007). Severe plastic deformation (SPD) and nanostructured materials by machining. *J. Mat. Sci.* **42**(5), 1529–1541.

5-15. Atkinson, H. V. and Davies, S. (2000). Fundamental aspects of hot isostatic pressing: An overview. *Metall. Mat. Trans. A* **31A**(12), 2981–3000.

5-16. Ye, J., Ajdelsztajn, L., and Schoenung, J. M. (2006). Bulk nanocrystalline aluminum 5083 alloy fabricated by a novel technique: Cryomilling and spark plasma sintering. *Metall. Mat. Trans. A* **37A**(8), 2569–2579.

5-17. Ajdelsztajn, L., Jodoin, B., Kim, G. E., and Schoenung, J. M. (2005). Cold spray deposition of nanocrystalline aluminum alloys. *Metall. Mat. Trans. A.* **36A**(3), 657–666.

5-18. Jin, Z. Q., et al. (2004). Shock compression response of magnetic nanocomposite powders. *Acta Materialia* **52**(8), 2147–2154.

5-19. Chen, T., Hampikian, J. M., and Thadhani, N. N. (1999). Synthesis and characterization of mechanically alloyed and shock-consolidated nanocrystalline NiAl intermetallic. *Acta Materialia* **47**(8), 2567–2579.

5-20. Morris, D. G. (1981). Melting and solidification during dynamic powder compaction of tool steel. *Metal Sci.* **15**(3), 116–124.

5-21. Gourdin, W. H. (1984). Energy deposition and microstructural modification in dynamically consolidated metal powders. *J. Appl. Phy.* **55**(1), 172–181.

5-22. Brochu, M., et al. (2007). Dynamic consolidation of nanostructured Al-7.5%Mg alloy powders. *Mat. Sci. Engineer. A* **466**(1–2), 84–89.

5-23. Raybould, D., Morris, D. G., and Cooper, G. A. (1979). A new powder metallurgy method. *J. Mat. Sci.* **14**(10), 2523–2526.

5-24. Nieh, T. G., et al. (1996). Dynamic compaction of aluminum nanocrystals. *Acta Materialia* **44**(9), 3781–3788.

5-25. Liao, X. Z., et al. (2003). Nanostructures and deformation mechanisms in a cryogenically ball-milled Al-Mg alloy. *Philosophical Magazine* **83**(26), 3065–3075.

5-26. Goods, S. H. and Brown, L. M. (1979). The nucleation of cavities by plastic deformation. *Acta Metallurgica* **27**(1), 1–15.

5-27. Fredenburg, D. A., Vogler, T. J., Saldana, C. J., and Thadhani, N. N. (2007). Shock consolidation of nanocrystalline aluminum for bulk component formation in Shock Compression of Condensed Matter – 2007. *Am. Inst. Phy.* **955**, 1029–1032.

5-28. Hugo, R. C. and Hoagland, R. G. (1998). *In-situ* TEM observation of aluminum embrittlement by liquid gallium. *Scripta Materialia* **38**(3), 523–529.

5-29. Lao, X. Z., et al. (2003). Deformation mechanisms in nanocrystalline Al: Partial dislocation slip. *Appl. Phy. Lett.* **83**(4), 632–634.

5-30. Warner, D. H., Sansoz, F., and Molinari, J. F. (2006). Atomistic based continuum investigation of plastic deformation in nanocrystalline copper. *Int. J. Plasticity* **22**(4), 754–774.

5-31. Derlet, P. M. and Van Swygenhoven, H. (2002). Length scale effects in the simulation of deformation properties of nanocrystalline metals. *Scripta Materialia* **47**(11), 719–724.

5-32. Zhu, Y. T. and Langdon, T. G. (2005). Influence of grain size on deformation mechanisms: An extension to nanocrystalline materials. *Mat. Sci. Engineer. A* **409**(1–2), 234–242.

5-33. Zhu, Y. T., et al. (2004). Nucleation and growth of deformation twins in nanocrystalline aluminum. *Appl. Phy. Lett.* **85**(21), 5049–5051.

5-34. Gray, G. T. (1988). Deformation twinning in al-4.8 wt% Mg. *Acta Metallurgica* **36**(7), 1745–1754.

5-35. Bae, D. H. and Ghosh, A. K. (2002). Cavity formation and early growth in a superplastic Al-Mg alloy. *Acta Materialia* **50**(3), 511–523.

6-1. Koch, C. C. (2007). *Nanostructured Materials: Processing, Properties, and Applications,* 2nd edition. William Andrews Publishing, New York, NY.

6-2. Ajdelsztajn, L., Jodoin, B., Kim, G. E., Schoenung, J. M., and Mondoux, J. (2005). Cold spray deposition of nanocrystalline aluminum alloys. *Metall. Mat.—Trans. A* **36A**, 657–666.

6-3. Ajdelsztajn, L., Jodoin, B., and Schoenung, J. M. (2006). Synthesis and mechanical properties of nanocrystalline Ni coatinga produced by cold gas dynamic spraying. *Surface and Coatings Technology* **201**, 1166–1172.

6-4. Kim, H. J., Lee, C. H., and Hwang, S. Y. (2005). Fabrication of WC-Co coatings by cold spray deposition. *Surface and Coatings Technology* **191**, 335–340.

6-5. Phani, P. S., Vishnukanthan, V., and Sundararajan, G (2007). Effect of heat treatment on properties of cold sprayed nanocrystalline copper alumina coatings. *Acta Materialia* **55**, 4741–4751.

6-6. Brochu, M., Zimmerly, T., Ajdelsztajn, L., Lavernia, E. J., and Kim, G. (2007). Dynamic consolidation of nanostructured Al-7.5%Mg alloy powders. *Mat. Sci. Engineer. A* **466**, 84–89.

6-7. Lee, Z., Witkin, D. B., Radmilovic, V., Lavernia, E. J., and Nutt, S. R. (2005). Bimodal microstructure and deformation of cryomilled bulk nanocrystalline Al-7.5Mg alloy. *Mat. Sci. Engineer. A* **410–411**, 462–467.

6-8. Chen, T., Hampikian, J. M., and Thadhani, N. N. (1999). Synthesis and characterization of mechanically alloyed and shock consolidated nanocrystalline NiAl intermetallic. *Acta Materialia* **47**, 2567–2579.

6-9. Tanimoto, H., Pasquini, L., Prummer, R., Knonmuller, H., and Schaefer, H.-E. (2000). Self-diffusion and magnetic properties in explosion densified nanocrystalline Fe. *Scripta Materialia* **42**, 961–966.

6-10. Chen, K. H., Jin, Z. Q., Li, J., Kennedy, G., Wang, Z. L., Thadhani, N. N., Zeng, H., Cheng, S.-F., and Liu, J. P. (2004). Bulk nanocomposite magnets produced by dynamic shock compaction. *J. Appl. Phy.* **96**, 1276–1278.

6-11. Hall, A. C., Brewer, L. N., and Roemer, T. J. (2008). Preparation of aluminum coatings containing homogeneous nanocrystalline microstructures using the cold spray process. *J. Ther. Spray Tech.* **17**, 352–359.

6-12. Hall, A., Yang, P., Brewer, L., Buchheit, T., and Roemer, T. Preparation and mechanical properties of cold sprayed nanocrystalline aluminum. *Expected submission to J. Ther. Spray Tech.*

6-13. Fredenburg, D. A., Vogler, T. J., Thadhani, N. N., and Dance, W. Shock consolidation of nano-crystalline 6061 aluminum powders. *Expected submission to Metall. Mat. Trans. A.*

6-14. Eckert, J. (1995). Relationships governing the grain size of nanocrystalline metals and alloys. *Nanostructured Materials* **6**, 413–416.

6-15. Dykhuizen, R. C., Smith, M. F., Gilmore, D. L., Neiser, R. A., Jiang, X., and Sampath, S. (1999). Impact of high velocity cold spray particles. *J. Ther. Spray Tech.* **8**, 559–564.

6-16. Nishijima, S., Ishii, A., Kanazawa, K., Matsuoka, S., and Masuda, C. (1989). Fundamental fatigue properties of JIS steels for machine structural use. *National Research Institute for Metals, Ibaraki*, 77.

6-17. Dieter, G. E. (1986). *Mechanical Metallurgy.* McGraw-Hill, New York, p. 189.

6-18. Elmustafa, A. A. and Stone, D. S. (2002). Indentation size effect in polycrystalline F.C.C. metals. *Acta Materialia* **50**, 3641–3650.

11

1. Kamenetsky, V. S., Davidson, P., Mernagh, T. P., et al. (2002). Fluid bubbles in melt inclusions and pillow-rim glasses: High-temperature precursors to hydrothermal fluids? *Chem. Geo.* **183**, 349–364.

2. Moore, J. G. and Calk, L. (1971). Sulfide spherules in vesicles of dredged pillow basalt. *Am. Mineralogist* **56**, 476–488.

3. Mathez, E. A. and Yeats, R. S. (1976). Magmatic sulfides in basalt glass from DSDP Hole 319A and Site 320, Nazca Plate. *Initial Reports of the Deep Sea Drilling Project.* US Government Printing Office, Washington, DC, USA.

4. Moore, J. G., Batchelder, J. N., and Cunningham, C. G. (1977). CO2-filled vesicles in mid-ocean basalt. *J. Volcan. Geother. Res.* **2**(4), 309–327.

5. Yeats, R. S. and Mathez, E. A. (1976). Decorated vesicles in deep-sea basalt glass, eastern pacific. *J. Geophy. Res.* **81**(23), 4277–4284.

6. Mungall, J. E., Bagdassarov, N. S., Romano, C., and Dingwell, D. B. (1996). Numerical modelling of stress generation and microfracturing of vesicle walls in glassy rocks. *J. Volcan. Geother. Res.* **73**(1–2), 33–46.

7. Furnes, H., Banerjee, N. R., Staudigel, H., et al. (2007). Comparing petrographic signatures of bioalteration in recent to Mesoarchean pillow lavas: Tracing subsurface life in oceanic igneous rocks. *Precambrian Res.* **158**(3–4), 156–176.

8. French, J. E., Muehlenbachs, K., Blake, D. F., and Banerjee, N. R. Identification of alpha-recoil and fission track etch-tunnels in mid-ocean ridge basaltic glass: Theoretical modelling of natural porosity. Submitted to *Geochimica et Cosmochimica Acta.*

9. Wada, Y. (1998). Possible application of micromachine technology for nanometer lithography. *Microelec. J.* **29**(9), 601–611.

10. Smith, H. I. (2001). Low cost nanolithography with nanoaccuracy. *Physica E* **11**(2–3), 104–109.

11. Shipboard Scientific Parties (1980). Site 418. *Initial Reports of the Deep Sea Drilling Project*. US Government Printing Office, Washington, DC, USA.

12. Arnéodo, A., Couder, Y., Grasseau, G., Hakim, V., and Rabaud, M. (1989). Uncovering the analytical Saffman-Taylor finger in unstable viscous fingering and diffusion-limited aggregation. *Phy. Rev. Lett.* **63**(9), 984–987.

13. Lajeunesse, E. and Couder, Y. (2000). On the tip-splitting instability of viscous fingers. *J. Fluid Mech.* **419**, 125–149.

14. Liang, S. (1986). Random-walk simulations of flow in Hele Shaw cells. *Phy. Rev. A* **33**(4), 2663–2674.

15. Turcotte, D. L. (1997). *Fractals and Chaos in Geology and Geophysics*, Vol. 398 and 2nd edition. Cambridge University Press, Cambridge, UK.

16. Turcotte, D. L. (2002). Fractals in petrology. *Lithos*, **65**(3–4), 261–271.

17. Witten Jr., T. A. and Sander, L. M. (1981). Diffusion-limited aggregation, a kinetic critical phenomenon. *Phy. Rev. Lett.* **47**(19), 1400–1403.

18. Glicksman, M. E. and Lupulescu, A. O. (2004). Dendritic crystal growth in pure materials. *J. Crystal Growth* **264**(4), 541–549.

19. Turcotte, D. L., Pelletier, J. D., and Newman, W. I. (1998). Networks with side branching in biology. *J. Theor. Bio.* **193**(4), 577–592.

20. Masek, J. G. and Turcotte, D. L. (1993). A diffusion-limited aggregation model for the evolution of drainage networks. *Earth and Planetary Sci. Lett.* **119**(3), 379–386.

21. Homsy, G. M. (1987). Viscous fingering in porous media. *Ann. Rev. Fluid Mech.* **19**, 271–311.

22. Bensimon, D., Kadanoff, L. P., Liang, S., Shraiman, B. I., and Tang, C. (1986). Viscous flows in two dimensions. *Rev. Modern Phy.* **58**(4), 977–999.

23. Saffman, P. G. (1986). Viscous fingering in Hele-Shaw cells. *J. Fluid Mech.* **173**, 73–94.

24. Kopf-Sill, A. R. and Homsy, G. M. (1988). Nonlinear unstable viscous fingers in Hele-Shaw flows. *Phy. Fluids* **31**, 242–249.

25. Nittmann, J., Daccord, G., and Stanley, H. E. (1985). Fractal growth viscous fingers: Quantitative characterization of a fluid instability phenomenon. *Nature*, **314**, 141–144.

26. Twiss, R. J. and Moores, E. M. (1992). *Structural Geology*, Vol. 532. W.H. Freeman, New York, NY, USA.

27. Auerbach, D. and Strobel, G. (1995). Fingering patterns and their fractal dimensions. *Chaos, Solitons & Fractals* **5**(10), 1765–1773.

28. Fenghour, A., Wakeham, W. A., and Vesovic, V. (1998). The viscosity of carbon dioxide. *J. Phy. Chem. Ref. Data* **27**(1), 31–44.

29. Gottsmann, J., Harris, A. J. L., and Dingwell, D. B. (2004). Thermal history of Hawaiian pāhoehoe lava crusts at the glass transition: Implications for flow rheology

and emplacement. *Earth and Planetary Sci. Lett.* **228**(3–4), 343–353.

30. Moore, J. G. (1979). Vesicularity and CO2 in mid-ocean ridge basalt. *Nature* **282**(5736), 250–253.

31. Burnard, P. (1999). The bubble-by-bubble volatile evolution of two mid-ocean ridge basalts. *Earth and Planetary Sci. Lett.* **174**(1–2), 199–211.

32. Lemaire, E., Levitz, P., Daccord, G., and Van Damme, H. (1991). From viscous fingering to viscoelastic fracturing in colloidal fluids. *Phy. Rev. Lett.* **67**(15), 2009–2012.

33. Javoy, M. and Pineau, F. (1991). The volatiles record of a "popping" rock from the Mid-Atlantic Ridge at 14°N: Chemical and isotopic composition of gas trapped in the vesicles. *Earth and Planetary Sci. Lett.* **107**(3–4), 598–611.

34. Zhang, J.-H. and Liu, Z.-H. (1998). Study of the relationship between fractal dimension and viscosity ratio for viscous fingering with a modified DLA model. *J. Petroleum Sci. Engineer.* **21**(1–2), 123–128.

35. Buka, A., Kertész, J., and Vicsek, T. (1986). Transitions of viscous fingering patterns in nematic liquid crystals. *Nature* **323**(6087), 424–425.

36. Fast, P., Kondic, L., Shelley, M. J., and Palffy-Muhoray, P. (2001). Pattern formation in non-Newtonian Hele-Shaw flow. *Phy. Fluids* **13**(5), 1191–1212.

37. Lindner, A., Bonn, D., Poiré, E. C., Amar, M. B., and Meunier, J. (2002). Viscous fingering in non-Newtonian fluids. *J. Fluid Mech.* **469**, 237–256.

38. Webb, S. L. and Dingwell, D. B. (1990). Non-Newtonian rheology of igneous melts at high stresses and strain rates: Experimental results for rhyolite, andesite, basalt, and nephelinite. *J. Geophy. Res.* **95**(B10), 15695–15701.

39. Nyamjav, D. and Ivanisevic, A. (2005). Templates for DNA-templated Fe3O4 nanoparticles. *Biomaterials* **26**(15), 2749–2757.

40. Wang, K., Yue, S., Wang, L., et al. (2006). Nanofluidic channels fabrication and manipulation of DNA molecules. *IEE Proc. Nanobiotech.* **153**(1), 11–15.

41. Hanson, G. W. (2008). *Fundamentals of Nanoelectronics*, Vol. 385. Pearson Prentice Hall, Upper Saddle River, NJ, USA.

42. Rabaey, J., Chandrakasan, A., and Nikolić, B. (2003). *Digital Integrated Circuits: A Design Perspective*, Vol. 761 and 2nd edition. Pearson Education, Upper Saddle River, NJ, USA.

43. Till, W. C. and Luxon, J. T. (1982). *Integrated Circuits: Materials, Devices, and Fabrication*, Vol. 462. Prentice-Hall, Englewood Cliffs, NJ, USA.

44. Ma, Y. and Seminario, J. M. (2007). Analysis of programmable molecular electronic systems. In *Molecular and Nano Electronics: Analysis, Design and Simulation*. J. M. Seminario (Ed.). Elsevier, Milan, Italy, pp. 96–140.

45. Huang, Y., Duan, X., Cui, Y., and Lieber, C. M. (2002). Gallium nitride nanowire nanodevices. *Nano Lett.* **2**(2), 101–104.

46. Cerofolini, G. F. and Mascolo, D. (2006). A hybrid route from CMOS to nano and molecular electronics. In *Nanotechnology for Electronic Materials and Devices*. A. Korkin, J. Labanowski, E. Gusev, and S. Luryi (Eds.). Springer, New York, NY, USA, pp. 1–65.

47. Byerly, G. R. and Sinton, J. M. (1980). Compositional trends in natural basaltic glasses from Deep Sea Drilling Project Holes 417D and 418A. *Initial Report Deep Sea Drilling Project*. US Government Printing Office, Washington, DC, USA.

48. Vieu, C., Carcenac, F., Pépin, A., et al. (2000). Electron beam lithography: Resolution limits and applications. *Appl. Surf. Sci.* **164**(1–4), 111–117.

49. Tamai, I. and Hasegawa, H. (2007). Selective MBE growth of hexagonal networks of trapezoidal and triangular GaAs nanowires on patterned (111)B substrates. *J. Crystal Growth* **301–302**, 857–861.

50. Kasai, S. and Hasegawa, H. (2002). GaAs and InGaAs single electron hexagonal nanowire circuits based on binary decision

diagram logic architecture. *Physica E* **13**(2–4), 925–929.

12

1. Jordan, A., Wust, P., Scholz, R., Tesche, B., Fähling, H., Mitrovics, T., Vogl, T., Cervós-Navarro, J., and Felix, R. (1996). Cellular uptake of magnetic fluid particles and their effects on human adenocarcinoma cells exposed to AC magnetic fields *in vitro*. *Int. J. Hyperthermia* **12**, 705722.

2. Zhao, D. L., Zhang, H. L., Zeng, X. W., Xia, Q. S., and Tang J. T. (2006). Inductive heat property of Fe3O4/polymer composite nanoparticles in an ac magnetic field for localized hyperthermia. *Biomed Mater* **1**, 198201.

3. Shiyan, Y., Dongsheng, Z. and Ning, G. (2005). Therapeutic effect of Fe_2O_3 nanoparticles combined with magnetic fluid hyperthermia on cultured liver cancer cells and xenograft liver cancers. *J. Nanosci. Nanotechnol.* **5**, 18.

4. Bode, A. M. and Dong, Z. (2002). The paradox of arsenic: Molecular mechanisms of cell transformation and chemotherapeutic effects. *Crit. Rev. Oncol. Hematol.* **42**, 524.

5. Soignet, S. L., Frankel, S. R., Douer, D., Tallman, M. S., Kantarjian, H., Calleja, E., Stone, R. M., Kalaycio, M., Scheinberg, D. A., Steinherz, P., Sievers, E. L., Coutré, S., Dahlberg, S., Ellison, R., and Warrell, Jr., R. P. (2001). United States multicenter study of arsenic trioxide in relapsed acute promyelocytic leukaemia. *J. Clin. Oncol.* **19**, 38523860.

6. Xiao, Y. F., Wu, D. D., Liu, S. X., Chen, X., and Ren, L. F. (2007). Effect of arsenic trioxide on vascular endothelial cell proliferation and expression of vascular endothelial growth factor receptors Flt-1 and KDR in gastric cancer in nude mice. *World J. Gastroenterol.* **13**, 64986505.

7. Florea, A. M., Splettstoesser, F., and Büsselberg, D. (2007). Arsenic trioxide (As2O3) induced calcium signals and cytotoxicity in two human cell lines: SY-5Y neuroblastoma and 293 embryonic kidney (HEK). *Toxicol. Appl. Pharmacol.* **220**, 292301.

8. Shen, Z. Y., Zhang, Y., Chen, J. Y., Chen, M. H., Shen, J., Luo, W. H., and Zeng, Y. (2004). The alternation of mitochondria is an early event of arsenic trioxide induced apoptosis in esophageal carcinoma cells. *Oncol. Rep.* **11**, 155159.

9. Seol, J. G., Park, W. H., Kim, E. S., Jung, C. W., Hyun, J. M., Lee, Y. Y., and Kim, B. K. (2001). Potential role of caspase-3 and -9 in arsenic trioxide-mediated apoptosis in PCI-1 head and neck cancer cells. *Int. J. Oncol.* **18**, 249255.

10. Yu, J., Qian, H., Li, Y., Wang, Y., Zhang, X., Liang, X., Fu, M., and Lin, C. (2007). Arsenic trioxide (As_2O_3) reduces the invasive and metastatic properties of cervical cancer cells *in vitro* and *in vivo*. *Gynecol. Oncol.* **106**, 400406.

11. Lin, T. H., Kuo, H. C., Chou, F. P., and Lu, F. J. (2008). Berberine enhances inhibition of glioma tumor cell migration and invasiveness mediated by arsenic trioxide. *BMC Cancer* **8**, 5872.

12. Du, C. W., Wen, B. G., Li, D. R., Peng, X., Hong, C. Q., Chen, J. Y., Lin, Z. Z., Hong, X., Lin, Y. C., Xie, L. X., Wu, M. Y., and Zhang, H. (2006). Arsenic trioxide reduces the invasive and metastatic properties of nasopharyngeal carcinoma cells *in vitro*. *Braz. J. Med. Biol. Res.* **39**, 677685.

13. Yiqun, D., Dongsheng, Z., and Haiyan, N. (2006). The therapeutic effect of nano-sized As2O3/Fe3O4 complex in combination with magnetic fluid hyperthermia (MFH) on cervical cancer. In *Proceedings of the 1st IEEE International Conference on Nano/Micro Engineered and Molecular Systems*. 1821 January 2006. L. Yung-Chun (Ed.). Zhu Hai, China, pp. 343348.

14. Goodison, S., Yoshida, K., Sugino, T., Woodman, A., Gorham, H., Bolodeoku, J., Kaufmann, M., and Tarin, D. (1997). Rapid analysis of distinctive CD44 RNA splicing preferences that characterize colonic tumors. *Cancer Res.* **572**, 31403144.

15. Ueda, M., Terai, Y., Yamashita, Y., Kumagai, K., Ueki, K., Yamaguchi, H., Akise, D., Hung, Y. C., and Ueki, M. (2002). Correlation between vascular endothelial growth factor-C and invasion phenotype in cervical carcinomas. *Int. J. Cancer* **98**, 335343.

16. Zheng, J., Deng, Y. P., Lin, C., Fu, M., Xiao, P. G., and Wu, M. (1999). Arsenic trioxide induces apoptosis of HPV16 DNA-immortalized human cervical epithelial cells and selectively inhibits gene expression. *Int. J. Cancer* **82**, 286292.

17. Yiqun, D. (2006). Research of preparation and biocompatibility evaluation of nanosized As2O3/Fe3O4 complex and its theraputic effect of thermochemotherapy on cervical cancer. In *Master Thesis*. Department of Pathology and Pathophysiology, Southeast University, China.

18. Tasci, T. O., Vargel, I., Arat, A., Guzel, E., Korkusuz, P., and Atalar, E. (2009). Focused RF hyperthermia using magnetic fluids. *Med. Phs.* **36**, 19061912.

19. Jordan, A., Scholz, R., Wust, P., Fähling, H., Krause, J., Wlodarczyk, W., Sander, B., Vogl, T., and Felix, R. (1997). Effects of magnetic fluid hyperthermia (MFH) on C3H mammary carcinoma *in vivo*. *Int. J. Hyperthermia* **13**, 587605.

20. Yiqun, D., Dongsheng, Z., Haiyan, N., Ning, G., Shiyan, Y., Qiusha, T., Liqiang, J., and Meiling, W. (2005). Preparation and characterization of Fe_3O_4 nanomagnetic particles for tumor hyperthermia. *J. Chin. Electr. Microsc. Soc.* **24**, 608612.

21. Jordan, A. and Maier-Hauff, K. (2007). Magnetic nanoparticles for intracranial thermotherapy. *J. Nanosci. Nanotechnol.* **7**, 46044606.

22. Maier-Hauff, K., Rothe, R., Scholz, R., Gneveckow, U., Wust, P., Thiesen, B., Feussner, A., von Deimling, A., Waldoefner, N., Felix, R., and Jordan, A. (2007). Intracranial thermotherapy using magnetic nanoparticles combined with external beam radiotherapy: Results of a feasibility study on patients with glioblastoma multiforme. *J. Neurooncol.* **81**, 5360.

23. Johannsen, M., Gneveckow, U., Eckelt, L., Feussner, A., Waldöfner, N., Scholz, R., Deger, S., Wust, P., Loening, S. A., and Jordan, A. (2005). Clinical hyperthermia of prostate cancer using magnetic nanoparticles: Presentation of a new interstitial technique. *J. Nanosci. Nanotechnol.* **21**, 637647.

24. Johannsen, M., Gneveckow, U., Taymoorian, K., Cho, C. H., Thiesen, B., Scholz, R., Waldöfner, N., Loening, S. A., Wust, P., and Jordan, A. (2007). Thermal therapy of prostate cancer using magnetic nanoparticles. *Actas. Urol. Esp.* **31**, 660667.

25. Wust, P., Gneveckow, U., Johannsen, M., Böhmer, D., Henkel, T., Kahmann, F., Sehouli, J., Felix, R., Ricke, J., and Jordan, A. (2006). Magnetic nanoparticles for interstitial thermotherapyFeasibility, tolerance and achieved temperatures. *J. Nanosci. Nanotechnol.* **22**, 673685.

26. Moroz, P., Jones, S. K., Winter, J., and Gray, B. N. (2001). Targeting liver tumors with hyperthermia: Ferromagnetic embolization in a rabbit liver tumor model. *J. Surg. Oncol.* **78**, 2229.

27. Moroz, P., Jones, S. K., and Gray, B. N. (2002). The effect of tumour size on ferromagnetic embolization hyperthermia in a rabbit liver tumour model. *Int. J. Hyperthermia.* **18**, 129140.

28. Moroz, P., Jones, S. K., Metcalf, C., and Gray, B. N. (2003). Hepatic clearance of arterially infused ferromagnetic particles. *Int. J. Hyperthermia* **19**, 2334.

29. Granov, A. M., Tiutin, L. A., Tarazov, P. G., and Granov, D. A. (2003). Modern technologies of diagnosis and combined surgical treatment in liver tumors. *Vestn. Ross. Akad. Med. Nauk.* **00**, 5154.

30. Yanase, M., Shinkai, M., Honda, H., Wakabayashi, T., Yoshida, J., and Kobayashi, T. (1998). Intracellular hyperthermia for cancer using magnetite cationic liposomes: An *in vivo* study. *Jpn. J. Cancer Res.* **89**, 463469.

31. Hong, S. C., Song, J. Y., Lee, J. K., Lee, N. W., Kim, S. H., Yeom, B. W., and Lee, K. W. (2006). Significance of CD44v6 expression in gynecologic malignancies. *J. Obstet. Gynaecol. Res.* **32**, 379386.

32. Tiefu, L., Binlin, C., Yuguang, G., and Tao, L. (2001). Study on relationship between expressions of CD44 andinhibition effect of arsenic trioxide on carcinoma. *J. Harbin med. University* **35**, 111112.

33. Mylona, E., Magkou, C., Gorantonakis, G., Giannopoulou, I., Nomikos, A., Zarogiannos, A., Zervas, A., and Nakopoulou, L. (2006). Evaluation of the vascular endothelial growth factor (VEGF)-C role in urothelial

carcinomas of the bladder. *Anticancer Res.* **26**, 35673571.

34. Roboz, G. J., Dias, S., Lam, G., Lane, W. J., Soignet, S. L., Warrell, Jr., R. P., and Rafii, S. (2000). Arsenic trioxide induces dose- and time-dependent apoptosis of endothelium and may exert an antileukemic effect via inhibition of angiogenesis. *Blood* **96**, 15251530.

35. Sawaji, Y., Sato, T., Takeuchi, A., Hirata, M., and Ito, A. (2002). Anti-angiogenic action of hyperthermia by suppressing gene expression and production of tumour-derived vascular endothelial growth factor *in vivo* and *in vitro*. *Br. J. Cancer* **86**, 15971603.

36. Tan Shelian, O. Y., Zhao, B., Fang, S., Yang, B., and Zhao, P. (2004). Effects of localized thermochemotherapy on angiogenes is and proliferation of C6 gliomas in rats. *Academic Second Military Medical University* **25**, 773776.

37. Okada, A. (1999). Roles of matrix metalloproteinases and tissue inhibitor of metalloproteinase (TIMP) in cancer invasion and metastasis. *Gan to Kagaku Ryoho* 26, 22472252.

Index

Printed and bound by CPI Group (UK) Ltd, Croydon, CR0 4YY

23/10/2024

01777675-0005